The Science of Light

SERGE HAROCHE
NOBEL PRIZE IN PHYSICS

The Science of Light

From Galileo's Telescope to Quantum Physics

Odile Jacob

Translated from the French by Daniel Becker and the Author.

The present English-language edition is published
by Odile Jacob Publishing.

Originally published in French as *La Lumière révélée*
by Serge Haroche.

www.odilejacob.com
www.odilejacobpublishing.com

ISBN : 978-2-4150-0003-5

Although the invention of plausible hypotheses, independent of any connection with experimental observations, can be of very little use in the promotion of natural knowledge; yet the discovery of simple uniform principles, by which a great number of apparently heterogeneous phenomena are reduced to coherent and universal laws, must ever be allowed to be of considerable importance towards the improvement of the human intellect.

Thomas Young, *Bakerian Lectures*, 1801.

For Elsa, Rachel, Samuel, and Vadim.

Contents

Foreword

Light has illuminated and fascinated mankind since the beginning of time. Only in the last four centuries, however, have we begun to reveal its secrets. Only much more recently have we domesticated it by way of the modern technologies that have revolutionized our lives. It has been little more than a hundred years since the discovery of microwaves, those cousins of visible light that are omnipresent in modern communication, navigation, and medical radiography devices. And it has barely been sixty years since we tamed visible light by inventing the laser. The extraordinary properties of its rays have allowed us to make fundamental discoveries and invent instruments that were unimaginable in my youth.

I have had the fortune of taking part in this adventure over the past half century. In describing the trajectory of a scientific life dedicated to light, I aim to share with the reader the pleasure a researcher feels each time he sees a new phenomenon that illuminates the world in an unexpected and surprising way. After many years of research, my team and I managed to trap a few microwave photons for over a tenth of a second inside a box with reflective walls. By causing these fragile and elusive particles of light to interact with atoms excited by laser beams, we have observed the simultaneous wave and corpuscular behavior of light in experiments that illustrate the strange properties of the quantum world. The pleasure of this discovery was enhanced by the exciting thought that this work might one day lead to new applications, even if it remains difficult to predict what they will actually be. Any researcher who has discovered something new and promising has experienced a kindred sense of pleasure and excitement.

At a time when the need for science is greater than ever, it is important that a non-specialist public be able to understand, through a personal account, what motivates a researcher, what inspires his curiosity, and what role luck plays in a process that never ceases to surprise. It is also essential to recall that research is above all a source of knowledge that enriches a cultural heritage assembled over the course of centuries. Researchers see the world from slightly higher up than others because, according to the famous remark attributed to Newton, they stand on the shoulders of the giants who came before them. From this privileged vantage, they are the transmitters, from one generation to the next, of the knowledge and the rational scientific methodology that are essential to our civilization.

By speaking of science—that which I have practiced myself, but also the work of others that has enriched me and made me look more deeply at the world—I wish to share my passion with young students and beginners in research, so that they might carry forward this endlessly renewed adventure. I also hope to interest those members of the general public who are curious to learn more about a story that has profoundly influenced how we see the world and given us powerful means of action and control over it. I hope no less to interest the reader who already knows the broad outlines of this story, by providing my own personal insights. In this book I have tried to describe what we know about light today and how we learned it, but I have also alluded to what we still don't know, what it will be up to future generations to elucidate.

It seemed impossible to me to discuss my research without placing it in the context of a rich saga of scholarship that spans several centuries. In addition to optics, this history has touched on all currents of knowledge. Discovering the nature of light involved physics, of course, but also had a profound influence on other sciences: astronomy, chemistry, biology, and the life sciences. In such undertakings as the exploration of our planet and the determination of its precise size and shape, advances in our understanding of light have played a major role. To speak of light, then, is to evoke all fields of knowledge.

The ability to measure with ever-increasing precision is an integral part of this story. Observation of nature became truly scientific with the invention of instruments that made it possible to quantify the phenomena under study and to describe them using numbers that could objectively and reproducibly measure first distances and time intervals, then subtler quantities such as forces, charges, and fields. Mathematics, geometry and algebra progressing in tandem, has connected these numbers through theoretical models that unify apparently disparate phenomena in a global explanatory framework. To evoke this history is to show how scientific knowledge has been built step by step, in a constant interaction between advances in instrumentation and those in methods of calculation. The craftsmen who cut the first lenses and combined them in magnifying glasses, along with the watchmakers who built the first precise pendulum clocks, are essential players in this story, as are the mathematicians who discovered complex numbers or the notions of derivatives and integral calculation.

Presenting science to a non-specialist audience is a challenging art. It is tempting to draw on images and metaphors, but these must not be misleading. Evoking quantum physics, essential to understanding what light is, risks casting us onto the slippery slope of mysticism. This physics is perplexing because we don't directly perceive it with our senses or our intuition of the macroscopic world, but there is in fact nothing mysterious about it. It has revealed itself to its discoverers logically, leading to a rigorous mathematical theory that allows us to calculate the observed phenomena with precision, leaving no room for esoteric blur.

Galileo was undoubtedly one of the first scientists to attempt to present his discoveries to a wide audience in an educational manner. In his *Dialogue Concerning the Two Chief World Systems,* he introduced his theory of relativity of motion to disbelieving and disoriented contemporaries. Renouncing the evidence of an immobile Earth at the center of the world was as difficult for the Renaissance man as it has been for the modern man to leave behind the classical images of Newtonian trajectories and embrace the description of an indeterministic world of atoms and photons. The risk Galileo faced was great because opposing the dogmas of religion was heresy, the worst crime under the Inquisition; today, the scientist who seeks to present the counterintuitive concepts of a physical science whose multiple applications have revolutionized our daily life obviously doesn't risk the same fate as the seventeenth-century scholar.

I am aware, nonetheless, of the dangers—less dramatic but very real—confronting a researcher who addresses non-specialists. He or she risks being either too technical or too simplistic. I have tried to avoid these pitfalls by progressively presenting the concepts of light, relativity, and quantum physics, without equations or formalism. I believe that following the genealogy of ideas and theories through the centuries, and evoking the questions they raised for the great thinkers of the past, will make them more familiar and easier to digest as the pages go by.

Going back in history to the origins of modern science has given me occasion to talk about my scientific heroes, from Galileo to Einstein, touching on the work of illustrious scholars whom everyone knows at least by name and others, more obscure, who contributed to this great adventure as well. This book is not the impartial work of a science historian: I may have gotten some details wrong in setting down a history of knowledge so rich in twists and turns. Rather, the following pages should be seen as my personal vision of the evolution over centuries of our knowledge about light, such as I have portrayed it to myself and as it has guided and inspired my own research.

This book intertwines that history of light with my personal experience. It is divided into two parts of roughly equal length. Three chapters—the first and the final two—cover the last fifty years, describing my research and that of those contemporaries whose discoveries I have witnessed. Readers who have some knowledge of the basics of physics and who are interested in modern developments in the science of light and lasers may begin with these chapters. The central part of the book, from Chapters 2 through 5, forms a backdrop presenting a panorama of light science from the seventeenth to twentieth centuries, in which I show how it has evolved and decisively influenced our view of the world. These chapters also illustrate the deep links that have been forged, since the very beginnings of modern science, between fundamental research motivated by pure curiosity and human activities such as the exploration of the Earth and the development of trade and industry. I hope it will interest the non-scientific reader, and remind specialists of some unknown or forgotten details in an exciting and surprising history.

I have illustrated this book with many figures but do not refer explicitly to them in the text, to ensure a more fluid reading experience. Some of these illustrations and their captions are complements that can be consulted independently. I have also tried to the best of my ability to make the chapters readable as standalones, focusing each one on different periods in the history of light or in my personal journey. A few callbacks from one chapter to another, however, link the ideas and concepts discussed throughout in different contexts. For the historical section, I have drawn inspiration from an abundant body of literature, presented in a few references in an appendix along with three publications from my own research group. An alphabetical index of the scientists involved in this history is given in the appendix as well: their biographies, easily accessible on Wikipedia, may be a useful complement to this book.

CHAPTER 1
The dawn of a vocation

In recent years, I have often been asked: "What made you a researcher? Where did your passion for science come from?" When I speak to high school or college students, I can no longer escape these questions, though I never got them when I was younger. When I would give a lecture twenty years ago, my listeners were more interested in my research than in my motivations. This new line of questioning is the product of age and the honors that have accompanied it. I don't shy away; I try to answer as honestly and precisely as I can, because even beyond my own case the topic is interesting: why does one become a researcher? What did science mean, sixty years ago, to a young person embarking on this adventure?

It is a nostalgic but invigorating exercise to recall my childhood and my adolescence before an audience of young people who live in a world very different from the one back then. The discussion that often follows shows me that, despite the passage of time, the curiosity of youth has remained the same. Our fundamental understanding of life and the universe has grown considerably, and our means of learning and gathering information about the world are vastly more powerful today, but the enthusiasm I see in the eyes of the students listening to me and the questions they ask are no different from those that inspired me at their age. The world in which they are growing up is merely more complex, more difficult to grasp, than the one in which I was fortunate enough to come of age.

In my youth during the *Trente Glorieuses*, despite the Cold War and the upheavals of decolonization, the hope reigned that the world was headed toward a future of progress and an ever more advanced and enlightened civilization. Young people drawn to research found it easier than it is today to follow their passion. Confidence in knowledge had not yet been undermined by the poison of post-truth, which today attacks the very values of science. André Malraux had declared that the twenty-first century would be religious or would not be at all, but we didn't really believe it; I never expected to live, as I do today, in a world so irrational that creationism is thriving and a non-trivial portion of the population believes the Earth is flat or vaccines are dangerous.

The students who ask me questions don't believe in this nonsense, of course, but then they are a select audience, ready to listen to me and to share the values of the scientific method. It is crucial that these values not be the province of an educated minority in the face of a dubious population, or one that allows itself to be influenced by lies. Our society needs science more than ever, and it is essential to speak of curiosity in general, and scientific curiosity in particular, and what awakens and sustains them.

This is the message I try to pass on to those who come to listen to me.

I tell them about the advances in knowledge that have impassioned me, and those I have witnessed firsthand, for more than half a century. In this way I hope to show them the beauty of the scientific approach and the strength of its values. Speaking to them about science leads me to the subtle and ever-evolving notion of what constitutes scientific truth. It is this search for truth by trial and error—a process that has its moments of doubt and self-questioning, but also moments of exaltation and dazzling triumph—that I wish to describe in this book.

First passions: from mathematics to astronomy

But let us return to the initial question: why did I become a researcher? For as long as I can recall, I have been attracted to numbers and passionate about measuring things. I remember counting, at a very early age, the number of ceramic tiles on the bathroom wall and the number of paving stones in the schoolyard. I would measure the length of the diagonal of a square or rectangle and compare it to the length of its sides. Without knowing it, I was doing trigonometry. The idea of classifying objects on the basis of precise measurements also led me to make a chart of metals in increasing order of density, from light aluminum to heavy uranium. There was no internet at the time, no Google, and I took all of this data from a popular dictionary, the *Petit Larousse illustré*. From a very young age, I loved to measure, to classify, to compare.

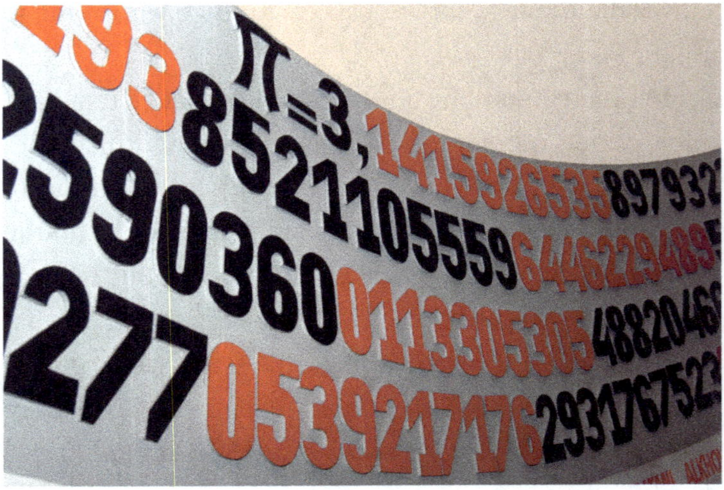

Figure I.1. The number π on the wall of the Palais de la Découverte (© Palais de la Découverte/C. Rousselin).

Geometry fascinated me as well. I soon began to draw circles with a compass, and ellipses by attaching a string to two pegs and tautening it with a pencil. From the age of ten or eleven on, I was fascinated by the number π. I remember its digits written in a long spiral on one of the walls of the Palais de la Découverte, which I visited as often as I could.

That this sequence goes on infinitely, without any regularity or recognizable repetition, fascinated me. How was it possible to determine it with such infinite precision when my own measurements of the figures I clumsily traced told me only that the ratio of a circle's circumference to its diameter was a little larger than 3?

This number's mystery did not end there. There was an interactive experiment at the Palais de la

Découverte that intrigued me: it involved dropping a pin onto a platform and counting the number of times it landed across two slats. The panel presenting this experiment explained that if the needle was equal in length to the width of the slats, the probability of this occurring was $2/\pi$, about 64 percent. Successive visitors would release the pin with the touch of a button, and the result was added to the statistics displayed on a counter.

The value of π, deduced after a few tens of thousands of throws, gave this number to two or three exact digits. Intrigued that this number could be determined by such an experiment, I began to grasp the notion of probability and see the link between probability and mathematics. Sometimes, when I got home, I would repeat the experiment with a handful of pencils that I threw on my bedroom floor. Only much later was I able to convince myself through reasoning that the value of π and the properties of the circle did indeed play a role in calculating the probability that a pencil would hit two slats at once.

The planetarium at the Palais de la Découverte drew me very early to astronomy. I remember the starry vault traversed by a ballet of zigzagging planets, and sunrises above the silhouettes of Parisian monuments depicted at the base of the planetarium's dome. The stars gradually faded out as triumphant music accompanied the new dawn and the spectators filed out, their dazzled eyes gradually reacclimating to the daylight.

Camille Flammarion's large folio *Popular Astronomy* allowed me at the age of twelve or thirteen to go deeper into what the planetarium had allowed me to glimpse. I lost the book long ago, but I remember its illustrations: photographs of the Moon and the planets, Jupiter and Saturn especially, viewed through telescopes—photos much less precise than the ones space probes have sent us since, but which fascinated me. The book also discussed the great discoveries that had shown man his place in the universe: Tycho Brahe measuring the position of the planets with the naked eye, Copernicus's heliocentric hypothesis, Kepler determining the shape of the orbits and the law of planetary motion, Galileo becoming the first person to point an astronomical telescope at the sky, and Newton explaining by mathematics that he had to invent everything his predecessors had discovered. Applying my mania for classification to the planets, I ordered them according to their sizes, their distance from the sun, and their periods of orbital rotation.

Popular Astronomy also talked about a character less well-known than the illustrious scientists I've just mentioned: a young astronomer named Urbain Le Verrier, who predicted, a hundred years before I was born, the existence of an unknown planet perturbing the orbit of Uranus. He had indicated the region of the sky where astronomers should point their telescopes to discover this planet, which was called Neptune. That one could predict new phenomena by calculation, and that the universe obeys the laws of mathematics, struck me thunderously at the time—and I must say it continues to amaze me today. The book also explained that Le Verrier had been in competition with an English astronomer, John Couch Adams, who also predicted Neptune's existence, though less accurately. This story gave me a first glimpse of a side of research often neglected by youthful idealism: the bitter competition between researchers, sometimes tinged with nationalist rivalry, for the credit of being first to a discovery.

A few years later, in tenth grade or so, I knew enough math to understand Newton's theory of gravitation and to follow the reasoning which, based on the law of universal attraction, leads to elliptical planetary orbits. I was struck by the fact that this law explains both the fall of bodies and the motion of the planets around the Sun. As Flammarion's book suggested, I was able to find the Moon's trajectory around the Earth by calculating the distance it would fall toward our planet in one second if

Figure I.2. *Popular Astronomy* and the scientists whose discoveries it recounted (*from left to right and top to bottom*): Tycho Brahe, Copernicus, Kepler, Galileo, Newton, Le Verrier.

it were not driven in a tangential direction by its orbital motion. This movement, combined with its fall, caused the Moon to plummet continuously without ever reaching the Earth. This explanation was a revelation to me!

My passion for astronomy was amplified by current events. I was in ninth grade in 1957, when the Russians launched the first artificial satellite, Sputnik, kicking off the space race between the USSR and the United States. I was very proud to be able to calculate, using the mathematics I had just learned, Sputnik's speed around the Earth and its period of revolution of roughly an hour and a half. I also calculated the escape velocity, about eleven kilometers per second, to which a rocket had to be accelerated in order to reach the Moon or leave the solar system. My taste for classification and comparison naturally led me to calculate the value of the same parameters for the Moon and the different planets, and to determine how much I would weigh on Mars or Jupiter.

My fascination with astronomy soon intersected with another subject that had impassioned me from a very young age: the history of the exploration of the Earth. I had read about the adventures of Columbus, Magellan, Cook, Bougainville, and Lapérouse. I had been moved by the epic of Captain Scott dying of exhaustion and cold in the Antarctic after losing in the conquest of the South Pole to the Norwegian Amundsen—another story of competition to get there first, this one more tragic than Le Verrier's rivalry with Adams. I had written to

Figure I.3. Sputnik I, the first artificial satellite, launched in 1957 (© NSSDC/NASA).

Paul-Émile Victor, the French explorer of the polar regions, to tell him about my passion, and was proud to receive a postcard with a handwritten note in reply. With the race to conquer the Moon, my two passions—astronomy and exploration—came together.

This is a reconstruction of my impressions and experiences as a high school student at the end of the 1950s. As you will have gathered, I was a good student, curious about science and mad about math. The conquest of space, with its element of adventure, added a touch of romance to my passion for numbers. Being able to use my limited academic knowledge of integral and differential calculus to calculate the motion of the satellites and rockets that the newspapers talked about each day gave me a feeling of enthusiasm and exhilaration that I still remember vividly.

That said, my curiosity and desire to explore and discover the world were nothing out of the ordinary. These are innate traits of children, nurtured in my case by loving, cultured parents and excellent teachers who were often passionate about what they taught me, be it history, literature, or mathematics. My taste for arithmetic, and the pleasure I took in solving algebra or geometry problems, oriented my natural curiosity toward science, and my first inclination was toward astronomy, which seemed to me the natural extension of exploring the Earth. I eagerly followed the Apollo program, which would send men to the Moon, and which I saw as a virtual adventure that could take me to the stars through observation and mathematics. Gradually, even more than Scott and Cook had been, Kepler, Galileo, and Newton became my heroes.

Math and physics programs have changed a great deal since I was a student. Today's high school students no longer know how to find the orbit of a satellite; their calculating skills no longer allow them to directly grasp the fundamental phenomena of classical mechanics. They often learn physics as an object lesson, a qualitative history in which the simple laws of mechanics and the most esoteric results of modern physics are presented on more or less the same footing, as properties of the world with which they should be vaguely familiar without truly understanding them. I wonder what this different approach would have changed for me: would I have been so determined to become a researcher if I hadn't had, so early on, the chance to glimpse directly through mathematics the richness of science, the exaltation of understanding, and the privilege of being able to follow the thinking of giants like Newton or Galileo?

After passing the baccalaureate, I began taking preparatory classes at the Lycée Louis-le-Grand, the typical route to the elite Grandes Écoles. For two years of intensive study, math took priority over physics. I learned the basics of what would later serve me in analysis, differential calculus and vector space algebra. These names, mysterious to the general public, refer to the mathematical methods physicists use daily to calculate the trajectories of classical objects subjected to different types of forces, but also wave propagation, the bizarre behavior of quantum systems, or the statistical properties of a set of particles. I learned, among other things, how to calculate the decimals of π that I had found so dizzying a few years earlier. I continued to enjoy the challenge of solving difficult problems, even if the need to prepare for competitive exams meant I spent more time on typical, often tedious ones.

I also realized, during these years of preparatory classes, that some of my classmates were stronger in pure mathematics than I was. They were better at conceptualizing abstract principles, more interested in the structure of mathematical theories and their axiomatics than in their utility for calculating concrete effects. This was the 1960s, the heyday of Bourbakism, an approach to mathematics named after Nicolas Bourbaki, an imaginary character that a group of francophone mathematicians had invented and waggishly declared their master. The Bourbakism of the 1960s was marked by a strong tendency to formalize mathematics, leading in particular to the systematic introduction of set theory into high school curricula.

I remember discussions in which a friend whom I admired for his intuition told me that the beauty of math lay in its perfect uselessness. I objected, citing the nobility of the physicist who cannot indulge in sterile games but must abide by the constraints of reality and find the mathematical formulas that physics obeys. The physicist, I argued, was the explorer of nature, and mathematics was his flagship. Like any good navigator preparing his voyage, he had to equip himself with the right theoretical background and maintain his mathematical knowledge in order to put it to use in this adventure of discovery. Sailing through formulas for the pleasure of it, as a pure mathematician did, was a gratuitous game, something that wasn't for me. Perhaps this somewhat grandiloquent rationalization was my attempt to console myself for my limitations in math.

My continued education would show me that my friend and I were both naive in our argument, and that it was impossible to separate useless math from useful math. More than once, abstract mathematical theories born of the imagination of pure mathematicians proved essential to the development of powerful physical theories.

Group theory, conceived in 1832 by Évariste Galois to solve algebraic equations, became a fundamental tool a century later for explaining the symmetries phenomena obey, especially in quantum physics.

Another example is vector analysis, which describes the manipulation of vectors defined in an abstract space. These vectors are represented by sequences of numbers—their coordinates in this space, which can have an arbitrary number of dimensions—and their transformations are described by arrays of numbers called *matrices*. These transformations can be rotations, translations, or even homotheties, in which the vectors are lengthened or shortened. Mathematics defines the algebra of these operators, *i.e.* the rules governing their combinations. Generally speaking, the product of two transformations, the result of the successive action of two operators, depends on the order in which the operations are performed. Performing transformation A and then transformation B on a vector gives a result different from the product in the reverse order (B then A).

This noncommutative property can be verified simply in the case of rotational operations in our ordinary space. To illustrate this, place a book flat on a table with the cover facing you. Call Ox the axis that runs along the book's binding in the plane of the table, Oz the axis perpendicular to this plane, and O the point of intersection of these two axes coinciding with the book's lower left corner. Rotate the book 90° around Ox, then the same angle around Oz. It is now standing on its binding. Then return it to its original position

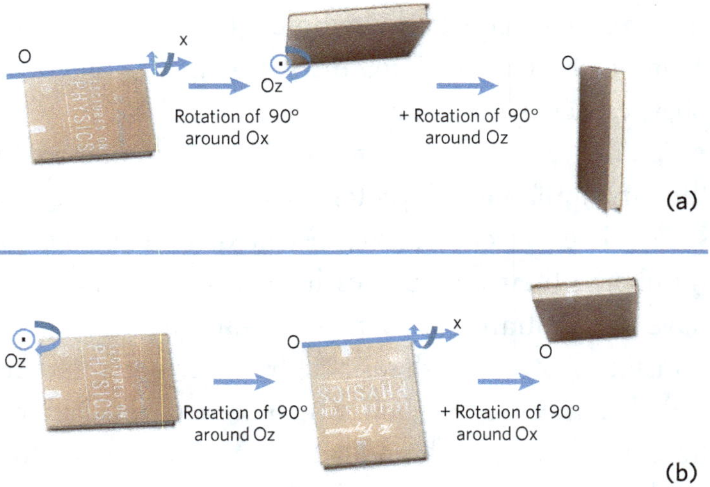

Figure I.4. The product of two rotations is noncommutative. Rotating a book around Ox, then around Oz (a) does not give the same final state as executing the rotations in the reverse order (b). The book used in this demonstration, Richard Feynman's *Lectures on Physics*, was one of my favorite readings when I began my university studies.

and rotate it in the reverse order: first around Oz, then around Ox. Ultimately the book is still standing, but this time on its smaller edge. The result of the product of the two operations therefore depends on their order.

This so-called *noncommutative algebra* is different from the algebra of ordinary numbers, in which multiplication is naturally independent of the order of factors. It was initially developed as an abstract mathematical game on matrices of numbers, and later proved essential in quantum physics for understanding the symmetry properties of quantum systems and the counterintuitive behavior of atoms. Separating mathematics in this way—between abstract games and ideas truly useful for understanding the real world—is difficult, even impossible. What arises as a game in the mind of an imaginative mathematician may later turn out to be a rule of nature. And so we find, at an even more fundamental level, what has fascinated me since my youth: the extraordinary equivalency between mathematics and the laws of physics.

In July 1963, I was accepted to the École Polytechnique (known as X) and the École Normale Supérieure (known as Normale Sup). I chose the latter school without hesitation: it was the natural path to the goal I had set myself, at a time when X was chiefly an engineering school and had not yet developed its research program. And it was in my years at Normale Sup that I truly learned what the life of a researcher is like. This led me to choose a different path from the one I had dreamed of in high school and in my preparatory classes. I would now like to discuss how this change came about, in order to illustrate the important role that chance encounters and simple luck can play in guiding a researcher's career.

Introduction to modern physics

The first year at Normale Sup is a time for recovery and relaxation, after the pressures of preparatory classes and intensive cramming for the Grandes Écoles' entrance exams. This was the year I met Claudine, a psychology and sociology student at the Sorbonne. We've been together ever since. Understandably, I took few courses that year, spending more time in the cinemas and cafés of the Latin Quarter than I did on the benches of lecture halls. I chose to take only two undergraduate courses, one on advanced mathematics for physics and one on the theory of relativity. My inclination toward astronomy required a deeper understanding of what gravitation was all about. I knew Einstein had dethroned Newton half a century earlier with a radically new description of gravitation, revolutionizing the concepts of space and time; I wanted to know more about this mysterious theory which everyone was talking about but none of my friends really understood.

I will speak later about relativity and the counterintuitive ideas it brought to physics; here I will only describe the effect this first contact with modern physics had on me. What I had learned to date was physics as it was known at the end of the nineteenth century, so-called *classical physics*. This included mechanics, the science of the motion of bodies subjected to forces, founded by Newton; electromagnetism, the science of electrical, magnetic, and optical phenomena, crowned a century before my entry into Normale Sup by the work of Maxwell; and thermodynamics, the science of exchanges between work and heat, or between order and disorder, which came into focus in the nineteenth century through the findings of a line of thinkers from Sadi Carnot to Ludwig Boltzmann.

In classical physics, time and space defined a universal and immutable stage on which a perfectly deterministic performance played out. In principle, you could calculate everything that would happen in an experiment if you knew the initial conditions. For simple situations, described by a small number of parameters, determinism could be applied easily and the system's future evolution calculated from knowledge of the present. For systems made up of a large number of particles, such as an atomic or molecular gas, knowledge was limited only by the impossibility of knowing all the parameters (the position and speed of all the particles) at a given moment, and physicists got away with codifying their ignorance thanks to the theory of probability, which allowed them to calculate average quantities accessible to measurement. These quantities were the only relevant parameters for the system concerned.

I had learned, during my final year of high school and in my preparatory classes at Louis-le-Grand, about two powerful theories—relativity and quantum physics, both born less than half a century before me—that had revolutionized those reassuring notions of absolute space and time and of an all-conquering determinism. But all of this remained mysterious, and so in my first year at Normale Sup I decided to tackle the enigma of relativity, which it seemed essential to understand if I wanted to do research in astrophysics.

This course in relativity was a revelation. From a simple principle—the independence of the speed of light from the reference point at which it is measured—everything followed in logical, inevitable fashion. If the speed of light is a constant, uniform for all observers, then it cannot be added to or subtracted from any other speed; therefore the law of composition of speeds, accepted since Galileo's and Newton's time, is no longer valid. In everyday life, for example, that law tells us that if we are driving on a road at speed v_1 and a moving object comes toward us at speed v_2, we see it approaching with speed $v_1 + v_2$. If, on the contrary, it passes us while driving in the same direction, we see it moving away from us at the speed $v_2 - v_1$. To establish this simple law, we accept the seemingly obvious fact that the distances between two points and the time interval between two instants are absolute data, valid for all observers. If the classical composition law for velocities is false when it comes to light, it's because this "obvious" fact is misleading, and we must renounce our intuitive notions of absolute time and space.

Einstein had described this revolution using simple images, the course explained. He had imagined thought experiments in which clocks located on station platforms or aboard trains compared their times by exchanging light signals. From these virtual experiments, we could establish in a few lines the relationships describing how lengths and intervals of time differ depending on whether you're on the platform or on the train. The differences are minimal for train speeds of a few tens or even hundreds of kilometers per hour, and the variation of lengths and times in these ordinary situations is negligible, which salvages Newtonian physics in our daily lives. But this is not the case at very high speeds: at that point, so-called *relativistic corrections* become necessary. More on those later. What fascinated and captivated me at the time was the inevitability of this theory, the fact that you could draw such fundamental conclusions from simple premises, and that, strange as they were, you had no choice but to accept them.

So far I have discussed only what is called *special relativity*. The course went on to sketch the theory of general relativity, which Einstein had developed to extend the concepts of his theory to accelerated motions. Instead of trains running at constant speed, these thought experiments involved rockets accelerating on takeoff or elevators in freefall. The trajectory of objects in these non-uniformly moving reference frames was compared to that of objects in the gravitational field of massive bodies, and from this comparison emerged the idea that masses deformed space itself, giving it a curvature that influenced the trajectory of objects moving in their presence. Thus the problem of gravitation became a problem of geometry in a curved space, a problem the mathematician Bernhard Riemann had treated as one of those "gratuitous" questions I mentioned earlier. Yet another example of the unexpected application to physics of so-called "useless" mathematics!

The equations of special relativity are elegant, with a symmetry reflecting the beauty and simplicity of the ideas that led to them. Starting from the relations that describe the transformations of space and time coordinates when we change our reference point, we can deduce in a few lines of calculation the famous equation $E = mc^2$, which expresses that any mass of matter m can potentially deliver, by annihilating itself, an energy E equal to this mass multiplied by the square of the speed of light c. This formula, probably the most widely known in all of science, has been called the precursor to the atomic bomb and nuclear power plants. Being able to re-derive this formula myself, grasp where it came from, and catch a glimpse of its implications was an extraordinarily exhilarating feeling to me.

The course on relativity was taught by a young professor barely thirty years old, Claude Cohen-Tannoudji, who laid out the physical ideas for us with great clarity, placing them in historical context, showing us the lineage from Galileo, who had established the principles of relativity of motion in mechanics, to Newton, who had drawn their mathematical consequences, and Einstein, who, by applying the same ideas to light phenomena, had revolutionized our concepts of space and time. Starting with the physical principle, Claude described Einstein's thought experiments and lucidly explained each step of the calculations that led to the theory's equations. Finally, he returned to physics to draw together all the consequences, describing the phenomena that proved the validity of the principles of special and general relativity. His clarity, the rigor of his reasoning, and the enthusiasm with which he shared his admiration for the giants of physics whose ideas he was teaching us were contagious. I left his classes with the impression that I had understood everything, even if I had to work hard to absorb all the subtleties.

I even organized a small seminar on relativity for my classmates who were not taking the course but who had been won over by my glowing commentary. I took it upon myself to pass on my newly acquired knowledge. We discussed the paradoxes of the theory and its more or less philosophical implications for our vision of the world. These impromptu lectures that I gave my classmates within Normale Sup's atmosphere of great intellectual freedom gave me a taste for teaching. I liked the challenge of making subtle ideas understandable, of finding the best way to convey a scientific truth and examine all its consequences. This pleasure has stayed with me ever since. The pedagogical effort required to clearly present a scientific question to my audience helps me first clarify it for myself, and has often given me ideas to further deepen my knowledge and follow new directions in my research. Since then I have known that research, for me, cannot be dissociated from teaching.

When I learned that Cohen-Tannoudji was teaching another class at Normale Sup on quantum mechanics, the other mysterious subject I understood only vaguely, I enrolled without hesitation in the graduate (DEA, or *diplôme d'études approfondies*) program where he taught. If he could explain quantum physics as well as he did relativity, I could satisfy my curiosity, no matter if the subject wasn't directly useful to astrophysical research. (On this point I was wrong, as quantum physics has since proven essential for the study of cosmology and the description of the origin of the universe.)

In any case, it was during this year of specialization, my second year at Normale Sup, that my passion for physics truly crystallized and I decided on the direction my research career would take. I owe this to an outstanding and charismatic teacher and his influence on the student I was at the time. The importance of human encounters and relationships in the origins and evolution of a life in research cannot be overstated. Claude Cohen-Tannoudji was my first university professor, then my thesis advisor, and then a reference and a guiding light for the rest of my scientific life.

The DEA course in quantum mechanics largely lived up to its promise. Claude put the same enthusiasm and lucidity into it as he did into teaching relativity. The basic postulates, which had been stated less than forty years prior by the Danish physicist Niels Bohr, were less simple than those that form the basis of relativity. But Claude, who had and shared with us a great admiration for Bohr, presented the quantum postulates with ample authority and clarity, and all their consequences followed easily. Gradually, the microscopic world of atoms revealed itself to us.

Quantum physics was born from the observation of atomic phenomena that could not be explained by nineteenth-century classical physics. Claude began his course by reminding us of the historical evolution of those ideas that had led to the quantum revolution. At first, I thought I was on more or less familiar ground: after all, hadn't I learned in high school and in preparatory classes that the atom was like a small solar system in which the nucleus played the role of the Sun and the electrons that of the planets? And didn't the law of electrical attraction between positively charged nuclei and negative electrons take the same form as the law of gravitation, with force decreasing as the inverse of the square of the distance between the nucleus and the electron? The magnitudes were very different, of course, but surely things were similar, and this analogy would allow us to approach the physics of atoms armed with familiar concepts.

However, this impression was soon to fade away. I had been taught in preparatory classes that the major difference between the classical physics obeyed by planets and that of the electrons in atoms was the quantization of electron orbits: the fact that for a reason that then seemed mysterious to me, the electrons in an atom can gravitate only on certain orbits of well-defined energy. This strange quantization, I had further learned, also applies to light, which is made up of discrete energy packets called *photons*. Each photon of a wave of frequency v carries an energy $E = hv$ where h is the famous constant that the German physicist Max Planck introduced to physics in 1900, symbolically marking the birth of quantum physics.

Atomic electrons can transit from one orbit to another only by instantaneous quantum jumps, emitting or absorbing a photon whose energy is equal to the difference in energies between the initial and final orbits. The frequency of light, inversely proportional to its wavelength, increases continuously—as we learn in middle school—when the color shifts from red toward blue. As a result, atoms emit or absorb only certain well-defined colors, those of photons whose energy is equal to that which the electrons lose or gain in the transition from one discrete level to another.

Flammarion's *Popular Astronomy* had already shown me the absorption spectra observed in the first half of the nineteenth century by the German physicist Joseph von Fraunhofer. Using a diffraction grating of his invention made of parallel thin metallic wires, he scattered white sunlight into a continuous spectrum from red to violet and saw very fine dark lines on the rainbow, corresponding to the frequencies that had been absorbed by atoms present in the solar atmosphere. These lines constitute a kind of universal barcode characterizing the different types of atoms everywhere in the universe. Hydrogen, carbon, and helium have different spectra, and their identification in the light coming from the atmosphere of stars or in interstellar gases attests to the presence of these elements in stellar atmospheres and intersidereal spaces.

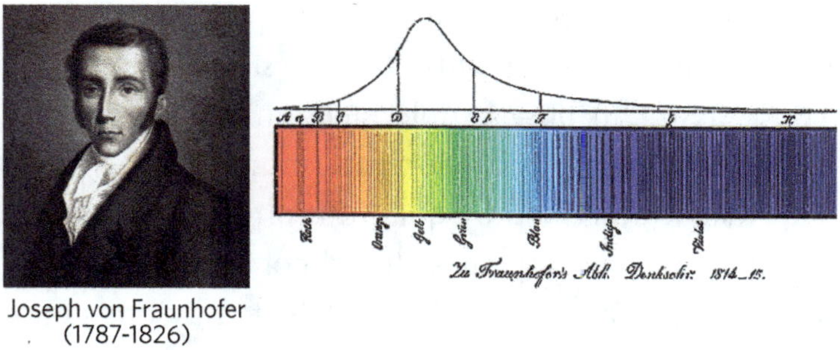

Figure I.5. Joseph von Fraunhofer and the dark atomic absorption lines in the spectrum of the Sun.

I was intrigued and fascinated to learn that light thus carries essential information about the composition of the world, that atoms have the same spectral signatures everywhere, and that we can locate their presence millions of light years away from Earth by observing and classifying spectra measured by instruments attached to our telescopes. But why this was so remained a mystery. What explained the discrete nature of these spectra, and how could we be sure of their universality? By what calculations could they be predicted, and was there a formula for atomic physics as simple and general as relativity's $E = mc^2$, whose demonstration had brought me so much joy the year before? I would get my answers to these questions by following the DEA's quantum mechanics curriculum.

The answers were astonishing, illustrating once again the deep and surprising link between abstract mathematics and real-world physics. Quantum theory assigns each state or configuration of a physical system a mathematical vector in an abstract space whose number of dimensions depends on the system under study. The system can be a simple electron, a molecule, or a solid consisting of an enormous number of particles. The system described by quantum physics can also be immaterial, for example visible light or an invisible electromagnetic wave like the ones detected by our radios or television sets (or modern cell phones, which didn't exist at the time). The quantum vector assigned to a given system has as many coordinates, as many dimensions in the abstract space in which it evolves, as there are possible configurations for the system. The classical three-dimensional space in which physical phenomena occur should not be confused with the abstract space of the quantum world, whose number of dimensions depends on the systems under study. For some systems, this number is finite ($D = 2$ in the simplest case, where the system under study evolves between only two quantum states); for others, the dimension of the state space is infinite, in which case the vector representing the quantum state has a continuous infinity of coordinates.

The theory also introduces operators acting on these vectors, which represent the transformations the system can undergo: its rotations or translations in space, or its evolution in time. These operators generally do not commute (see pages 33–34 for an introduction to this property), and it is this algebraic property of non-commutation that leads to the discretization of the values taken by some of the quantities measured in these systems. We thus understand the quantization of atomic energy states and that of the electromagnetic field itself, comprising components of different frequencies called *field modes*. Each mode has a scale of discrete energy states, separated by equidistant rungs. Each rung corresponds to a precise number of photons in the field mode under consideration.

It is not my intent here to give a detailed description of quantum theory, nor to analyze all of its strange and counterintuitive properties. I will return to some of these aspects later. I am merely trying to relive my state of mind in that 1964–1965 academic year, as I was learning the rules of the quantum world and starting to apply them. The indispensable mathematical tool was the vector algebra of the *Hilbert space*, named after the German mathematician and contemporary of Einstein's who had established its rules. This was the space of states in which any quantum system evolves; I had known this algebra since my preparatory classes, but I could now bring it to life by using it to analyze the phenomena of the atomic world, finally understanding how to calculate the frequencies of the spectrum emitted or absorbed by atoms. Planck's simple formula connecting energies to frequencies, $E = h\nu$, was omnipresent in the theory: this was, so to speak, the quantum physics counterpart of relativity's $E = mc^2$.

Shut up and calculate!

It's worth remembering that the quantum formalism I was learning at the time had been discovered less than forty years prior. I will return in other chapters to the debates between Einstein, Bohr, and other players in the birth of quantum physics. Interpretations of the formalism and attempts to pin baffling quantum ideas to classical conceptions of the world were at first a dominant aspect in the way physicists thought. But soon, beginning in the early 1930s, the scientific community devoted itself almost exclusively to exploiting the formalism in order to understand the properties of the microscopic world and of light. Questions of interpretation, seen as sterile at the time, were relegated to the background. This attitude was justified largely by the success of this practical approach, which quickly revealed the mysteries of the atom and its nucleus as well as those of condensed matter. It also made it possible to elucidate the fundamental properties of the interaction between electrons and photons, leading to the powerful theory of quantum electrodynamics. And during my years at Normale Sup and those immediately after, this pragmatic approach led to field theory and the Standard Model of elementary particles, which to this day represents the most advanced physical theory of nature. The dominant approach was to use quantum theory and not ask too many philosophical questions about its interpretation; those who continued to ask questions—most notably Einstein, then in the United States, and Louis de Broglie in France—were advised to "shut up and calculate," a refrain coined by the American physicist David Mermin.

Despite his admiration for Einstein, Claude Cohen-Tannoudji largely adhered to this pragmatic point of view. Without evading the problems involved in interpreting the theory, which he quickly exposed, he spent most of his lectures giving students the tools to become good calculators of the quantum world. His attitude, I believe, was that it's good to observe that nature obeys laws that may seem strange, that we may be surprised at this, but that there's no point in contesting it and looking elsewhere until you've determined all the consequences of the theory by confronting it with increasingly precise experimental results.

About twenty years later, in the 1980s, attitudes about quantum mechanics changed. Questions about theoretical interpretation returned to the forefront—without, however, undermining the pragmatic approach, which continues to prove its efficacy. The reason for this turn back toward questions of interpretation comes from the fact that the strange aspects of quantum physics, those that are most counterintuitive to us observers of the classical world of macroscopic objects, can now be seen in spectacular fashion through experiments in which we manipulate atoms, molecules, or isolated photons.

Quantum strangeness, long hidden in objects made up of a large number of particles, can now reveal itself before our eyes, so to speak. I will say more later about this resurgence of questions of interpretation in quantum physics, but I am grateful to have learned this physics in Claude Cohen-Tannoudji's DEA course without having to ask them of myself. And I wonder about the current trend of introducing quantum physics to young students by bringing these questions to the fore. I think it is no doubt more efficient to start with "shut up and calculate" before tackling questions that our brains, wired by Darwinian evolution to intuitively understand the world of macroscopic objects rather than that of atoms or photons, may never definitively answer.

When atoms and photons are spinning tops: optical pumping

The quantum mechanics course was not the only one that year to teach us about atoms. Claude's two mentors, Alfred Kastler and Jean Brossel, also taught at Normale Sup. Together, in the 1950s, they had invented the *optical pumping* method: a process that allows the small magnets carried by the atoms of a gas contained in a glass cell to be oriented in the same direction by illuminating them with light. This *atomic magnetization* is intimately linked to the rotational motions of the electric charges contained in the atoms. To explain optical pumping, which played an essential role in my training as a researcher, I must therefore speak of spinning electrons, atoms, and photons.

Rotations are everywhere around us. To begin with the universe, whose study was my first passion, the planets spin on their axes and revolve around the Sun, and the entire solar system is pulled, along with the other stars of our Milky Way, into an overall motion of galactic rotation. The same is true of the billions of other galaxies, grouped in huge clusters, that populate the universe. The eddies of stars in these galaxies often form spiral arms that sprawl over hundreds of thousands of light years. These spiral structures provide direct evidence of the gigantic rotational motions at work in the universe.

Back on Earth, the carriage wheel and the spinning wheel were the first human inventions to exploit the phenomenon of rotation. Modern industrial revolutions have multiplied their use through dynamos, engine turbines, windmills, and so on. Finally, on a completely different scale, quantum physics has revealed to us the importance of the gyratory phenomena hidden in matter and light. These phenomena are exploited today in various devices, some of them based on optical pumping.

An object's rotation is characterized by its frequency, often indicated by the Greek letter ν, which is equal to the number of revolutions it makes per second. The unit of frequency is the *hertz* (Hz), named after a late-nineteenth-century German physicist whom we will meet again. Rotation can also be expressed by its angular velocity. Angles of rotation are generally counted in *radians*, the angle of about 57° that projects from the center of a circle to form an arc equal in length to the circle's radius. By definition, one full circle is 2π radians. The angular velocity—usually written as ω—of an object rotating at frequency ν is therefore equal to $2\pi\nu$ radians per second.

The rotations observed in nature and in the objects of our industrial civilization cover an enormous spectrum of frequencies. A galactic rotation, whose period can extend over billions of years, corresponds to a frequency on the order of 10^{-17} Hz (a fraction of a hertz equal to 1 divided by 100 million billion!). At the opposite extreme, an electron can be classically represented as a particle rotating on its orbit around an atomic nucleus with a typical frequency of 10^{15} Hz (one million billion revolutions per second). Between these two limits, we find the orders of magnitude of the planets' rotation frequencies around the Sun (roughly 3×10^{-8} Hz for the Earth) and their diurnal spinning rotations (one day corresponding to 1/86,400 Hz). Higher up, approximately equidistant (counting in powers of ten) between the frequencies of the galaxies and those of the electrons, we find the frequencies of clock hands, wheels, spinning wheels, windmill and turbine blades—in short, of all the objects our civilization has invented (from 1/3,600 Hz for a watch's minute hand to a few hundred hertz for turbines and sirens).

That these common frequencies fall in the middle of the logarithmic scale of frequencies is no coincidence. It is due, of course, to the value of the unit of time, the second as it has been defined since the Babylonian era, which measures a short interval on the order of a heartbeat, directly perceptible to our senses. It is therefore unsurprising that the frequencies of the phenomena of our daily life are expressed in numbers of hertz neither too small nor too large, while those of cosmic or atomic phenomena, less directly accessible to our senses, are expressed with very high positive or negative powers of ten.

The measure of the speed or frequency of an object's rotation is called *kinematics*. At a more fundamental level, we must understand the laws governing the object's motion: how it can be rotated, how its angular velocity can change, how its energy depends on its velocity, and which laws of conservation its motion obeys. This is no longer a question of kinematics but one of dynamics, a science created by Newton in the seventeenth century. It is suitable for describing the motion of objects at the scale of our daily experience, but had to be modified to account for phenomena at the scale of the universe (taking into account the relativistic curvature of space-time) and that of atoms (for which we must adopt quantum theory). But Newton's classical physics, as it is taught in high school, introduces general concepts of dynamics that must be grasped before we can understand how quantum thinking has modified them.

Let us start, then, by describing the rotation of the macroscopic objects around us. As anyone who has played with a spinning top knows, making a solid spin requires that we apply a force F_t perpendicular to the axis around which it rotates, in the direction tangential to the motion. We can do this, for example, by pulling on a string wound around the top's axis. The efficiency of this tangential force depends on the distance r to the axis of its point of application. We can define the torque exerted by this force as the product of its intensity F_t and the distance r. The effect of this torque on the solid's motion depends on its moment of inertia I, a parameter that describes the solid's resistance to rotation. The more massive the object is and the farther its mass is distributed from the axis of rotation, the greater the moment of inertia and the greater the torque that must be applied to give it a given spinning frequency.

To define the "rotational quantity" contained in a body rotating around an axis, Newtonian dynamics introduces the concept of *angular momentum*, a vector quantity aligned along the axis of rotation whose modulus $L = I\omega$ is the product of the moment of inertia and the angular velocity ω. For rotations, this angular momentum is the equivalent of the momentum $p = mv$ which measures the "quantity of motion" associated with the translation of a solid of mass m moving in space at velocity v. Just as the change in momentum per unit time of a particle is equal to the force applied to it (which expresses the fundamental law of Newtonian mechanics), so the change per unit time of a rotating body's angular momentum is equal to the torque applied to it.

In the absence of external influence (*i.e.*, applied torque), the angular momentum is invariant. This is the generalization to rotational dynamics of Newton's law of inertia, which says that, in the absence of force, a moving body continues its course at uniform speed, with no acceleration. This law of conservation

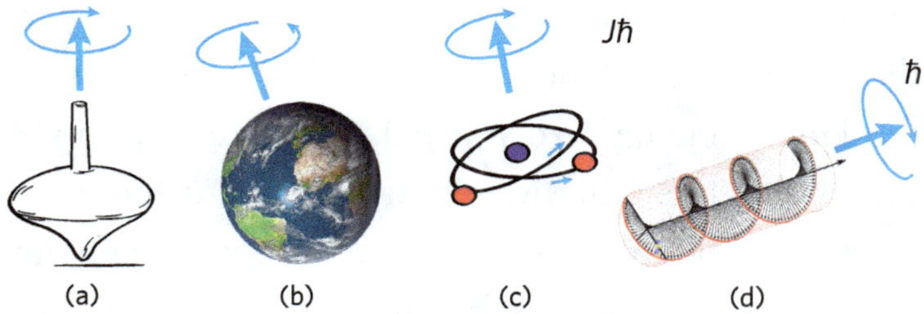

Figure I.6. Various angular momenta: (a) a spinning top rotating on a vertical axis; (b) the Earth rotating on its axis inclined on the plane of its orbit; (c) atoms consisting of moving charges whose total angular momentum $J\hbar$ results from the combination of orbital angular momenta and spins of electrons and nucleus; (d) a circularly polarized photon of angular momentum $\hbar\omega$, the quantum of a light wave whose electric field tip describes a helical motion around the direction of light propagation.

of angular momentum explains, for example, why a swiveling skater who suddenly brings her arms closer to her body will begin to turn faster: she decreases her moment of inertia, and since the product of this quantity and her angular velocity must remain constant, the spinning frequency increases. On an astronomical scale, this same law of conservation explains why the period of the Earth's diurnal rotation remains constant over time.

The connection between the dynamics of rotation and that of translation is found in the expression of their respective kinetic energies. A mass with linear velocity v carries the energy $E_c = mv^2/2 = pv/2$. Similarly, a rotating solid with moment of inertia I has a rotational kinetic energy $E_c = I\omega^2/2 = L\omega/2$. This energy, proportional to the square of the angular velocity, is returned to the environment when the solid stops spinning. In the case of the rotation of a car wheel, for example, the kinetic energy is dissipated as heat on the brake drums that block the wheel, or partially, if the car is electric, as energy recharging the battery. The energy stored in a rotating body is equal to half of its angular momentum multiplied by its angular velocity.

If a rotating object carries electric charges, its rotation is accompanied by magnetic phenomena. A charge rotating around an axis generates a magnetic field whose spatial distribution is the same as that of a magnet aligned along that axis. The intensity of this magnetization is measured by the *magnetic moment* of the rotating object, a quantity proportional to its angular momentum (the ratio of the two is called the *gyromagnetic ratio*). The greater the electrical charge and the faster it rotates, the greater the magnetic moment and its associated magnetization. The universally known example of this correspondence between rotation and magnetism is the Earth, which has an intrinsic magnetic moment roughly aligned with its polar axis of rotation. The Earth's magnetic field results from this. Its origin lies in hydrodynamic phenomena in the underground magma, which contains moving charges. The situation is more complex than that of a rotating point charge, but the general idea is the same: the rotational and magnetic properties of bodies are intimately linked.

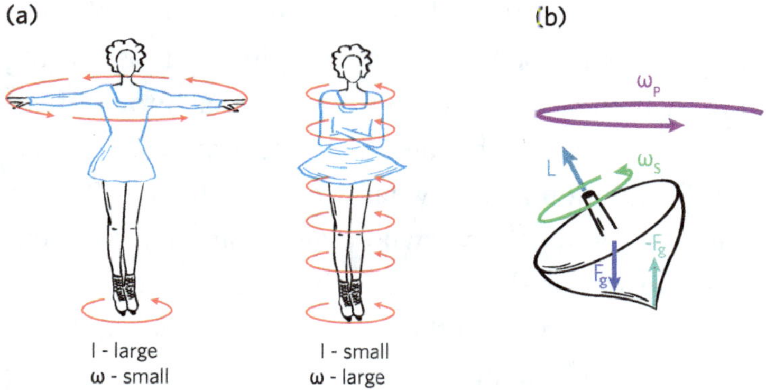

Figure I.7. Conservation of angular momentum: (a) a skater spins faster by bringing her arms closer to her body. Since no torque is exerted on her, her angular momentum $L = I\omega$ remains constant, ω increasing as I decreases; (b) a spinning top rotating with angular velocity ω_S on an inclined axis precesses around the vertical direction (precessional angular velocity ω_P). The weight F_g and ground reaction $-F_g$, two opposing forces applied on the spinning top's axis of rotation, exert zero torque with respect to this axis: ω_S is therefore constant. The vertical component of the top's angular momentum L is also invariant because the torque of the forces F_g and $-F_g$ is zero with respect to the vertical. The modulus of L and its projection on the vertical being preserved, the axis of the spinning top rotates on a cone, keeping a constant angle of inclination with the vertical.

These classical phenomena can be generalized to atoms, with some modifications imposed by quantum physics. The state of an atom is not only defined by its internal energy (that of the electrons orbiting the nucleus and that contained inside the nucleus as a result of nuclear forces). The atom possesses an angular momentum as well, a combination of the angular momenta associated with the electrons' orbital rotations around the nucleus and the intrinsic rotations of these particles, which behave like small spinning tops twirling around themselves. The intrinsic angular momenta of the electrons and of the atomic nucleus are called *spins*.

This situation is somewhat similar to that of the solar system described above, whose global angular momentum consists of three terms: the spinning motion of the Sun (analogous to the spin of the atomic nucleus); the angular momenta of the planets' diurnal spinning motions (analogous to the electrons' spins); and the planets' orbital rotations around the Sun (analogous to the electrons' orbital angular momenta). As in classical physics, all the components of atomic angular momentum have associated magnetic moments. Quantum mechanics describes how these moments interact with each other, how their presence slightly modifies the atom's energy levels, and how they combine to form a global angular momentum associated with an overall atomic magnetic moment.

As with energy, the angular momenta of the electrons, that of the nucleus, and the atom's overall angular momentum are quantized and vary in discrete increments. If we wish to make a classical representation, we can view a quantum angular momentum as a vector in ordinary space whose components along the three spatial coordinate axes are expressed in units of $h/2\pi$, where h is Planck's constant. We use a crossed-out h ($\hbar$, pronounced "h-bar") to indicate the quantum of angular momentum $h/2\pi$. Thus, to describe quantum angular momentum, we find the constant that links the energy absorbed or emitted by an atom to the frequency of the photon that brings or takes away that energy.

We can intuit this result qualitatively if we observe that angular momentum is classically expressed as an energy divided by an angular velocity. If the energy of an atom varies in discrete packets $h\nu = \hbar\omega$ when an electron jumps from one orbit to another while its angular velocity is on the order of ω, we understand that the angular momentum changes in increments on the order of $\hbar$, the result of dividing $\hbar\omega$ by ω.

This intuitive argument is not, of course, a demonstration. The rigorous quantum theory that Claude taught me in 1964 assigns to the spatial components of angular momentum operators that describe how the quantum states of a system are transformed in Hilbert space when the system is subjected to rotations in ordinary space. The non-commutation of these operations, described above, leads to the non-commutation of the operators describing the components of angular momentum along the three spatial coordinate axes. In a few lines of computation, Claude showed us that this non-commutation accounts for the quantization of the values, by jumps of $\hbar$ that an atom's angular momentum or an electron's spin can take on when measured in a given direction.

The simplest angular momentum is that of the electron's spin, which can take on only two values, $+\hbar/2$ or $-\hbar/2$, when its component is measured along any axis. The spin is aligned either parallel or antiparallel to this axis, and is symbolically represented as a small vector pointing up or down (*spin up* or *spin down*). This is called a *spin-1/2*. This property is shared by certain nuclei, such as that of the hydrogen atom, whose proton also carries a spin-1/2. The evolution of these two-state quantum systems is described in a two-dimensional Hilbert space. The study of their dynamics in a magnetic field is thus a very simple quantum problem, which exemplifies and models many situations where the system under study evolves essentially between two states.

The value of an atom's global angular momentum depends on the configuration of the electron orbitals and the spin of the nucleus. More precisely, the component of an atom's angular momentum along an arbitrary axis can take the $2J+1$ equidistant values between $J\hbar$ and $-J\hbar$, these two extreme values corresponding to cases where the global angular momentum is exactly parallel or antiparallel to the axis in question. The intermediate values between the extremes correspond classically to cases where the angular momentum is oriented transversely, in a direction making a greater or lesser angle with the axis. The value of J, which can be an integer or a half-integer (1/2, 1, 3/2, 2, etc.), depends on the atom under consideration and on the energy level of its electrons (lowest-energy ground state or higher-energy excited states).

In a magnetic solid, the moments of the different atoms are aligned with each other and the whole solid has a macroscopic magnetic moment. This is how the magnetism of some materials, for example iron oxides such as magnetite, can be explained as a function of the properties of the atoms that constitute them. A heated magnet loses its magnetism at a certain temperature when thermal agitation overcomes the forces that tend to align the moments of neighboring atoms. In a diluted gas, the interatomic forces are in a first approximation negligible because of the relatively large average distance between the atoms, and thermal agitation causes the angular momentum components of the atoms to be nearly equally distributed among all possible values. In this case the gas's overall angular momentum and magnetic moment are zero, even though each atom in the medium carries an individual magnetic moment.

Light, too, can carry angular momentum. The electric and magnetic fields associated with a light beam oscillate in a plane perpendicular to its direction of propagation. When light is circularly polarized—this is achieved by passing it through a transparent material plate with special optical properties—the emerging electric field rotates at the frequency of the wave, in a plane transverse to the direction of the beam. This rotational motion lends angular momentum to the field. Each of its photons, of energy $\hbar\omega$, then carries a quantity $+\hbar$ or $-\hbar$ (depending on whether the electric field rotates clockwise or counterclockwise) of angular momentum aligned along the light beam. The spin of the photon, twice as large as that of the electron, is said to be equal to 1.

We can now understand qualitatively what happens when an atomic gas interacts with a circularly polarized beam of light. If its photons have the energy $h\nu$ corresponding to the energy interval between the atom's ground state and an excited state, the light will cause the atoms to undergo repeated cycles of absorption and emission, obeying the principle of conservation of angular momentum. When an atom absorbs a photon from the incident field and is brought to an excited state, the overall angular momentum of matter and field is conserved and the atom's angular momentum increases by $\hbar$ while that of the light, which loses a photon, decreases by the same amount. The atom takes in one unit of angular momentum borrowed from the light.

After absorption, the atom returns to its ground state by spontaneously emitting a photon in a random direction. The angular momentum of this photon along the direction of propagation of the incident light is generally smaller than $\hbar$; it is equal to that value only in the event that the re-emitted photon is found to have the same direction and the same polarization as the absorbed photon. The outcome of the exchange of angular momentum between the atom and the field after an absorption followed by an emission cycle is thus, on average, favorable to the atom. The field loses angular momentum along its direction of propagation and the atom gains it.

After a few cycles of this process of light absorption followed by emission, the atoms find themselves in the state of maximum angular momentum (and therefore maximum magnetic moment), in which all of the moments are oriented in the direction of light propagation. The

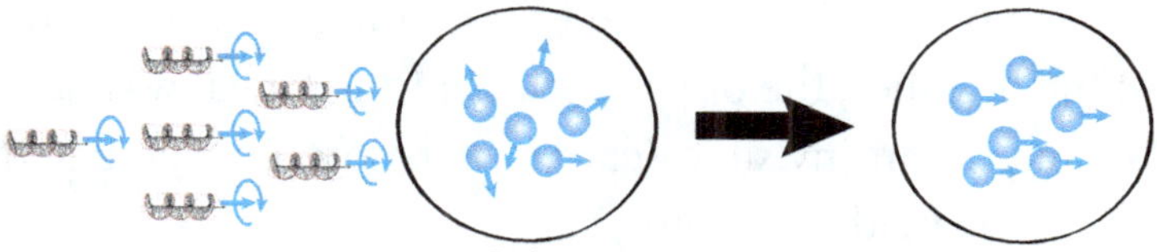

Figure I.8. Optical pumping. The exchange of angular momentum between atoms and circularly polarized light orients the magnetic moments of all atoms in the direction of the light beam, by virtue of the conservation of the total angular momentum of the atom-light system: the gas whose atomic magnetic moments are initially randomly distributed (*before*) becomes a magnetized sample (*after*).

atomic sample, optically "pumped" into this state out of thermodynamic equilibrium, becomes like a macroscopic magnet whose atoms all have their magnetic moment pointing in the same direction.

The light beam that performs the pumping can be used at the same time to detect the atoms' orientation. When pumped in the maximum component state of angular momentum, the atoms can no longer absorb light and the gas becomes transparent. The amount of light transmitted by the pumping beam thus serves as a measure of the gas's orientation state.

I have described the principle of optical pumping in detail because it played an essential role in guiding my vocation as a researcher. I remember as though they were yesterday the lectures of Alfred Kastler, who made us visualize the dance of atomic angular momenta illuminated by light. He gave vivid, sometimes poetic descriptions of atoms and their interactions with photons. We could tell he had in him something of the dreamer, the one who had intuited what was happening on an invisible scale long before he calculated its effects mathematically.

Kastler was, in some ways, a scientist of the old, classical world. He had learned physics in an era that preceded the emergence of the quantum theory to which Claude was exposing us in his lectures. Kastler, born in pre-war Prussian Alsace, told us about his studies in German, and how he had been taught the rudiments of what was then called quantum mechanics from the famous volume by the German physicist Arnold Sommerfeld, *Atombau und Spektrallinien* (*Atomic Structure and Spectral Lines*). In this book, written in 1919—six years before the advent of modern quantum physics—Sommerfeld described atomic angular momenta and their evolutions using simple images, representing them as vectors pointing in space, not unlike the way I have tried to explain them here. These images were how Kastler, still clinging to classical physics, saw these phenomena as well. The quantum theory that Claude was teaching us in parallel broke free from this necessarily approximate vision and described the components of angular momentum as operators acting in the abstract space of quantum states, operators whose non-commutation accounted for all the quantum properties I have described above.

To see the world as something rich and strange

Taking Claude's and Kastler's courses simultaneously, I understood that mathematics and formalism were not everything. Discovering or inventing something new would require more than mathematical rigor, more than the ability to solve equations. It required an intuition that went further, an imagination that made it possible to dream about what went on in a world not directly accessible to the senses. Kastler had that imagination. It let him invent a new way of seeing the interaction between matter and light, opening up fields of research that were inconceivable to us at the time.

After Kastler, the poet and dreamer, I must mention Jean Brossel, the man of details with a passion for precision. Brossel was Kastler's collaborator, the one who tempered the dreamer's ideas with reality, creating the glass cells to contain the optically pumped atomic gases. Their paraffin coatings, which only Brossel knew how to apply to the glass cell walls, ensured that the atoms' oriented magnetic moments did not lose their direction after bouncing off of the walls. This property was essential for some gases to be oriented efficiently by light. Brossel also filled the small glass lamps with atoms that, under the effect of an electric discharge, emitted the photons that illuminated the pumping cells. Juggling with isotopic shifts, the fact that atoms of the same element but whose nuclei contain different numbers of neutrons radiate at slightly different frequencies, he was able to produce light sources that resonated exactly with the atoms in the cells, or emitted photons slightly out of resonance.

Brossel taught us all of this in his DEA course. In a physics experiment, he showed us, the devil is in the details. An idea, even a great one, is not enough; to implement it, you must also consider all the real-life factors that complicate the ideal scheme and disrupt the conditions of the thought experiment. Only when these disturbances are eliminated or compensated for does nature allow us to see in the laboratory that which intuition initially imagined. Anticipating all the difficulties and finding ways to avoid them often requires an imagination and creativity as great as those which led to the idea of the experiment in the first place.

Taking Kastler's and Brossel's courses, I could see this essential complementarity in experimental research: having an idea and knowing how to implement it in the lab. These two sides of the researcher's work were illustrated beautifully in the comparison between these two scholars' classes.

Brossel didn't stop at teaching us about optical pumping. He also described the great experiments in atomic physics of the previous half-century, which had increasingly validated quantum concepts and established quantum electrodynamics, the theory that explains with extraordinary precision the phenomena of interaction between atoms and light. I remember his description of an experiment by Otto Stern and Walther Gerlach, two German physicists who discovered the electron spin in 1922: a remarkable example of an unexpected finding, heralding numerous applications and inventions that have had a huge impact on modern technology.

Intending to measure the magnetic moment of silver atoms, Stern and Gerlach had experimented with a rudimentary atomic beam propagating in a vacuum, sending the atoms from a furnace through the air gap of a magnet. When they collected the atoms on a glass plate after passing them through the magnet, they expected to see an oblong mark caused by the deflection of the atoms at different angles, based on the orientation of the small magnets they carried. Instead of this single mark, Stern and Gerlach were surprised to find a double trace, the first direct manifestation of the spatial quantization of atomic angular momenta.

An analysis of the experiment showed that this angular momentum was due not to the orbital rotation of the silver atom's external electron but to its intrinsic spin, which, pointing upward or downward, sent the atoms on two distinct trajectories inside the magnet's air gap—explaining the two marks observed. Brossel described this experiment in detail and showed us how it led Stern's student Isidor Rabi to refine the atomic and molecular beam method and develop the magnetic resonance method for precisely measuring the magnetic moments of many atomic nuclei. These historic experiments laid the groundwork for all of the advances in physics that would lead to the invention of atomic clocks, lasers, magnetic resonance imaging (MRI) machines, and many other devices.

Brossel also spent some time in his course describing the 1947 experiment by the American Willis Lamb, a student of Rabi's, who measured the shift of atomic energy levels that now bears his name: the *Lamb shift*. I must now say a few words about this famous shift and the important role it played in atomic physics. Since the 1920s, quantum mechanics had given physicists a mathematical tool to accurately calculate the energies of atoms in their ground state as well as in states corresponding to excited electronic orbits. The theory's accuracy was first tested on the simplest atom, hydrogen, which has only one electron. The dominant effect, the electrical attraction of the electron by the proton, treated in a non-relativistic way in this first approach, gives a good approximation of the atom's energy levels in its ground and excited states. For this we use either Schrödinger's equation or Heisenberg's, which present non-relativistic quantum mechanics in two different but completely equivalent ways.

We must then take into account the magnetic effects mentioned above, the coupling of the electron spin with its orbital motion, which shifts the energy levels via an effect called *fine structure*. A complete description of these effects must account for relativity, which provides a coherent framework incorporating all magnetic phenomena. Relativity also introduces, even in the absence of spin, orbital corrections related to the fact that one cannot fully neglect the electron's velocity on its orbit compared to the speed of light. This velocity *v* is roughly a hundred times smaller than *c*, which produces relativistic corrections of a few parts per ten thousand, of the second order in *v/c*, which are on the same order as the magnetic corrections. As it turns out, this coincidence does not occur by chance: magnetic phenomena are in fact relativistic effects, and all of these corrections appear together once we no longer neglect the electronic velocities of the objects under study in comparison with the speed of light *c*. When we incorporate all of these corrections, as the English physicist Paul Dirac did only one year after Schrödinger and Heisenberg, we can calculate with great precision the energy levels of hydrogen. And we find that the first two excited levels, which have an electronic angular momentum of 1/2 (noted $S_{1/2}$ and $P_{1/2}$ in spectroscopy jargon), should have exactly the same energy—should be *degenerate*, as we say.

Now, Lamb's experiment showed that these two levels were separated by a small energy gap on the order of one ten-millionth of the gap separating these first two excited states from the ground state. This very delicate spectroscopic experiment consisted in irradiating a hydrogen beam with a radiofrequency field resonating with the transition between the two levels and detecting the transfer from one to the other by exploiting the fact that they fall back toward the ground level at very different rates. Brossel explained all of this to us, detailing the problems Lamb had encountered and how he solved them.

This experiment was crucial in that it revealed the importance of the vacuum in quantum electrodynamics, and in physics more generally. What Dirac had neglected in his equation was that all particles in the universe, in particular the electron in the hydrogen atom, are bathed in a vacuum traversed by quantum fluctuations. Even in the absence of light sources, the field in space cannot be zero. There are small fluctuations, such as radiation wavelets that ultimately result from the non-commutation of the quantum operators describing the electromagnetic field. These fluctuations can be described as the appearance of so-called *virtual photons* that constantly pop up and disappear randomly in empty space. These photons can in turn create pairs of an electron and its antiparticle, the positron, whose existence Dirac's theory had also predicted. These electron-positron pairs annihilate after a very short time, but their fleeting appearance with the virtual photons populates the quantum vacuum and changes the dynamics of the electron orbiting around the atom's nucleus.

These vacuum fluctuations perturb the two states $S_{1/2}$ and $P_{1/2}$ of hydrogen in different ways, lifting the degeneracy between them. Lamb's measurement, in perfect harmony with the theory of quantum electrodynamics, played a crucial role in the latter's acceptance in the 1940s and 1950s. Physics has since shown us that alongside the electromagnetic vacuum there are vacuums of other quantum fields associated with particles besides photons, electrons, and positrons.

Lamb's experiment, as described in Brossel's course, taught me another unforgettable lesson, one that reminded me of what I had read several years prior in Camille Flammarion's *Popular Astronomy*: that measuring with ever greater precision can lead to a great discovery. This is, after all, what allowed Le Verrier to discover Neptune. The precise calculation of Uranus's orbit had taken into account the disturbance brought to its motion by the presence of the other known planets, primarily Saturn and Jupiter, but there remained a small difference between the observed orbit and the orbit calculated by taking these perturbations into account. This minute difference was caused by the presence of a new, theretofore unknown planet: Neptune. So it was for the *Lamb shift*. Once all known magnetic and relativistic perturbations of the hydrogen spectrum had been accounted for, there remained a small difference that led to the discovery of the physical effect of the vacuum. Accuracy in measurement is not merely a frivolous obsession among researchers, a compulsion driving them to add more and more decimals to the quantities they measure. What motivates this obsession is that there is always, in the quest for precision, the possibility of discovering something unexpected, and sometimes of fundamental importance.

Wherever I turned my attention during this second year at Normale Sup, I saw the importance of quantum physics. This was especially apparent in Brossel's and Kastler's classes, which focused on the properties of isolated atoms that could be studied by emerging experimental techniques—but it was also present in my other DEA courses. I remember in particular Pierre-Gilles de Gennes's class, which focused chiefly on superconductivity, the property certain metals have at very low temperatures whereby they can conduct electricity without any loss. This effect, discovered in 1911, had not been explained theoretically until a few years before my arrival at Normale Sup in 1957. The theory accounted for the collective properties of the electrons of these metals that, below a critical temperature, behave like a fluid that flows without resistance. Quantum mechanics is essential to understand how this works. De Gennes told us about the Josephson effect, a strange phenomenon of electron flow through a thin insulating wall separating two superconducting samples.

Here again, quantum physics could be felt vividly. De Gennes said that Josephson, who had just given his name to this effect, was a young English doctoral student about our age. I still knew nothing about the true nature of research, but I could glimpse the possibility for even a young person, still in learning mode, to contribute to the advancement of knowledge. This made me all the more impatient to begin real research work.

With this in mind, I turned to Claude, who had just created his own group in Kastler's and Brossel's lab. It was his courses in relativity and quantum mechanics that had impressed me most. They didn't cover astronomy or astrophysics, the fields that had initially oriented me toward science, but I had realized that physics is an intrinsic whole, that the problems encountered in studying the infinitely large and the infinitesimally small have many points in common, and that what had attracted me to astrophysics, the possibility of applying mathematics to the understanding of the world, was no less true in the exploration of the microscopic world.

Moreover, atomic physics, as Kastler and Brossel described it, had the advantage of being carried out in a laboratory on a human scale, without the need for large and distant instruments. The atoms and their dance within magnetic fields were within reach, so to speak, and great discoveries could be expected. I am trying to rationalize my decision, but the truth is there was no hesitation. This choice seemed obvious to me, and in the fall of 1965 I began a master's thesis under Claude's direction.

I had a glass cell, prepared by Brossel, containing a drop of liquid mercury. Heating the cell released a gas of mercury atoms into it. Like the proton, the chosen isotope, mercury 199, has a nucleus with a nuclear spin-1/2, which constitutes the atom's global angular momentum. In this atom's ground state, the contributions from electronic magnetism are canceled out exactly by the quantum rules governing the combination of angular momenta. I was thus able to experiment with a very simple system, a set of spins-1/2 that were practically independent of each other and which I could submit to the light of the mercury lamp (also made by Brossel) that emitted by construction at the wavelength required to excite the atoms of my gas. Very quickly, I observed the optical pumping signals Kastler and Brossel had described in their lectures and was able to begin experimenting myself.

I won't describe my master's thesis experiments here, but I would like to evoke one of them by way of describing the impression they made on me and my state of mind when I observed an atomic phenomenon by myself for the first time. The principle of this experiment is very simple: once the spins were oriented by optical pumping, I abruptly applied a magnetic field perpendicular to that orientation. One might naively imagine that, like compasses, the small atomic magnets would detect the field and align themselves along its direction by rotating 90°. I knew enough physics, however, to predict that the spins' evolution would be different, somehow counterintuitive.

Indeed, each atom is not only a magnet but also a small spinning top, and the effect of a force on a top defies naive intuition. If a top rotates around an axis that forms an angle with the vertical, we observe that instead of falling under the effect of gravity and tilting more and more, it starts—in addition to its rapid rotation around its axis—to revolve slower around the vertical while maintaining a constant angle with this direction. This is one consequence of the conservation of the vertical component of angular momentum. The force of gravity and the reaction of the ground exert a zero torque with respect to the vertical and the projection of the angular momentum along this direction must remain constant, leading to the precession motion of the spinning top's axis.

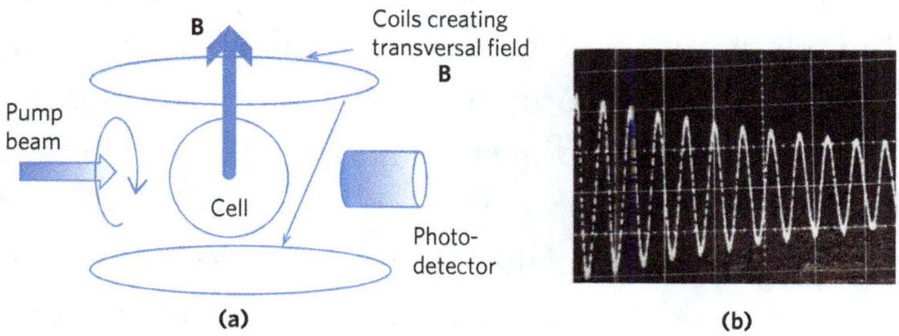

Figure I.9. Larmor precession experiment on atoms in a transverse magnetic field B. (a) Schematic diagram of the experiment: the resonance cell containing the mercury vapor is optically pumped by the horizontal beam of circularly polarized light. The orientation of the atoms is detected by a photodetector measuring the intensity of the light transmitted by the cell. A vertical magnetic field B, generated by current coils, is applied abruptly to the atoms. (b) Oscilloscope signal revealing the Larmor precession of the atomic magnetization (one square represents 0.2 seconds).

A similar effect occurs on the small magnetic spinning tops that we call atoms. Instead of aligning themselves with the field, they start to rotate around it at a frequency proportional to the field amplitude. This is called the *Larmor precession*, after the physicist who first explained it. The magnetization component of the gas in the direction transverse to the field along the optical pumping beam thus begins to oscillate, resulting in a modulation of the intensity of the light transmitted by the sample, which I detected easily. This oscillation was gradually damped over time, as the collisions between the atoms and the cell walls destroyed the sample's initial orientation.

This was one of the first experiments I performed. In itself, it was nothing novel and did not deserve publication—but it gave me the chance to realize for the first time that I could steer atoms. Even if I couldn't see them with my own eyes, by observing the oscillation that retraced the spins' rotation on the screen of my oscilloscope, I could be sure that very close to me, invisible to the eye but undoubtedly present, the atoms were doing what I had predicted. That green trace on the screen, which I could repeat as many times as I wanted, whose frequency I could change at will by modifying the amplitude of the magnetic field, fascinated me.

Many years later I found, in the Nobel Lecture given by Edward Purcell in 1952, a sentence that expresses what I felt then better than I ever could. Purcell, along with Felix Bloch, discovered the phenomenon of nuclear magnetic resonance in solids and liquids, known as NMR, which led to the invention of magnetic resonance imaging. My experiment on mercury was nothing more than an NMR experiment performed in a diluted gas, not in a solid or liquid. The nucleus rotating in my experiment was that of mercury, not the proton of the hydrogen atom whose precession Purcell had observed for the first time in 1945, but the physics was the same. To express what he experienced at the time, Purcell wrote:

> Commonplace as such experiments have become in our laboratories, I have not yet lost a feeling of wonder, and of delight, that this delicate motion [of atomic nuclei] should

> reside in all the ordinary things around us, revealing itself only to him who looks for it. I remember, in the winter of our first experiments, just seven years ago, looking on snow with new eyes. There the snow lay around my doorstep—great heaps of protons quietly precessing in the earth's magnetic field. To see the world for a moment as something rich and strange is the private reward of many a discovery.

I had done only a simple demonstration, but I felt the same wonder. It seemed to me that I had touched a deep and hidden truth. Atoms were real, albeit invisible in the cell, ready to lend themselves to the experiments I would attempt. There were billions of atoms spinning together, but as they all evolved in concert, the signal I observed revealed the behavior of each one, a microscopic entity spinning in the magnetic field independently of all the others. It would take another thirty years before I could truly manipulate and observe individual atoms or photons, but already I felt I was at the beginning of a great adventure.

An apprenticeship in trust and freedom

I had just started my fourth year at Normale Sup in 1966 when it was announced that Kastler had been awarded the Nobel Prize for the invention of optical pumping. It was an October day, an early snow was falling on Paris, and the revolution in the laboratory was instantaneous: journalists flocked, champagne flowed, photos were shot in the lab rooms with researchers, students, technicians, and the lab secretary all sharing in the same joy. I felt immense pleasure and thought how lucky I was to have found this laboratory, to be working in the field of research that had just received such worldwide recognition, alongside those who had made the discovery being consecrated.

There was just one hitch: the Nobel Committee had rewarded Kastler the dreamer, who had thought up the method, but not Brossel, the lover of details with a passion for precision, who had made the dream possible. The photos from that day show Brossel smiling but somewhat absent. I can only imagine the disappointment that mingled with his joy, the bittersweet taste it gave the day. Kastler himself has repeatedly expressed regret that he did not share the prize with his colleague, but Brossel never showed the slightest bitterness. All of us in the lab asked ourselves what the reason was for such an omission; more than fifty years later, I still wonder.

This Nobel Prize underscored the importance of optical pumping, the first method to use light to manipulate atoms, to make them evolve in a precise and controlled manner, with new possibilities for applications. At the time it was the angular momenta of the atoms that were at stake, but later, as we will see, the method would be extended to the external variables of motion, the velocities of atoms. Light would be used to cool, trap, and manipulate them.

Figure I.10. At the ENS Spectroscopy Laboratory on the day Kastler's Nobel Prize was announced in October 1966. *Left to right*: Franck Laloë, Claude Cohen-Tannoudji, Alfred Kastler, me, Jean Brossel, and Alain Omont, a young researcher a few years older than me.

Kastler's Nobel Prize marked the rebirth of postwar French research. It was the first Nobel awarded to a French physicist since Louis de Broglie's in 1929. The creation of the National Center for Scientific Research (CNRS) in 1939 and, especially after the war, the research development initiated by the government of the Fourth Republic at the particular instigation of Pierre Mendès France, and then by the nascent Fifth Republic at the impetus of General de Gaulle, had fostered high-quality, internationally competitive research in France. In the years that followed, seven other Nobel Prizes in Physics would be awarded to

French researchers. That said, these awards, which testify to the efforts made in support of research from the 1960s through the 1980s, should not distract us from the crisis faced by research in France today, on which more later.

As a young student, I enjoyed exceptional working conditions, quite different from those that greet a beginning researcher in France today. I worked under the supervision of a young, enthusiastic mentor who didn't have to worry constantly about seeking funding for his research or justifying it in terms of its potential utility. Instead of writing proposals and reports, Claude was able to devote all his time to doing research, mentoring young students, and writing the articles that shared our results. I was learning from a master whose mind was wholly invested in research and teaching, something I would surely not have received from a thesis director today, beset as he would be by administrative tasks, heavy teaching loads, and all the constraints of an entrepreneur constantly preoccupied with the need to ensure funding for his team. And I had no trouble, in 1967, before I had finished my doctorate, getting hired as a researcher by the CNRS, securing my career and my ability to devote myself freely to that which excited me.

I remember fondly the atmosphere that prevailed in the Kastler and Brossel lab at the time, the freedom afforded to young researchers to choose their research subject from among the infinite possibilities that the ideas of optical pumping opened toward a deeper knowledge of atoms and the interaction between matter and radiation. Once we had proven ourselves by passing exams and demonstrating our passion for research, we were trusted. If we had an idea that seemed promising, all we had to do was present it succinctly to Brossel and he would give us the means to carry it out, sparing us all the red tape and bureaucracy that dog the work of researchers today. In spite of all the vicissitudes of the present, this spirit has survived at least partially in what is now called the Kastler Brossel Laboratory, or LKB. It was a privilege to be able to carry out my life's work there. I know how lucky I was to have chosen it in the first place, simply because a course taught by an enthusiastic young professor, in which I had enrolled almost by chance, attracted me to it.

Promises of the laser

I was also lucky to have come to research shortly after the advent of the laser. This extraordinary light source, which appeared in laboratories in 1960, would open up immense possibilities in the fundamental and applied sciences. Today we are familiar with its countless uses in our daily lives, from reading CDs and DVDs to fiber-optic communication and the internet, from the ultra-precise cutting of materials to ocular surgery, from scanning barcodes at store checkout to the rangefinders used on all construction sites. But the general public knows less about the fundamental role the laser has played over the last half-century in basic physics, chemistry, biology, and astronomy research.

The first lasers were beginning to appear in laboratories during my years of training at Normale Sup. I remember the amazement caused by these thin beams of intense blue, red, or green light that propagated over long distances without diverging. Their impact on a white wall or a sheet of paper would form a shimmering spot of light with strange streaks never produced by light from conventional lamps.

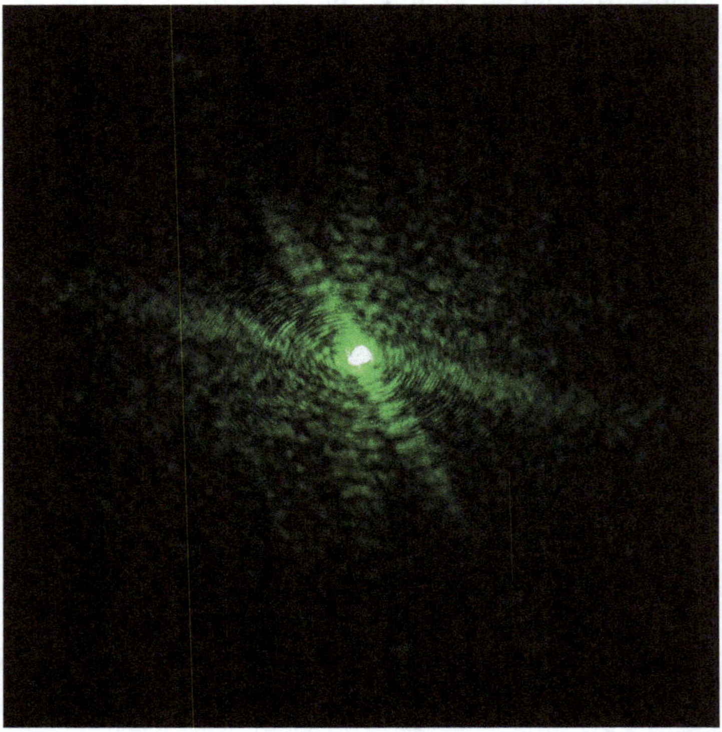

Figure I.11. Fluctuating spots, or *speckles*, produced by a laser beam on a screen (Patrick Imbert, Collège de France photo service).

These spots, known as *speckles*, are light fluctuations scattered by the irregularities of the surface illuminated by the laser beam; they are due to the frequency and phase stability of its light, which gives rise to interference phenomena when scattered by microscopic roughnesses. This stability permitted enormous progress in spectroscopy in the 1970s, but for now lasers emitted only at fixed frequencies, and we had to take advantage of rare coincidences with atomic or molecular transitions to use them in atomic physics experiments. The experiments I did for my thesis could not yet exploit this new light source, and I had to make do with the relatively weak light of Brossel's classical spectral lamps. However, Claude and I dreamed of all we could do if we had tunable lasers, whose color could be changed with the turn of a knob so as to scan the spectral lines of the atoms we were optically pumping.

In 1968, a visionary article by Arthur Ashkin, a physicist from the famous Bell Laboratories in the United States, attracted our attention. Ashkin proposed using the radiation pressure of lasers to control the motion of atoms in space. In fact, he suggested doing to the atoms' linear momentum what optical pumping did to their angular momentum. Could light be used to control not only the axis of rotation of atomic spinning tops, but also their velocities?

The idea seemed crazy: lasers were still far from having the qualities required for such experiments. Twenty years later, however, it became a reality. The progress of lasers made it possible to stop atoms in their path, to cancel their thermal motion almost completely, and even to trap them in light boxes made with laser beams. These *laser cooling and trapping* experiments revolutionized atomic physics and earned Claude and two American physicists, Steve Chu and William Phillips, a Nobel Prize in 1997. Arthur Ashkin had to wait fifty years for his prophetic idea to be rewarded with the same prize in 2018, when he was ninety-seven. This somewhat extreme delay between discovery and recognition tells us something about research: about how long it takes for the fundamental idea to pass to application, and also how there are now so many researchers in the world who have done important work in so many fields that it can sometimes take many years before a great discovery is recognized by the Nobel Prize.

In 1968, even in our wildest dreams we couldn't imagine how much lasers would contribute to fundamental research. Without laser light, I would not have been able to perform any of the experiments that allowed me to explore the quantum world. Beyond using it in my own research, I have also had the privilege of witnessing the extraordinary advances it has enabled in many other fields. Although we were far from predicting the extent of these advances, we intuited even in the 1960s that these new light sources would open up new and still unexplored fields of research in atomic physics.

Some of my fellow Normaliens had chosen a different path. Theoretical physics and elementary particle physics were undergoing a period of great development. As I have said, these were the years when the Standard Model of elementary interactions was being elaborated in an ongoing exchange between theorists and experimenters working on large particle accelerators. Beyond atomic physics, the goal was to understand the nature of nuclei and their constituents. We were beginning to talk about quarks, the elementary particles of which protons and neutrons are made. It was clearly, and rightly, a popular and fashionable subject that attracted many brilliant students.

My friends who had embarked on this path looked condescendingly at the one I had chosen. Atomic physics was a bit of a dusty field, they told me. The laws of quantum mechanics obeyed by atoms had been well known for almost half a century; I was just double-checking the obvious. Recording atomic spectra with increasing accuracy was, in their view, more the job of an archivist than that of a true physicist. In moments of doubt, I sometimes wondered whether I was succumbing to my childhood mania for measuring things for the sake of accuracy and classification. In short, according to my classmates, I was wasting my time. They would have been right had the laser not opened up so many new possibilities. By trusting in the future of atomic physics, I saw—subconsciously, no doubt—that there were still great things to be discovered over the horizon, thanks to this light source and its fantastic properties.

Beginnings in research

In any case, I was captivated by the work I was doing for my master's thesis, then for my PhD. I hadn't lost the sense of wonder I derived from observing the signals atoms sent me through the devices I used to detect them. These experiments were much simpler than the ones today's laser-juggling students must master: all I needed were one or two lamps, one or two cells containing the atoms, a photomultiplier, and a pen recorder or an oscilloscope. We were a far cry from the hundreds of mirrors, lenses, semi-reflecting beam splitters, and crystals through which laser beams must now pass on large optical tables before interacting with atoms.

I rounded out this elementary equipment with a cylindrical shield made of mu-metal, a nickel-iron alloy that cancels the Earth's magnetic field and greatly reduces ambient magnetic noise. In this protected environment, I developed an optically pumped rubidium atom magnetometer. I did this in collaboration with Jacques Dupont-Roc, another student of Claude's. We detected very small changes in the magnetic field by measuring the variations they induced on the intensity of the light transmitted by the optically pumped cell. The spins of rubidium atoms, whose angular momentum is of electronic origin, have magnetic moments about a thousand times larger than the nuclear spins of the mercury I studied in my first experiments. This made these atoms much more sensitive to variations of the magnetic field.

This showed me for the first time the possible link between basic and applied research. We even registered a patent for this magnetometer. It hasn't earned us any money, even though barely different versions of our device are now used in medicine and research to record magnetocardiograms and magnetoencephalograms measuring the tiny fluctuations in the magnetic field produced by heartbeats and neuronic brain currents.

In order to demonstrate the sensitivity of our magnetometer, we decided to use it to observe the precession of helium-3 nuclear spins. This was thus a double optical pumping experiment: on the helium-3, which had to be oriented first, and then on the rubidium of our magnetometer. The helium-3 nucleus is composed of two protons and one neutron, and has a spin-1/2 like mercury-199. This spin carries, in helium as in mercury, all the atomic magnetism. Orienting it requires a more complex procedure than the one described above, which I will not specify here. It was implemented by another student, Franck Laloë, who was working on a thesis with helium atoms at the time. Jacques, Franck, and I joined forces for this experiment.

Once the helium spins were oriented, they were subjected to a very small transverse magnetic field on the order of one ten-thousandth of the Earth's field. They then began to rotate around the field at a rate of about one revolution in two and a half minutes. The cell of the rubidium magnetometer was placed in the vicinity of the helium cell. The precession of the helium nuclei produced a small rotating magnetic field whose periodic variations were detected a few centimeters away by the rubidium magnetometer. The modulation of the light transmitted by the cell of this magnetometer was recorded in the form of a slow sinusoid inscribed in red ink by a pen moving up and down on a long, slowly unwinding strip of millimeter-sized paper. The helium spins remained oriented for hours and I watched, hypnotized, for more than a day as hundreds

(a)

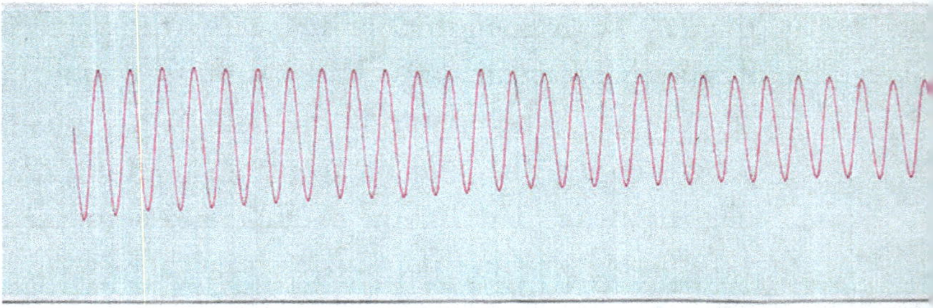

of slow oscillations gradually damped. Behind the walls of mu-metal, the helium and rubidium atoms spoke to each other through tiny variations in the magnetic field whose amplitudes were on the order of a millionth of the Earth's magnetic field. The information was transmitted from helium to rubidium, then from rubidium to light, and finally from light to the pen of the recorder.

There was a playful side to this experiment: once the spins were revolving like microscopic spinning tops, I could leave them to it. One day I quietly closed the door to the lab and left a note—DO NOT DISTURB, EXPERIMENT IN PROGRESS—and went to see a movie in the Latin Quarter with Claudine. I forgot my spins for two hours; when we came back, they were still spinning. I kept this long strip of paper in a drawer, as a record of the precession of the helium atoms: the only trace I've kept of this experiment. Today it would have been digitized directly and stored in the memory of a computer. There was no such equipment at the time, and I'm not sure the durability of this fragile signal would have been guaranteed any better by electronic memory than it was by the antiquated method of archiving on paper.

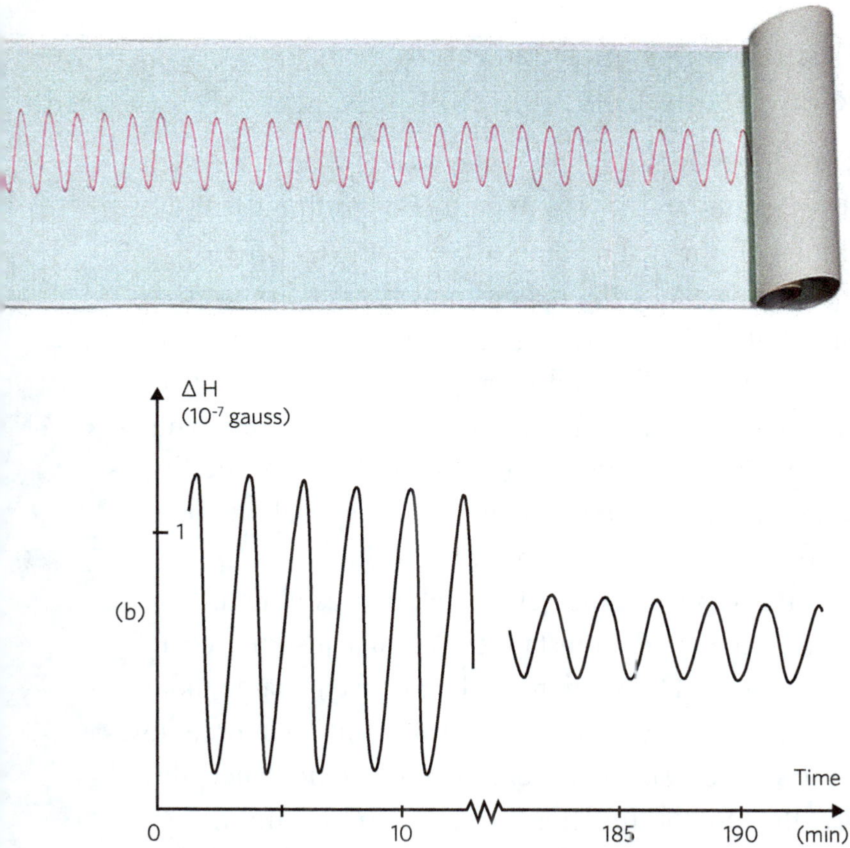

Figure I.12. (a) Observation via rubidium magnetometer of the slow precession of helium-3 nuclei in a magnetic field of 2 microgauss (2×10^{-10} tesla). More than a hundred periods were recorded on this paper strip running at a speed of one centimeter per minute. The precession period was two and a half minutes. The signal faded slowly and remained visible after ten hours. (b) The same signal as it was presented in our paper, published in *Physical Review Letters* in 1969 (unable to show the entire signal, we focused the figure on two portions of the recording: at the beginning, and after three hours).

First trip to America and return to my first passion

The same year we did this experiment, Claude decided that Jacques and I were mature enough to present the results of our magnetometry work at a conference in Canada. We took the opportunity to extend our trip to the United States, visiting research laboratories where Brossel and Kastler had referred us to their colleagues and friends. Claudine and Jacques's wife, Roselyne, came with us. It was my first trip to America, the first in a long series of transatlantic crossings that are today de rigueur for all researchers.

I still recall that first jet lag, which was difficult due to a particular circumstance. We had just arrived in Washington, D.C., where we had booked a large room with a color television (a novelty at the time) in order to witness an exceptional event. By pure coincidence, this was the day the first men were to land on the Moon. The historic moonwalk was expected to take place at 9 p.m. on July 21, 1969. I still had my passion for space exploration and did not want to miss this event at any cost. To ward off the sleepiness that threatened to overcome us that afternoon, we forced ourselves to visit the National Gallery of Art, the great painting museum in the American capital. The splendor of its collection and the museum fever it inspired only made us more exhausted. Back at the hotel, the four of us fell asleep at 7 p.m., missing the suspenseful leadup to the Apollo 11 landing. I woke up two hours later, just as the ghostly figure of Neil Armstrong set foot on the lunar surface and said the historic words: "That's one small step for man, one giant leap for mankind." I had quite a hard time waking Claudine and our friends and getting them to share my enthusiasm.

I relate this anecdote because it brings us back to the laser. Three days later, we arrived in Boulder, Colorado, where Brossel had advised us to go see his colleague Peter Bender at JILA, the Joint Institute for Laboratory Astrophysics. As its name implies, JILA was a renowned center for astrophysics, but was also—and still is—a center for cutting-edge research in atomic physics and optics. Our visit was unforgettable. Peter was responsible for the Apollo program's lunar telemetry experiment. The night we had such difficulty keeping our eyes open, Armstrong and his colleague Buzz Aldrin had placed a reflector, a panel tiled with small reflective cube corners, on the Moon; Peter and his team in Boulder had directed a powerful pulsed laser beam of light to it through a telescope. The light reflected from the cube corners was sent back to the Earth. By detecting it and measuring the delay between the departure of each light pulse and its return, an interval of about 2.5 seconds, Peter and his colleagues were able to directly and accurately measure an astronomical distance for the first time in history. I will return later to the long and fascinating history that, since Galileo, has linked the question of light speed to astronomy.

Figure I.13. The catadioptric laser reflector deposited by the Apollo 11 mission on the Moon. Peter Bender and his team had just detected the light of a laser reflected by this device when I visited them in July 1969 (© NASA).

This method of astronomical telemetry has been carried on and refined over the past fifty years using the same lunar reflector, along with two others deposited by the Apollo 14 and 15 missions, and has made it possible to determine the lunar orbit with extraordinary precision, measuring the Earth–Moon distance at any time to within a few millimeters. These observations confirm the predictions of Einstein's theory of general relativity.

From the outset, this experiment was a brilliant demonstration of the power of the laser, opening new windows on the world in a concrete way. It was a moment of great joy and emotion to witness live, so to speak, such a crucial scientific event. That Peter Bender, despite the thrill of the discovery, was willing to spend an entire afternoon explaining to us the details of the experiment added to my excitement. If he considered me worthy of hearing his explanations and sharing his enthusiasm, I had truly become a researcher.

"Blue sky" research

Half a century has passed since then, and I've never stopped marveling at the complex beauty of nature with the same sense of exhilaration and profound joy. Each time I have observed something new, I have seen the world as something rich and strange, to borrow Purcell's beautiful words. Juggling atoms and photons with instruments that technology has continued to refine, I have also enjoyed myself tremendously, as a new experiment is always a fun challenge one sets for oneself.

I have been fortunate to be able to create and lead a research group of students and colleagues who share the same passion. The feeling of belonging to a community of scientists, driven across borders and cultures by the same curiosity to explore nature, has not left me. From generation to generation, this community connects us to all those who have come before us for centuries, paving the path of knowledge that we continue to travel.

A scientist's privilege is not only to contribute to this gathering of knowledge, but also to witness the discoveries made by his colleagues around the world, and to be able to understand them and appreciate their beauty.

My work has always been motivated by pure curiosity. It has always been an attempt to answer a fundamental question: "How will an atom and a photon behave if I put them in this or that situation?" And also: "How could I observe this with minimal disturbance to the system?" True, sometimes the potential practical application of a result can be naturally envisioned: for example, the optically pumped magnetometer. But I myself have never pursued the line from an idea to the production of an instrument, let alone to its market launch. That requires other qualities and other skills.

The same is true of the fundamental research my team has done on the manipulation of individual quantum systems, which was recognized by the Nobel Prize in Physics. We did not undertake this research in order to build a *quantum computer*, despite what many journalists seem to believe. In fact, we were initially unaware of the theoretical quantum computing algorithms that mathematicians and computer scientists began developing in the 1980s. We were simply curious to know whether, by manipulating atoms and photons protected from their environment, we could bring to light the strange quantum logic that allows an atom or an electromagnetic field to be in many places at once, or to occupy several energy states simultaneously. And we wanted to explore all the implications of this logic.

The English have a name for this "free" fundamental research, free in the sense that it seeks not economic profit but only a better understanding of the world: *blue sky research*. I like this name because it calls to mind the sky, as well as astronomy and astrophysics, my first passion in physics. Aren't these sciences of the universe by nature "useless," since they're not linked to any direct application? Knowing why the sky is blue, or why it turns red at sunrise and sunset, is apparently useless except to satisfy our curiosity. And yet, by elucidating the light-scattering properties of gas molecules in the atmosphere, we have discovered phenomena that are now exploited by many optical instruments. So it will probably be one day for our experiments on atoms and photons, whether they lead to a quantum computer or to something else.

When I lecture about my research, after the question about why I became a physicist and what led me to my field of study, I can rarely escape another question that seems to follow inevitably: *What is the purpose of your work?* We can generalize this question by asking why we do fundamental research. Why blue sky science? This book is my attempt to answer.

CHAPTER 2

Reflections in the Observatory square

Each day for more than forty years, I have walked across the square of the Paris Observatory to get to my laboratory. This lovely grassy path, lined with chestnut trees, connects the observatory built under Louis XIV to the Luxembourg Gardens, following the line that marks the Paris meridian. It is the arc of this meridian, which runs from Dunkirk to Barcelona, that the astronomers Delambre and Méchain measured during the French Revolution in order to define the meter. Crossing the square, I walk eastward over the imaginary line that marked the origin of longitudes on French world maps until 1880, when they adopted the Greenwich meridian, which passes some 2° farther west.

At the end of the path to the south—on my right—I can admire the white dome of the observatory, where, in 1676, the Danish astronomer Römer made the observations of Jupiter's satellites that enabled the first measurement of the speed of light. On my left, I see the Palais du Luxembourg, built under Marie de Medici. In 1808, the engineer Étienne Louis Malus, observing from his house the rays from the sun reflected by the palace's windows, discovered the properties of the polarization of light, which as we have seen played an important role in the optical pumping experiments of my first years as a researcher.

Thus, I often have occasion to meditate on the long history of the science of light. It is a history whose protagonists are physicists, of course, but also mathematicians, astronomers, engineers, and navigators. Their memory calls to mind all the passions of my youth. I like the story of light because it illustrates how far the different fields of knowledge have progressed together over the centuries, revealing in ever greater detail the rich and strange beauty of the world. In order to unravel the mysteries of light, it was necessary to learn to observe the sky and the Earth with ever greater acuity, to invent new instruments of measurement, and to develop powerful methods of calculation.

Passion for precision has played an essential role in this adventure of knowledge. It was by measuring time and distance with increasing care that we were able to verify the laws of Newtonian mechanics, determining the shape of the Earth and the dimensions of the solar system. And it was by pushing precision still further that we were later led to radically revise the classical worldview when we discovered the laws of relativity and of quantum physics. It is this scientific history, which truly begins with Galileo and brings us to Einstein and the emergence of modern physics, that I will discuss in this chapter and the following three. Questions about light are the common thread, but we will take some detours along the way, so full has this story been of surprises and unexpected revelations.

Mankind has been fascinated by light since the dawn of time. As opposed to darkness, a source of terror and a symbol of death, light was a sign of life and rebirth. The sun, which illuminates and gives off warmth, was the object of worship in all primitive religions. In one way or another, the celebration of the seasonal cycle was part of these expressions of faith. Light has also been a symbol of knowledge, as it brings us most of the information we receive from the world—from the light that comes from the sky, which tells us about the universe, to that which comes from the things and beings around us, allowing us to orient ourselves in our environment and take heed of the dangers that may threaten us.

The Ancients' immemorial veneration of light was not rational. To its own questions, humanity proposed mythical or religious answers that shed no light, as it were, on the mysteries of light. Does it fill all space instantaneously or does it propagate at a finite speed? Is it of the same nature as matter, or is it of a different essence? Why are some environments transparent and others opaque? There were no answers to these questions.

The laws of optics governing the propagation of light rays had certainly been discussed in Antiquity and the Middle Ages. The Basra mathematician and astronomer Ibn al-Haytham, known in the West as Alhazen, understood circa the year 1000 that light does not emanate from the eye or from objects around us, but that these reflect toward our pupils rays from light sources like the sun and lamps. But this was still very qualitative. It was not until the birth of the scientific method in the seventeenth century that light's secrets were truly revealed. This was when observation, experimentation, quantitative measurement, and the development of mathematical theories took precedence over mythical storytelling. Fundamental knowledge and instrumentation progressed in concert, illustrating even then the symbiosis of research inspired by pure curiosity—*blue sky research*—and technology.

Two instruments at the origins of the scientific revolution: the refracting telescope and the pendulum clock

Let's begin with the question of the speed of light. Against the ancient belief that radiation filled all space instantaneously, Galileo was the first scientist of the modern era to intuit that light, like sound, should propagate at a finite speed. It is said that he even tried to measure this speed with one of his assistants, each holding a lantern while they stood a few kilometers apart on two Tuscan hills. The two lanterns were initially masked; Galileo uncovered his and asked his assistant to do the same as soon as he saw the light. By seeing the signal his assistant sent back, Galileo hoped to measure a delay which, based on the distance between the two hills, would allow him to estimate the speed of the light rays.

The result was disappointing. There was a fraction of a second's delay, barely detectable, between the emission of the signal and its return—but it was independent of the distance between the two men, and in fact revealed only the response time of their brains. We know now that the time interval they were trying to measure was merely a few millionths of a second: far too short for our senses to register, and something no instrument at the time could measure. It would be two and a half centuries before Galileo's experiment, enhanced by technological advances, gave a positive and accurate result. What was clear at the beginning of the seventeenth century was that light was very fast, and that if one wanted to estimate its speed one had to learn how to precisely measure time intervals and very long distances.

Galileo himself tackled both of these problems, though without any particular focus on the speed of light, which seems to have stopped preoccupying him after the failure of his naive experiment. It was by studying the motion of pendulums, and using the first magnifying telescope to observe the sky, that he made two major discoveries that would, half a century later, make possible the first estimation of the speed of light rays.

Until the seventeenth century, time was measured with primitive instruments. Clocks in church steeples counted the twisting oscillations of a long string attached to a horizontal wooden arm called a foliot; the inaccuracy could reach a quarter of an hour per day. Shorter intervals were measured by the human pulse or by water clepsydra. Galileo is said to have measured the time it took a marble to go down a slope by weighing the amount of water that flowed from the clepsydra between when the marble was set in motion and when it reached the bottom. Here again, the precision was poor.

Galileo's experiments on the point pendulum, a small sphere attached to a wire oscillating in the earth's gravitational field, opened up new perspectives. He found that the pendulum's swing period was independent of the suspended mass and the amplitude of its oscillation. In fact, this latter property, known as the *isochronism* of oscillations, is verified only for motions whose angular amplitude is less than a few degrees. The pendulum's period then depends only on the length of the suspension wire; it is two seconds for a wire of about one meter long (Galileo did not use the meter, of course, but the units of the time, the cubit or the toise). It would still take enormous progress to go from a simple quasi-point pendulum beating approximately once in two seconds to a clock that counted time autonomously and accurately, but the idea of basing the measurement of time on the regular motion of an oscillator was already there.

It was Christiaan Huygens, half a century later, who transformed Galileo's pendulum into a real clock, making it an instrument for the precise measurement of time. Huygens demonstrated the formula that all high school students have learned, the one that expresses the period of a point pendulum as 2π times the square root of the ratio of the length l of the suspension wire to the gravitational acceleration. The latter, often written as g, is about 9.8 m/s^2; it expresses the change in speed per second of a mass falling into the Earth's gravitational field. Huygens also showed that if the pendulum was not a point but an object of any shape oscillating around a horizontal axis, calculating its period required defining its moment of inertia with respect to this axis, a quantity I described in the previous chapter. Knowing the mass, the moment of inertia, and the distance from the object's center of gravity to its point of suspension,

(a)

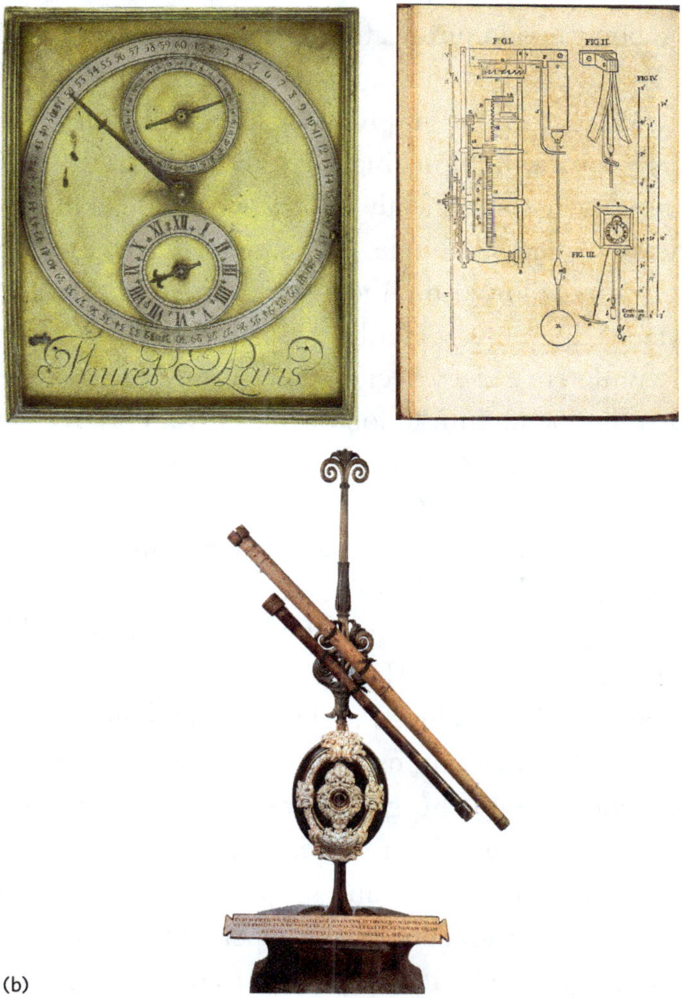

(b)

Figure II.1. The two instruments at the origin of the birth of modern physics. (a) The Huygens clock: (*left*) the dial of the clock made by the clockmaker Isaac Thuret from plans by Huygens (Boerhaave Museum, Leiden); (*right*) Huygens's drawing detailing the clock's mechanism. (b) Galileo's telescope, which allowed him to observe the Moon as well as Jupiter and its satellites (Galileo Museum, Florence) (© akg-images).

he could calculate the period and thus the length of the point pendulum beating at the same rate.

He also sought to eliminate altogether the period's dependence on the amplitude of the oscillation, which was a source of imprecision. He showed that the oscillations were not perfectly isochronous, independent of the oscillation amplitude, unless the pendulum's center of gravity followed not the arc of a circle but that of a cycloid, the curve describing a point's motion on the circumference of a wheel that rolls without slipping. To force the pendulum to follow this curve, Huygens had to conceive an ingenious system of rails placed near the axis around which the pendulum was oscillating. These rails changed the effective length of the pendulum during the oscillation.

Alongside these theoretical advances and the subtle calculations involved, Huygens was also a remarkable engineer. He developed a clever escapement mechanism with a dual purpose: by coupling it with a weight descending into the gravitational field, it delivered energy to the pendulum to compensate for the work of the forces of friction and keep it in a steady state of oscillation. In tandem, the escapement recorded the passage of time by moving hands forward along a dial to the rhythm of the pendulum's beats, marking seconds, minutes, and hours. It was no longer necessary for an assistant to count the oscillations: the machine did it automatically. It took great ingenuity to develop an escapement mechanism that did not alter the frequency stability of the oscillator, which could be set to beat the second. Huygens's clocks deviated by only ten to fifteen seconds a day from the astronomical time based on the observation of the Earth's diurnal rotation.

Galileo's other decisive contribution to the measurement of the speed of light was to refine the refracting telescope, a tube enclosed by two lenses, allowing the formation of enlarged images of observed objects. With this instrument—whose first prototypes had come from Holland, and whose magnification he increased from 3× to 30×—Galileo became, in 1610, the first person to look truly scientifically at the sky. He observed the Moon, made out the rings of Saturn, and above all—most importantly for our story—discovered four satellites of Jupiter. For the first time, a planetary system obeying Kepler's laws was visible directly to the eye, entrenching these systems' existence in the universe and confirming the Copernican model.

Jupiter and its procession of satellites were also fated to become an astronomical clock. According to Kepler's empirical laws, which would be justified at the end of the century by Newton's theory of gravitation, the period of Jupiter's satellites had to be stable and provide a measure of time. This astronomical time was accessible wherever Jupiter could be observed, offering a method of universal time synchronization in an era when our present-day standard of communication and synchronization via exchange of radio signals could not even be imagined.

Figure II.2. Jupiter and the four "Medicean" satellites discovered by Galileo, as he saw them in 1610 with his telescope, which included magnification between 20× and 30×. Io is the closest satellite to the planet, making one revolution every 42 hours, 27 minutes, and 21 seconds. This photo was taken by a simple digital camera on a tripod (© Igor Dotsenko).

Measuring the speed of light to survey the universe

This natural clock attracted the interest of astronomers. In 1676, Römer, a young Danish astronomer working at the newly inaugurated Paris Observatory, was tasked with tabulating the reappearances of Io, the satellite closest to Jupiter, coming out from the planet's shadow. Römer quickly noted that the time interval between two successive reappearances of Io was not fixed, as Kepler's laws had predicted, but that it lengthened for six months of the year and shortened for the following six. To measure this, he used the precise clocks Huygens had developed. Römer understood this phenomenon to be due to the finite speed of light.

As the Earth in its orbit moved away from Jupiter, the distance between the two planets increased while Io made a revolution. The time interval between two of Io's successive reappearances was then increased by the time it took for light to pass through the distance the Earth had traveled while Io completed a rotation on its orbit.

During the six months when the Earth was approaching Jupiter, the opposite phenomenon occurred: Io reappeared earlier each time. The advance or delay, on the order of a few seconds for each revolution of Io, reached about twenty minutes after six months. Römer concluded that this was the time light took to travel the diameter of the Earth's orbit, twice the distance from the Earth to the Sun. But what was this distance, which we now call the astronomical unit?

Another astronomer from the Observatory, Jean Richer, had been sent to French Guiana four years prior to answer this question. His task was to measure, from Cayenne, the parallax of the planet Mars at the favorable moment when it was closest to the Earth, and to compare this parallax to its simultaneous measurement—by the astronomer Jean-Dominique Cassini, Römer's boss—from the Paris Observatory. A very simple experiment will suffice to demonstrate parallax: extend your hand in front of you and look at your index finger pointing vertically, while fixing in the same axis a distant marker (such as a doorpost or a window about ten meters away). Make this observation by closing your eyes alternately. Depending on which eye is open,

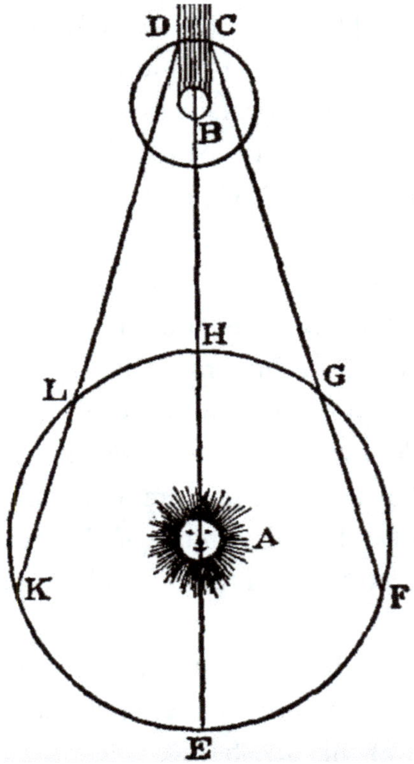

Figure II.3. Römer's drawing, published in the *Journal des savants on* December 7, 1676, illustrating the principle of his estimation of the speed of light: the lower circle with the Sun in A represents the Earth's clockwise orbit. The upper circle describes the orbit of the satellite Io around Jupiter, located in B. Io enters the shadow of Jupiter at C and exits at D. Meanwhile, the Earth moves from K to L (approaching Jupiter) at certain times of the year, and from G to F (moving away from it) six months later. (The distance traveled by Earth during Io's eclipse time is greatly exaggerated.) Deducing from Io's reappearance times the value of the speed of light required knowing the diameter EH of the Earth's orbit.

you will see your index finger at a different angle, either to the right or to the left of the marker. This change in perspective is the parallax effect. Now bend your arm to bring your index finger closer to your eyes and repeat the experiment: the effect is greater. The closer the object, the greater the difference in the angles at which it is observed by both eyes. For a very distant object (theoretically at infinity), the parallax is zero; this object can therefore be used as a benchmark to measure the parallax of closer objects.

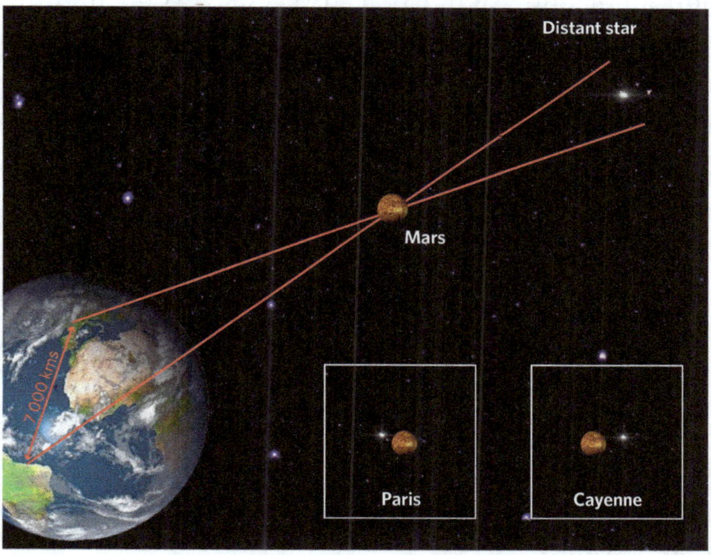

Figure II.4. The parallax of Mars measured by Richer and Cassini. At the same moment, the planet appears to the right of a distant star when seen from Paris and to its left when seen from Cayenne. The effect is highly exaggerated in the drawing because the distances are not to scale (at this time Mars was about four thousand diameters of Earth away from Earth, not at one or two as shown here).

In the 1672 experiment, Richer was one eye and Cassini the other, Mars was the index, and the marker was a star fixed in Mars's vicinity that night. The simultaneity of the measurements was approximate. It was enough to make sure that between the two measurements the Earth had not rotated more than a small fraction of the distance between Cayenne and Paris; a difference of a few minutes between observations was acceptable. The experiment gave a parallax difference of twenty-five arcseconds between Cassini's measurement and Richer's (it also took several months, of course, for Richer to return to Paris to make this comparison). A simple triangulation made it possible to deduce that Mars, that night, had been at about 7,500 times the distance from Paris to Cayenne, known to be approximately 1,700 leagues (7,000 kilometers). The Earth–Mars distance at the time of opposition (Mars, the Earth, and the Sun being aligned in that order) was thus about 13 million leagues (53 million kilometers). Knowing the laws governing the planets' orbits, this measurement was sufficient to deduce the parameters of Mars's and Earth's orbits, in particular the distance between the Earth and the Sun—the astronomical unit.

For this distance, Cassini and Richer found an approximate value of 30 million leagues; one could thus deduce from the delay of the Jovian satellite's reappearances that the speed of light was on the order of 50,000 leagues per second. The precision was not excellent (the real value, we now know, is very close to

300,000 km/s, or 75,000 leagues per second), but it was the first determination to ascribe this quantity the right order of magnitude. Curiously, Römer himself did not give any value for the speed of light; the result deduced from his observations was calculated and published by Huygens alone.

Making this first determination of an essential parameter of physics required using the universe as a laboratory, with instruments whose precision had improved considerably since Galileo's day. The refracting telescope was mounted on sextants that could be oriented by a very precise mechanism of screws and verniers, and fine reticles in the focal plane of its lens made it possible to observe the position of the stars with a precision of a few arcseconds.

With these experiments, the universe in all its immensity revealed itself to observation. The gigantic distances involved helped put the Earth and mankind in their place, continuing the reappraisal that had begun with the Copernican discovery. Not only were we not at the center of the world; it also turned out that our Earth was mere dust in infinite space. The size of our planet, determined by the explorations of Renaissance navigators, had provided the 7,000-kilometer toise between Paris and Cayenne that the astronomers of the Paris Observatory used to measure the size of the solar system. We now had a unit of measurement twenty thousand times larger—the radius of the Earth's orbit—which would serve in turn as a unit for determining the distance to the nearest stars.

Their parallax, measured six months apart at two diametrically opposite points on the Earth's orbit, is only a fraction of an arcsecond. It was not until a century and a half later that advances in astronomical telescopes made it possible to determine these tiny angles and place these stars at a few tens of thousands of astronomical units away. It took several years for their light to reach us—compared to eight and a half minutes for light from the Sun.

The finite speed of light, which Galileo had dreamed of measuring, added an essential parameter to astronomy: time. The light we see today left a nearby star years ago. From farther stars, whose distance from the Earth could not be estimated until much later and by other methods, light has traveled for millions or even billions of years before reaching us today. By receiving it, we can decipher information about the distant past of the universe. Astronomy, born in ancient times as a science of space, would become astrophysics, a science of space and time, whose ultimate goal is to study cosmology, the history of the universe. This was still a long way off in Cassini's and Römer's day, but it's worth remembering that this science was already germinating in the astronomical observations of Louis XIV's Grand Siècle.

The progress of observational astronomy in the seventeenth century was a result of the perfection of lens-based optical instruments. Their size, which for centuries had obeyed empirical recipes, became better controlled and more precise once laws were established to describe the propagation of light rays in the air and in transparent matter. This was how the scientific method, based on the precise observation of natural phenomena and leading to precise mathematical formulation, found one of its first applications.

The science of light becomes quantitative: Descartes and *Dioptrics*

The first person to systematically apply this method to the study of light propagation was Descartes. Following his *Discourse on the Method*, he wrote a treatise on optics called *Dioptrics*, published in 1637, in which he presented his views on the nature of light and explained the laws of reflection and refraction of light rays. Unlike Galileo, he assumed that these rays propagate with infinite speed (this was before Römer's observations), and did not say anything particularly clear about their nature. He simply assumed that the seemingly empty space was filled with rigid particles of a different nature from those that formed ordinary matter, in which light bathed. In this he was describing one of the first versions of the *aether*, that mysterious medium that would intrigue physicists for another two and a half centuries. These aether particles instantly transmitted, from sources of light to the material objects that reflected or refracted it, or even to the eye of an observer, a "tendency to motion" that was light. The instantaneity of light propagation was explained by the infinite rigidity of the particles that made up this aether. Although for Descartes light had an infinite speed and was not associated with motion, he was obliged to understand the path of light rays by way of an analogy to the trajectories of bodies actually moving in space. He likened the bounce of a ball on a flat floor to the reflection of light on a mirror or a flat surface separating air from a transparent medium, known as a *flat diopter*. He thus affirmed the law, known since antiquity, whereby light is reflected in the plane defined by the incident ray and the normal to the diopter, and the angle of reflection is equal to the angle of incidence.

Descartes then tackled the problem of refraction, the fact that when a light ray encounters a diopter part of the light enters at a different angle from the angle of incidence on the medium. The angle of the refracted beam is given by the famous *law of sines*, also called *Snell's law*, after the Dutch physicist who established it a few years before, apparently unbeknownst to Descartes. For readers with only a vague recollection of Pythagoras and his geometry of the triangle, let us recall that in a right triangle the sine of the angle formed by the hypotenuse and one side of the right angle is obtained by dividing the length of the other side by the hypotenuse. This number varies from 0 to 1 as the angle increases from 0 to 90°.

Snell's law says that the sine ratio of the angles of incidence and refraction of light at the passage of a plane diopter between two media is a fixed number, independent of the direction of incidence and depending only on the nature of the two media (*e.g.* air and water or air and glass). The angles in question are those the rays form with the normal to the plane of the diopter. When light passes from air to a denser transparent medium, the light ray is deflected toward the normal and the angle of refraction is smaller than the angle of incidence. The sine ratio is then a number greater than 1—in the language of modern physics, this is called the medium's *refractive index* relative to air.

The opposite occurs when the ray leaves the densest medium and enters air. The sine ratio of the incident angle to the exit angle in air is less than 1 and the ray exits at an angle with the normal greater than the angle of incidence. For a certain critical angle of incidence, the sine of the exit angle reaches 1, and the exit ray, making a 90° angle with the surface normal, emerges parallel to the diopter. If the angle of incidence increases further, there is no more emerging ray. The ray reaching the diopter is totally reflected inside the medium.

These results had been observed empirically since antiquity for small slopes of the rays relative to the normal to the diopter. In these cases, the sines are approximately proportional to the angles. Finding Snell's law, Descartes extended the formula to any slope, replacing the angles with their sines. To justify this result by empirical observation, he reused the analogy of a ball hitting a flat surface, considering the component of its motion

René Descartes
(1596-1650)

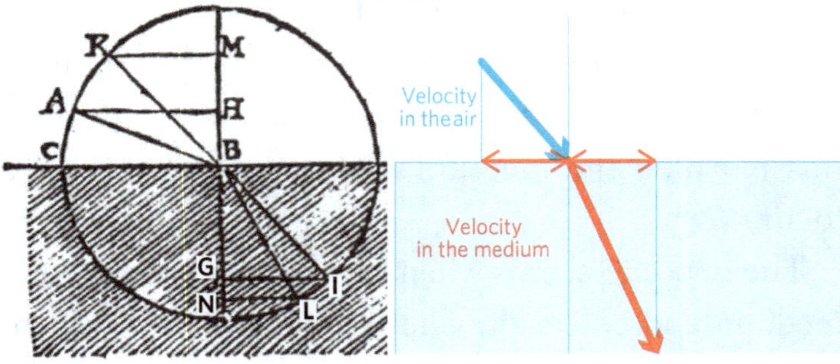

(a) (b)

Figure II.5. (a) The law of refraction illustrated in *Dioptrics*: the radii AB and KB are refracted respectively in BI and BL. Taking the radius of the circle BA = BK as the unit, the sines of the angles of incidence and refraction are given by the lengths AH and GI for the radius from A, and by KM and NL for the radius from K. Snell's law expresses that the ratios AH/GI and KM/NL are equal. (b) A diagram showing that a bullet propagates faster in the transparent medium than in air if its trajectory bends toward the normal to the diopter while keeping its speed's horizontal projection constant. The ratio of the speed in the medium and the speed in air is then equal to the ratio, greater than 1, of the sine angles of incidence and refraction.

parallel to the surface to be unmodified. He also hypothesized that, as a result of its interaction with the medium, the speed of the virtual ball (representing light) changes in a constant ratio independent of the angle of incidence. Here he reiterated the law of sines, but his mechanical analogy led to a paradoxical result.

If the ball's velocity decreases as it passes through air toward the denser transparent medium, as one would intuitively expect, it should move away from the perpendicular to the diopter while the light ray moves closer to it. He thus had to admit that light, unlike a material projectile, penetrates water more easily than it does air. This is how Descartes introduced the notion of light's greater ease in penetrating matter as a kind of attraction between light and the transparent medium.

This model of balls propagating with finite velocity to represent what he saw as a tendency to instantaneous motion presented an obvious internal contradiction. Newton would break free of it a few years later by asserting that light was made up of real particles propagating at finite speed. The law of sines implied that these particles had to go faster in water than in air, just like the virtual balls in Descartes's reasoning.

Relying on analogies and fuzzy concepts such as light's ease in penetrating a medium, Descartes manifested the perplexity of his time regarding the real nature of light, which, though we were beginning to understand the laws of its propagation in air and in transparent media, remained mysterious. The contribution of Snell and Descartes was to deduce, through observation of the optical phenomena of reflection and refraction, a quantitative mathematical law that solidly accounted for the observed phenomena. However, a satisfactory theory to justify this law was missing.

Nature works by the shortest and simplest ways: Fermat's principle

Pierre de Fermat was one of the first thinkers to point out, in the 1650s, the contradictions of *Dioptrics*. They could be resolved, he understood, if we admitted that light had a finite speed—this was still before Römer—and that in a material medium such as water or glass this speed was less than in air. The sine ratio was that of the corresponding speeds in both media. To reach this result, it was necessary to abandon the analogy between light and material projectiles. Without further questioning the nature of light rays, Fermat showed that the sine law, understood as a law of the ratios of the speed of light in the media through which it passed, could be expressed in an even simpler and more fundamental way:

> To go from one point to another, a light beam follows a path that takes a minimal time compared to all neighboring paths.

This is, expressed in modern language, the famous Fermat's principle, which has remained of considerable importance in optics, and in physics in general, to this day. To justify it, Fermat invoked a principle of economy whereby "nature always acts by the shortest and simplest ways." In a homogeneous medium, the principle explains, light propagates in a straight line.

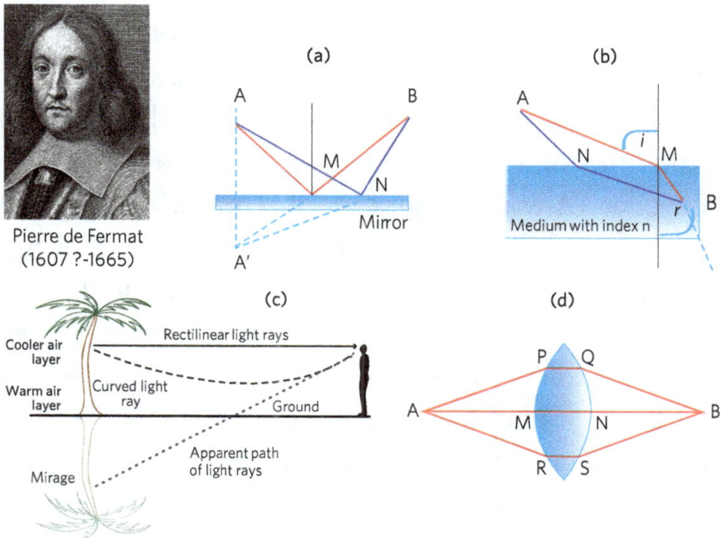

Figure II.6. Fermat's principle. (a) The law of reflection: to go from A to B, light is reflected in M such that AMB is of minimum length compared to all other paths (including *e.g.* ANB). M is the intersection of the mirror with the line joining B to the image A' of A. The angle of incidence is then equal to the angle of reflection. (b) The law of refraction: to go from A in air to B in a medium with index $n > 1$, the light follows the path AMB, which makes it go farther in air, where it travels faster. The point M of contact with the diopter is the one where the sine ratio of the angles of incidence i and reflection r is equal to n. On the non-minimal path ANB, the light gains time in air (AN < AM) but loses more time in the medium where it is slower (NB > MB). (c) Mirages: light curves toward the warm ground, where it goes faster, and the ray that reaches the eye gives the impression that the palm tree is reflected in the ground. (d) Focusing glass lenses: to go from A to B, light follows all paths of equal time AMNB, APQB, ARSB… The more the rays are inclined on the normal to the lens, the longer the paths in air become, but the delay on these paths is made up by the reduced travel time in the lens.

This also accounts for the law of reflection, because it's clear that if a ray is moving from one point to another and meets a mirror on its path, the shortest path is the one that meets the obstacle at the point for which the angle of incidence is equal to the angle of reflection. Any other point on the mirror corresponds to a longer path, as we can see from a simple drawing.

The same principle makes it possible to find the law of refraction. If light starting in the air is moving toward a point in the water situated obliquely with respect to the boundary between these two media, it will take a minimum time if it travels more through the air, where it goes faster, and less through the water, where it is slowed down. The optimal path corresponds precisely to Snell's law, with an angle of incidence in air greater than the angle of refraction in water. A common analogy is that of a lifeguard who sees a swimmer drowning: in order to minimize the time it takes to rescue the swimmer, she must run along the beach at an angle until she is close to the plumb line of the swimmer in distress. The point at which she enters the water is given by the law of sines, whose ratio, greater than 1, is equal to that of the speeds of the rescuer running on the beach and swimming in the water.

Fermat's principle is generalized to situations where light passes through all kinds of diopters or non-homogeneous media where the index can vary continuously. It explains in particular how mirages are formed. In the vicinity of a surface of land heated by the sun, warm air is less dense than the air farther from the ground, and light propagates there a little faster. Instead of going in a straight line, the light rays that go toward your eye from the top of a tree or a church's bell tower will seek the path of minimum time by following a curved trajectory that brings them close to the ground before rising toward you. By extrapolating as a straight line the direction of the light rays reaching you, you get the impression that they're coming from images on the other side of the ground. The layer of warm air behaves like the surface of a lake or a mirror, reflecting the sky and the objects in the landscape around you.

Fermat's principle also makes it possible to explain simply the properties of convergent lenses that form a near point image from a point source. Descartes had offered an explanation of this phenomenon essential to the functioning of optical instruments—telescopes and microscopes, which had just been invented—based on his law of sines. Fermat's principle gives a simpler intuitive image, if we modify the statement slightly.

Indeed, a more accurate formulation of this principle says that light follows paths where the travel time is steady with respect to small variations on either side of these paths. This is generally the case of a path with a minimum travel time, but there may be situations, on the contrary, where this time is maximal. There are also situations where the travel time between two points stays the same for a large number of neighboring paths connecting the points. This is precisely the case for the rays connecting the object and image points located on the axis of a glass converging lens, whose surfaces are convex spherical caps. The geometrically shortest ray from the object to the image is the straight line segment passing through the center of the lens, which crosses its thickest section.

Oblique rays that deviate from this geometric path of minimum length pass through a larger region in air but a smaller one in the thinner parts at the lens's edges. The time light loses in air, where it goes faster, is made up for by the shorter time it takes to pass through the lens, where it goes slower. Light's travel times are thus the same for all paths joining the object and image points following more or less oblique trajectories in the air. All of these paths obey Fermat's principle in the broadest sense, and contribute to form the image at the point of convergence. This requires a well-defined relationship between the distances of the object and image from the center of the lens and a parameter of the lens called its *focal length*, which depends on the geometry of the spherical caps that constitute it. This focal length, which measures the lens's convergence power, is also a function of the ratio between the speeds of light in the air and in the glass, which in modern language is called the *refractive index* of the transparent medium. These rules of geometrical optics, analyzed in Descartes's *Dioptrics* with the help of the sine formula, find a particularly pertinent geometric interpretation thanks to Fermat's principle.

Along with Newton's laws of mechanics, this principle constitutes what can be considered one of the very first quantitative theories of physics. Starting from an empirical observation—that the law of sines is satisfied by the refracted rays in a material medium—it posits a simple law of light's steady travel time. This principle makes it possible to explain other phenomena—mirages, the laws of lens convergence—and to predict something that was not yet experimentally verified: that light travels slower in a transparent medium than in air or in a vacuum. Crucially, this theory is minimal, in the sense that it makes no gratuitous hypotheses about that which was unknown, be it the nature of light, the medium it passes through, or its greater or lesser tendency to pass from one medium to another.

In discussing Fermat's principle I used the anthropomorphic image of light "seeking" the minimum time path. Does this expression make any sense? How does light perceive whether the path it is following is of minimal or steady time? We know now that yes, in fact, light does have a way of searching for this path: its wave-like character allows it to spread over small distances around this path, on the order of what we now call its wavelength. But it would take almost a century and a half after the statement of Fermat's principle for the notion of wavelength to be clearly established. Nonetheless, we see the beginnings of this concept in the *Treatise on Light* that Huygens published in 1690.

Huygens and the wave theory of light

That book, which Huygens began while he was a member of the Royal Academy of Sciences, is written in French, but—a sign of the troubled times—was not published until after its author, a Protestant, was forced to return to Holland, driven out of France by the revocation of the Edict of Nantes. The *Treatise of Light* is one of the great books in the history of science, offering a fresh look at light and establishing principles that would withstand the test of time. Huygens abandons Descartes's ball analogy in favor of another one: sound. Light, he says, like sound, is a wave propagating in space. But whereas sound is a vibration of air, light is the vibration of a hypothetical medium—the aether—whose properties he seeks to describe.

For Huygens, then, light is a wave comprising a succession of vibrations moving in space. The speed of this wave had just been determined by Römer at the Royal Observatory of Paris, where Huygens had himself observed Saturn and its rings in the 1670s. In his *Treatise*, Huygens uses Römer's observations to estimate the speed of light at six hundred thousand times that of sound (the exact value is closer to a million, but he had the right order of magnitude). Without giving his opinion on the exact nature of the aether, he imagines, in the manner of Descartes, that it consists of hard particles that transmit light vibrations by elastic shocks.

The analogy is made to sound propagated by the compression and expansion effects of air, but Huygens points out that the aether cannot be air, since light propagates in an empty volume from which air has been pumped or in transparent media into which air does not penetrate. He insists that it is not aether particles that move at the enormous speed of light: they merely move around their equilibrium position as the wave's jolt gradually reaches them. The same is true of air particles that do not move at the speed of sound but vibrate in place, or of water that rises and falls in place when a wave ripples the surface of a pond.

As in these simple examples, the light waves emitted by point sources in a homogeneous medium are spherical, moving in concentric circles away from their origin. These waves may encounter obstacles, propagating in environments in which the speed of light changes. To determine how a light wave propagates from one place to another, Huygens states a simple principle. He posits that each point illuminated at a given time becomes a virtual point source, emitting a spherical wavelet whose amplitude is proportional to that of the wave that has reached that point, and whose state of vibration is identical to that of that wave. The light radiated beyond a given surface results from the addition of all the auxiliary wavelets emitted by the virtual sources distributed on this surface. Therefore, if we know the state of the light on the surface, we can deduce how radiation will propagate beyond it without having to know the properties of the real physical sources located behind the surface.

According to Huygens, the light wave was thus a succession of gradually propagating aether tremors. The waves' surfaces are, as we now know, all points where the wave vibrates with the same phase. For Huygens, these were the envelopes of the wavelets emitted by the auxiliary sources gradually defined on the surfaces the wave reached at equal time intervals. The light ray passing through a point was the line connecting this point, considered a secondary wavelet source, with the point of contact between this wavelet and the envelope of all wavelets coming from this point and from neighboring ones.

Huygens tries to use this simple idea to explain why light in a homogeneous medium propagates in a straight line with rectilinear rays. If a screen with an opening punched in it is interposed in front of a point source of light, everything happens behind the screen as if it were the virtual sources distributed on the surface of the hole that were radiating.

These secondary sources all emit at the same time if the screen is in the plane of the incident wave, perpendicular to the line joining the point source to the center of the hole. The radiations of the auxiliary sources combine in this direction and weaken in the others, creating a rectilinear beam of light. The light ray propagates according to the normal to the wave front on which the virtual auxiliary sources are emitting at the

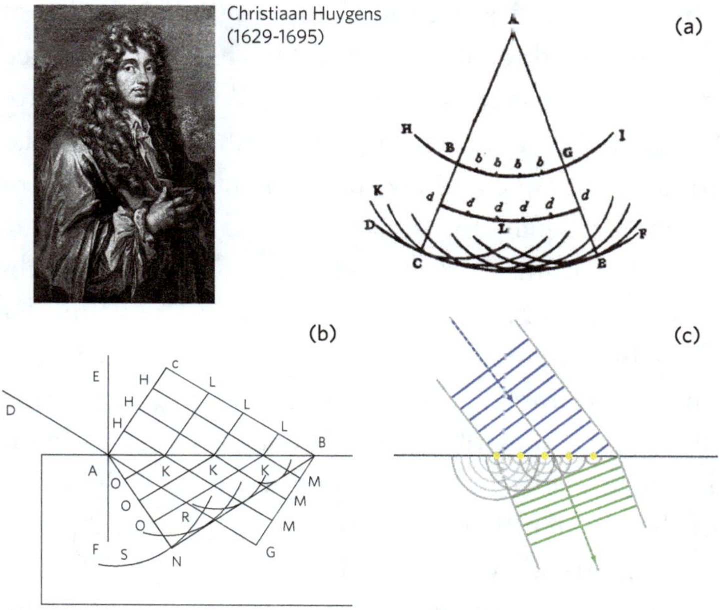

Figure II.7. (a) A figure from the *Treatise on Light* illustrating Huygens's principle: the wave emitted by source A beyond a spherical wave front appears as the envelope of auxiliary waves emitted by virtual sources distributed on this wave front. (b) A figure from the *Treatise* illustrating the law of refraction of light passing from air (*top*) to a medium of index $n > 1$ (*bottom*): the planar wave fronts inclined on the plane of the diopter reach it at points K, at instants following each other at regular intervals. Each point radiates a wave in the lower medium. The envelope of these waves forms parallel planes that are spaced more closely than in air. In the medium, the light ray normal to these planes corresponds to a refracted angle smaller than the angle of incidence. (c) A simplified representation explicitly showing the secondary spherical waves emitted by each point of the diopter, reached at successive instants by the incident planar waves (Wikimedia Commons).

same time. This principle of auxiliary waves, stated by Huygens and revisited and clarified at the beginning of the nineteenth century by Augustin Fresnel, has since been named after both physicists. When the electromagnetic theory of light was established, it lost its status as a principle and became a direct consequence of Maxwell's equations.

Based on this principle, Huygens was able to establish the laws of reflection and refraction. Let's consider the latter phenomenon only. Suppose light arrives at an oblique angle of incidence on a diopter separating air from a block of glass: let's analyze what happens in the plane of incidence defined by the direction of the incident beam and the normal to the diopter. The spherical waves radiated into the glass by the secondary sources distributed along the diopter line are spaced more closely in the glass than in air if the light propagates more slowly in the medium. These waves admit an envelope propagating in a different direction from that of the incident wave front, making a smaller angle with the normal to the diopter. A simple geometrical analysis brings us back to the law of sines, whose ratio is that of the speeds of light in the two media. This result is consistent with what Fermat thought.

The reader familiar with optics will be struck by the fact that I have thus far avoided mentioning the concepts of wavelength or interference. We know now that light decomposes into monochromatic waves, oscillating with a well-defined period and propagating in space with undulations whose successive peaks are separated by a distance defining what we call wavelength. But this notion would not be introduced until more than a century later, by Young and Fresnel.

It is striking, then, that although he speaks of waves for light, Huygens never mentions the spatial periodicity of the radiation. The explanation is that optical wavelengths are too small, on the order of a fraction of a micron, to be resolved in the simple experiments he was able to do. The wavelength is the distance traveled by light in one period. Despite the enormous value of the speed of light, its product with the period of light oscillation is tinier than could be measured in the seventeenth century. This reflects the fact that the period of a light wave vibration is extremely short, on the order of a millionth of a billionth of a second. Huygens and his contemporaries could not imagine distances and times as small as wavelengths and optical periods; it was not until the nineteenth century that such mysteries began to be explored.

Among the effects Huygens did not describe was the diffraction of light by small obstacles, which had already been studied by the priest Grimaldi. The reasoning I recalled above to explain light's rectilinear propagation from a diaphragm is valid only if its diameter is large compared to the light's wavelength. Light radiated through very small holes diverges at the output and has an intensity distribution with alternating bright areas and dark fringes. This phenomenon, already studied

Figure II.8. Double refraction of an Icelandic spar crystal: each of the crosses drawn on the black sheet gives two right-angled polarized images in transmission.

by Grimaldi, would be described again a few years later by Newton, but it would not be truly understood until the nineteenth century.

In his *Treatise on Light*, Huygens also gives the first description of the strange double image formation phenomenon observed when light passes through the transparent crystals of a substance that sailors and traders brought back from faraway seas: Icelandic spar. This is a crystal of rhombohedral shape, a kind of cube or parallelepiped that looks like it was flattened by compressing two opposite tops. If we were to look at an object through this crystal, for example a cross drawn on a sheet of paper, we would see it appear twice. If we turned the crystal, one of the images would remain fixed and the other would revolve around it. A ray arriving normally at the entrance face thus gives two rays: a non-deviated one that follows the sine rule (the ordinary ray), and one that deviates from it (the so-called *extraordinary ray*). This has since been called the phenomenon of *birefringence*.

These crystals have another strange property. If we followed the ordinary ray after it had passed through a first crystal and presented it to a second spar crystal, identical and oriented in the same way, the ray would pass through the second crystal without being deflected. But if the second crystal were rotated 90° around the direction of the ray, it would be deflected and become the extraordinary ray. The experiment thus revealed a characteristic of light related to its non-invariant behavior by rotation around its direction of propagation. This has come to be called the phenomenon of *polarization*. We know now that light is a transverse vibration, corresponding to the oscillation of an electric field in a direction normal to that of the wave's propagation. The spar crystal has the property of spatially separating waves whose electric field vibrates in orthogonal directions. Huygens's experiment revealed this phenomenon for the first time and demonstrated the spar crystal's polarizer and analyzer properties, which were not well understood until much later. He did not realize that the strange behavior of the light he had glimpsed was related to the fact that the light wave was a vibration transverse to the direction of the rays, but he was able to give a qualitatively correct interpretation of birefringence.

The propagation of the ordinary ray was due to the fact that it generated spherical wavelets on the crystal's entrance surface, leading to a refraction obeying the law of sines. The extraordinary ray produced wavelets with an ellipsoidal wave surface, indicating that the light associated with these waves propagated at different speeds depending on its direction. The extraordinary ray followed the direction defined by the center of each wavelet on the surface of the diopter and the point of contact in the crystal of the envelope of these oblong non-spherical waves. This direction was abnormal with respect to the law of sines—hence the name given to the ray.

These experiments further illustrated Huygens's experimental genius, which he had already revealed in his mechanical studies of the pendulum. His experiments on spar crystal showed for the first time the *anisotropy* of certain transparent media: the fact that the light passing through them can propagate at different speeds depending on the direction of its rays. That an incident ray can give two refracted rays also implicitly indicated a principle of superposition. Natural light had to have two superimposed components in the incident light ray, one giving rise to spherical secondary waves on the crystal's surface and the other to ellipsoidal waves. These two components are associated, as we now know, with two orthogonal directions of light's transverse polarization. It was not until the beginning of the nineteenth century, with Malus's experiments observing the sunlight reflected by the windows of the Palais du Luxembourg, that the phenomenon was fully understood and the link between polarization and birefringence established. But Huygens had set the stage by attempting to precisely describe the bizarre optical properties of a small piece of rock brought back from a distant expedition.

It is also remarkable that Huygens was able, despite the aforementioned limitations, to lay the foundations for what would become, a century later, the interference experiments I will describe elsewhere in this book. These experiments illustrate the fundamental principle of optics (and later of electromagnetism), which I have already mentioned in reference to birefringence: wave superposition. Huygens had another intuition of this superposition when he noticed that several waves can propagate in a medium without disturbing each other, combining their effects on the aether, which could respond independently to solicitations from different light sources. Huygens noted that light rays from different sources crossed in space without colliding, unlike what material particles would do. Two observers can see different objects even if the trajectories of the light rays forming these objects' images overlap. What one sees is in no way disturbed by what the other sees. This principle of superposition would be generalized much later to quantum physics, with strange consequences.

It is worth recalling that the wave hypothesis of light had been considered by other scientists before Huygens, notably the Italian Grimaldi, the Englishman Hooke, and Father Pardies, a French physicist who died in 1673 before he could publish the results of his work. But Huygens was the first to build a coherent wave theory explaining a large number of luminous phenomena in a general framework. His primacy and the importance of the principle he discovered were summarized by Leibniz in a letter to Huygens a few months before the latter's death in June 1694:

> Certainly Mr. Hooke and Father Pardies were careful not to arrive at the explanation of the laws of refraction by the thoughts they had about the ripples. It all consists in the way in which you have taken it upon yourself to consider each point of the ray as radiating and to compose a general wave of these auxiliary waves.

Newton, light particles, and color

Huygens's *Treatise on Light* says nothing about color. We know now that this essential property of radiation is associated with the frequency of light vibrations, the colors of the rainbow corresponding to radiations whose wavelengths range from 0.4 microns (blue) to 0.7 microns (red). Since he could neither estimate nor measure such dimensions, Huygens did not give an interpretation of colors in his treatise. Paradoxically, it was Newton who, in his great *Opticks*, published in 1704, addressed this question even though he did not believe in the wave nature of radiation. Newton describes his famous experiments on the dispersion of white light, carried out in the 1670s, of which Huygens must have been aware.

The subtitle of Newton's book—*Treatise on the Reflections, Refractions, Inflections, and Colors of Light*—summarizes all the questions he addresses. I will skip over the part of the book that deals with the laws of ray propagation in transparent media; adopting a point of view similar to Descartes's, Newton derives the sine laws by advancing the corpuscular hypothesis that light can be described as a set of particles moving in a straight line in a homogeneous medium and undergoing influences that change their trajectories at the interfaces. This model, in which light particles are thought to be attracted as they enter a transparent medium by a force normal to the diopter that separates it from the air, led Newton to admit—contrary to Fermat and Huygens—that light propagates faster in transparent media than in air.

Its analysis of light dispersion makes *Opticks* one of the great works of physics. Newton details the famous experiments in which he used prisms to break down sunlight into a spectrum ranging from red to violet. He describes how he obtained white light by recombining the radiations of different colors, and correctly explains the dispersive properties of the prism by assuming that the different colors correspond to different values of the sine ratio of incidence and refraction. In other words, the speed of light in a material medium depends on its color. According to Newton, blue light, which undergoes greater deflection, should propagate through glass faster than red light. The wave theory, of course, tells us the opposite.

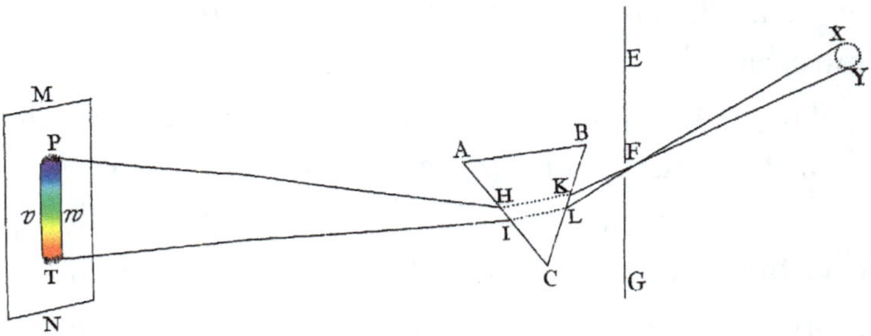

Figure II.9. Drawing 13 of Chapter 1 of Newton's *Opticks*, showing his first light dispersion device. Light rays emitted by the sun (XY disk) enter his darkroom through the F hole drilled in the EG shutter. The stain of width vw and length TP, observed after crossing the prism ABC on the sheet of paper MN, is elongated and colored. The spectrum has been added. Newton, whose figures were ink drawings, simply gives a qualitative description of this spectrum by specifying that the edge T (the least deviated) is red and the edge P (the most deviated) purple.

Newton also shows how colors can appear in the iridescence of soap bubbles. He describes, accurately and in detail, the colored rings he observed when white sunlight passed through a spherical lens on top of a flat glass slide, known as *Newton's rings*. He also mentions the colors appearing in the diffraction of light when it passes in the vicinity of small obstacles, noting for example the iridescence of spider webs illuminated at a certain angle by sunlight. He also takes an interest in the colors of peacock feathers. We know now that these are all interference and diffraction phenomena that manifest the undulatory character of the radiation. The book's paradox is that Newton describes in precise detail, through beautiful experiments that were great feats for their time, phenomena that his corpuscular theory of light couldn't really explain.

Although he was unable to present a convincing theory of colors, it should be noted that Newton's extraordinary physical intuition made him understand that the colors observed in his thin-sheet experiments must have depended on a phenomenon that made the light corpuscles "feel" the distance between the surfaces they crossed. Reaching these surfaces, he imagined, the light particles induced a disturbance in the transparent medium, which he called a *fit*. This perturbation propagated from one diopter to the other and induced, depending on the separation of the two surfaces, a greater tendency for the particle to be transmitted or to be reflected on the input diopter. This Newtonian *fit* can appear as a kind of vibration provoking a phenomenon of reinforcement or attenuation, pushing the particles to be reflected or transmitted at the level of the wave's entrance and exit diopters. Newtonian theory thus considers light to be a flow of particles propagating in space and transparent media, but inducing phenomena of a vibratory nature in the matter through which they pass, and also in the eye that perceives them in the form of different colors.

Some have said Newton was the first to glimpse, in this very qualitative model, the wave–particle dualism of quantum physics. This is quite an exaggeration. The historical truth is that this notion of *fit* remained vague, and that Newton simply anchored in the minds of most scientists of the century to come the idea that light was made up of discrete particles.

Nonetheless, Newton's work on color led to real revolutions in optical instrumentation. His experiments on the prism were the starting point for spectroscopy, the science that analyzes the colors of light absorbed and emitted by matter. The Fraunhofer spectra mentioned in the previous chapter, which revealed the presence of the elements in the universe, were recorded in the nineteenth century by devices that Newton's prisms anticipated.

After observing that the lenses of refracting telescopes produced images with colored edges because of the dispersion of colors in the glass, known as *chromatic aberrations*, Newton decided to replace the lens on these instruments with a concave mirror, forming the image of stars and planets at its focus by reflection rather than refraction. Since the laws of reflection are totally achromatic—the angle of reflection is equal to the angle of incidence regardless of color—he obtained sharper images.

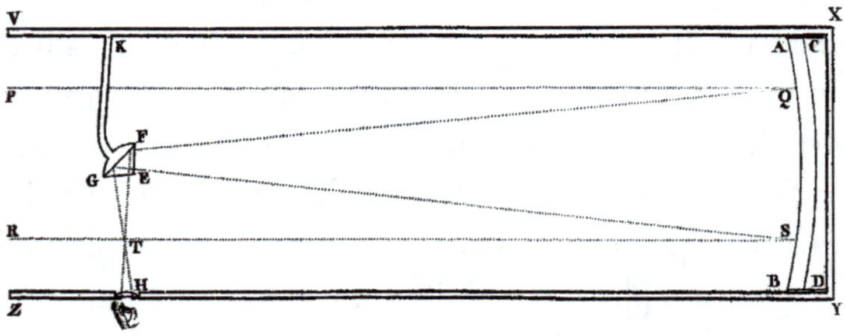

Figure II.10. Drawing 29 of Chapter 1 of *Opticks*, showing the schematic diagram of the parabolic mirror telescope. The light of a parallel beam is focused at the focus and reflected at right angles by a small mirror FG. The eyepiece is placed at point H.

Mirrors or lenses whose surfaces are spherical caps focus light only imperfectly, forming the image of a point in a small volume of finite dimension. To eliminate these sphericity aberrations and obtain a precise focal point, Newton understood, the mirror of his telescope had to be parabolic in shape. This is another consequence of Fermat's principle. The surface of revolution generated by a parabola rotating around its axis is indeed the set of points the sum of whose distances to the focus and to a plane normal to the axis is constant. All the rays coming from an infinitely distant star in the direction of this axis converge, according to Fermat's principle, exactly at the focus. Thus did Newton invent the optical reflecting telescope, which gradually replaced refracting telescopes and, two centuries later, became the observation instrument adopted by all major observatories around the world.

A century after Galileo and the beginnings of observational astronomy, then, optics had made enormous progress. Conceptually speaking, two theories of light were facing off, defended respectively by two giants of science, Newton and Huygens: a corpuscular theory—also called *emission theory* at the time—and the wave theory. Newton's greater notoriety, crowned by the success of his theory of universal gravitation, attracted the majority of physicists to his point of view.

One criticism made of wave theory, by Newton and his eighteenth-century successors, was its difficulty in explaining the straight-line propagation of light. Sound, which is a wave, can bypass obstacles and be heard even if walls separate the listener's ear from the source of the noise. But this is not the case for light. Obstacles seem to define specific areas of shadow. Huygens had tried to show with his principle of secondary waves that these only strengthen in the direction of the rays, but his explanation lacked precision because he didn't have the notion of wavelength and so couldn't really understand the phenomena of diffraction. In fact, light does penetrate well into shadow regions, but analyzing the phenomenon requires delicate quantitative experiments that neither Huygens nor Newton were able to realize. These experiments would have to wait another century for Young and Fresnel.

Let us conclude this journey through the history of light in the Grand Siècle with an astronomical advance that clarified Römer's estimate of the speed of light: Bradley's 1727 measurement of the aberration of stars. The direction in which a star appears to us depends on the respective directions of the light that reaches us and of the speed of the Earth in its orbit around the Sun. To estimate this effect in a simple case, let's assume that the observed star is at its zenith, in a direction orthogonal to the Earth's orbital speed. This speed v is on the order of 30 km/s, about one ten-thousandth of the speed of light c. The light ray coming from the star will appear inclined with respect to the direction of the zenith by a small angle proportional to the ratio v/c, about twenty arcseconds.

This deflection oscillates during the year, following the direction of the Earth's motion around the Sun, and thus makes the stars move in an apparent circular motion on the celestial vault. This observation yields a value for the speed of light in a vacuum very close to the one we know today.

This aberration effect of the position of the stars is similar to a phenomenon we have all observed. If you are traveling in a convertible under rain falling vertically, the drops will come toward you at an angle, a greater angle to the vertical the faster you drive. The windshield will protect you up to a certain point, even if you haven't had time to raise the top! For a while, Bradley's observation appeared to support the corpuscular hypothesis of light, which in effect equates light particles with raindrops in this analogy. In fact, it can be shown that the aberration of stars is also compatible with a wavelike interpretation of radiation. It should be noted that the angular aberration of starlight is not a parallax effect due to the change of the Earth's position with respect to the stars. Parallax, unlike aberration, depends on the stars' *distance* from the Earth, whereas the aberration effect depends only on the *speed* of our planet and is the same for near or distant stars observed in the same direction. The parallax effect is, as we have seen, at most a fraction of an arcsecond for the nearest stars and totally undetectable for distant ones.

Since we have just evoked the speed of light again, let us note that in order to choose between the corpuscular and wave theories it seemed an experiment was necessary. Measurement of the speed of light in a material medium must have been a critical test, since the two competing theories made opposite predictions. Such an experiment could not be carried out by astronomical observation; it had to be done in a laboratory. It took one hundred and fifty years to carry out, but by the time it was done, as we shall see, other crucial experiments had already settled Huygens's bet with Newton.

Beyond the controversy between the two men, other mysteries remained unsolved. If light was made of particles, what was their nature, and how did they carry the different colors? If it was a wave, to which properties of this wave did we owe the colors? And what made up this mysterious aether in which it propagated? It had to be infinitely rigid to transmit the shocks produced by luminous vibrations at such high speed, and at the same time formed of infinitely tenuous elements since it penetrated all transparent material media. Nor did we understand in detail the birefringence of light and the notion of its polarization.

In telling this story I have shown how different areas of science are interrelated, depending on one another for progress. There was no concept of interdisciplinarity at the time, but it was practiced naturally. This was a time when scholars like Galileo, Huygens, and Newton could grasp a wide spectrum of disciplines—practically all the scientific knowledge of the time. It is also fascinating to see how one observation, one experiment, led to another, often motivated by the discovery of something unexpected—and to realize the degree to which science was driven by curiosity, by men seeking to understand nature and its laws ever more deeply.

As we continue to follow the lineage of experiments on light in the following chapters, I will describe how the mysteries mentioned above were solved, and how this led to other, even deeper questions that would in turn lead to modern physics. Before returning to this road, however, let us stay a moment longer in the Grand Siècle and the Age of Enlightenment to talk about an adventure that may seem far off from the one I'm describing in this book, but will ultimately bring us back to it. I mean the measurement of the Earth, which during the second half of the seventeenth century and throughout the eighteenth occupied physicists, astronomers, mathematicians, and explorers, some of whom had already played an important role in the history of light.

Measuring the shape of the Earth

Jean Richer, who observed the parallax of Mars from Cayenne, carried with him a precise clock that beat to the second, undoubtedly to synchronize his observations with Cassini's. He had noted that the instrument was slower in Cayenne than in Paris, accumulating about two and a half minutes of delay per day—a difference ten times greater than the uncertainty of a clock incorporating Huygens's latest technological advances. Richer had thus discovered, likely by chance, that gravity was weaker near the equator than in temperate regions.

The explanation for this phenomenon came from Huygens again, who in the 1670s discovered the notion of centrifugal force. In order to understand this, we must recall the ideas Galileo had introduced at the beginning of the century. The Italian physicist understood that no mechanical experiment could tell whether one was stationary on Earth or moving at constant speed, for example aboard a ship sailing on a calm sea. Drop a ball on the ship and it will fall at your feet under the sway of gravity, no different than if you were at a stationary landmark. If you throw the ball, it will follow the same parabolic trajectory on the ship as if you had thrown it from the stationary landmark on Earth.

Here is Galileo, in his *Dialogue Concerning the Two Chief World Systems* in 1632:

> Shut yourself up with some friend in the main cabin below decks on some large [stationary] ship. [...] hang up a bottle that empties drop by drop into a wide vessel beneath it [...] the drops fall into the vessel beneath; and, in throwing something to your friend, you need throw it no more strongly in one direction than another, the distances being equal; jumping with your feet together, you pass equal spaces in every direction. When you have observed all these things carefully [...] have the ship proceed with any speed you like, so long as the motion is uniform and not fluctuating this way and that. You will discover not the least change in all the effects named, nor could you

tell from any of them whether the ship was moving or standing still.

Indeed, the notion of immobility has no absolute meaning. Two friends on dry land or two others on Galileo's ship are equally entitled to consider themselves "motionless"; it's a simple question of perspective. To define the situation more precisely, they are all in an inertial—a so-called *Galilean*—frame of reference, wherein a body on which no force is exerted remains in uniform rectilinear translational motion. If the Earth is an inertial frame of reference, the same applies to the ship so long as it is on a calm sea and does not undergo any change of speed relative to the Earth. If, on the other hand, the ship accelerates or slows down, it is no longer inertial and the behavior of the launched objects will be affected. If you drop a ball while the ship is accelerating or slowing down, it will follow a parabolic arc backward or forward, as if you had exerted a horizontal force on it in the direction opposite the acceleration of the landmark. You will feel this force yourself as you are thrown backward or forward as the vessel pitches. These inertial forces—forces that do not result from the interaction of the objects under study with other bodies—intervene to account for the movement of objects in a non-inertial frame of reference undergoing a change of speed, an acceleration relative to a Galilean frame of reference.

But what frames of reference can really be described as inertial? Until Copernicus, the answer had been clear throughout the history of mankind: it was "obvious" that the Earth was motionless and that the whole universe revolved around it. When this certainty disappeared, defining the Earth as an inertial landmark became more difficult. As Galileo said: *E puere se move* (and yet it moves). The Earth revolves around the Sun, which gives a small acceleration to all the reference points associated with it, but this acceleration is so small that we can neglect it (as was certainly the case during the time in question). What interested Huygens, who had just studied the forces exerted on rotating bodies, is that the Earth rotates on itself, so that any reference frame fixed relative to the ground, in Cayenne as in Paris, is in uniform circular motion corresponding to a centripetal acceleration, oriented toward the polar axis of the Earth's rotation. This acceleration, which Huygens found to be proportional to the distance from the axis of rotation and to the square of the angular velocity, is zero at the pole and maximal at the equator.

In any rotating terrestrial frame of reference, except at the pole, bodies are thus subject to a centrifugal inertial force, which must be taken into account when calculating their motions in relation to the reference frame. To get an intuitive idea of this force, imagine yourself on a rotating merry-go-round. You feel that the centrifugal force tends to pull you toward the outside of your seat, especially as the carousel rotates faster. Astronauts who practice in rapid centrifuges feel this force as much greater than their own weight.

In the case of the Earth, this rotation is very slow, one revolution in twenty-four hours, and we don't feel it because it's so small compared to the gravitational force that attracts us toward the center of the planet. The Huygens clock Richer took to Cayenne, however, was sensitive enough to feel this force in an experiment that was the first to detect the Earth's rotation by a method other than astronomical observation. Richer and Cassini, counting the oscillations of two identical clocks in Cayenne and Paris, were in the same situation as the two friends in Galileo's example. Even without looking at the outside world (the sky and the stars), they were able to realize by mechanical experimentation that the Earth was not a Galilean frame of reference.

Centrifugal acceleration decréases the acceleration of gravity near the equator by about 0.34 percent, and by only 0.15 percent at the latitude of Paris. There was thus a difference of 0.19 percent, which explains a 0.095 percent variation of the pendulum's frequency, proportional to the square root of the gravitational acceleration. This accounted for a shift in the pendulum of about eighty seconds per day between Cayenne and Paris, while the observed difference was a little less than double.

Was the centrifugal force of inertia only part of the explanation? Huygens realized in 1687 that this force intervened in another, more indirect way in the observed effect. Under the influence of its rotation, the Earth must not have been a perfect sphere, but rather an ellipsoid flattened at the poles and slightly swollen at the equator. Observation of Jupiter, which rotates rapidly on itself and has a distinctly oblate shape, showed clearly that the centrifugal force acting on a planet could deform it in such a way. The existence of fluid magma inside the Earth, which rises to the surface during volcanic eruptions, justified considering the planet to be made of malleable matter whose shape would depend on the balance between the centripetal effects of gravity and the centrifugal acceleration of diurnal rotation. Huygens built a model showing that this equilibrium led to an Earth that was ellipsoidal and flattened at the poles. Newton applied his theory of gravitation to the same problem in 1690 and found a greater flattening than Huygens's.

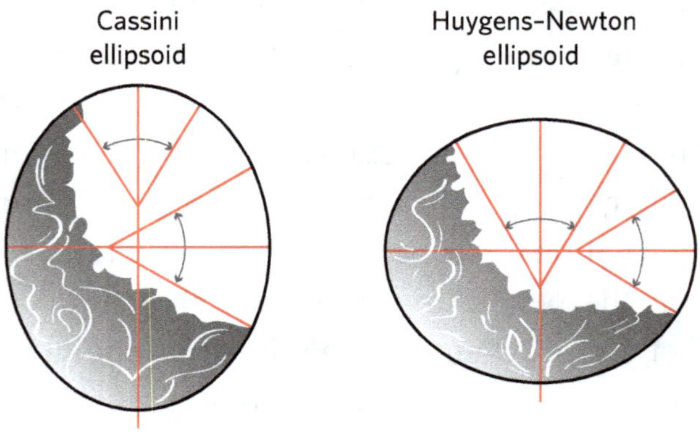

Figure II.11. The Cassini and Huygens–Newton models of the Earth's shape.

Such a deformation, in any case, must have added its effect to that of the centrifugal force acting directly on the pendulum. Flattening caused a point at the equator to be farther from the center of the Earth than a point at a temperate latitude. The gravitational force at the equator therefore had to be reduced in relation to its value in Paris; this effect was added to the difference in the centrifugal forces acting directly on the pendulum. The effect was a step toward explaining the discrepancy between the pendulum shift observed and the weaker one produced by the sole inertial force of the Earth's rotation.

The question of measuring the Earth and estimating its exact shape was thus of great scientific relevance at the beginning of the eighteenth century. Whether the Earth was really flattened became a controversial question after a large triangulation campaign was carried out to establish an accurate map of France. Louis XIV asked his Academicians and the astronomers of the Paris Observatory to remap his territory using the most efficient technologies of the time; the project continued under Louis XV. The precise refracting telescopes that had made possible the astronomical observations mentioned above could be adapted to precisely locate monuments or natural promontories in the French countryside, then make ultra-precise angular measurements. One could thus define triangles tessellating entire areas of the country, first along the Paris meridian and then in other regions. The side of one such triangle could be measured using a carefully plotted measuring rod, which could then be combined with trigonometric tables to obtain all other distances.

Coupling these measurements with astronomical surveys made it possible to measure the length of one degree of latitude along the Paris meridian. These measurements, continued by Jacques Cassini—son of Jean-Dominique, who had measured the parallax of Mars with Richer—seemed to show that one degree of latitude in the north of France was a bit smaller than one degree in the south, which led to the conclusion that instead of being oblate like an orange the Earth was oblong like a lemon, elongated toward the poles and narrowed at the equator. This went against the results of Richer's experiment and Huygens's and Newton's calculations, but Cassini had great influence and the question remained unresolved. The Royal Academy of Sciences decided in 1735 to resolve it by an indisputable measure. Let's give the floor to the Academician Maupertuis:

> The Academy was thus divided; its own lights had made it uncertain, when the King wanted to make a decision on this great question, which was not of those vain speculations which the idleness or useless subtlety of philosophers sometimes deals with, but which must have real influences on Astronomy and Navigation. In order to determine the figure of the Earth, it was necessary to compare together two degrees of the Meridian which were as different in latitude as was possible; because if these degrees go increasing or decreasing from the Equator to the Pole, the too-small difference between neighboring degrees could be confused with the errors of the observations, instead of the fact that if the degrees we count are at great distances from each other, this difference, being repeated as many

times as there are intermediate degrees, will make too considerable a sum to escape the observers.

At the Academy's suggestion, then, Louis XV sent out two expeditions—one to the polar regions, the other to Quito, near the equinoctial zone (as it was then called) in the viceroyalty of Spanish Peru—to measure a degree of latitude in two zones as far from one another as possible. Maupertuis himself led the expedition to Lapland in 1736, and the Academicians La Condamine and Bouguer led the one to Peru. The latter had extraordinary adventures that lasted more than a decade. Maupertuis reported his measurements to Paris in 1737; their precision was sufficient to show clearly that the degree of latitude at the pole was greater than the one measured in Paris. He considered this result sufficient to affirm, without waiting for La Condamine's return, that the Earth was indeed a flattened ellipsoid, contrary to the conviction of Cassini father and son. Upon hearing the news, Voltaire allegedly congratulated Maupertuis on having flattened the Earth... and the Cassinis.

When Bouguer returned to Paris in 1744, before his companion, his report to the Academy of Sciences confirmed Maupertuis's results. We now know from satellite observations, much more precise than their eighteenth-century counterparts, that the difference between the equatorial and polar radii of the Earth is 21.3 kilometers, for an average radius of 6,367 kilometers, or an ellipticity of 1/298—a value closer to the one calculated by Newton (1/230) than to the one predicted by Huygens (1/578).

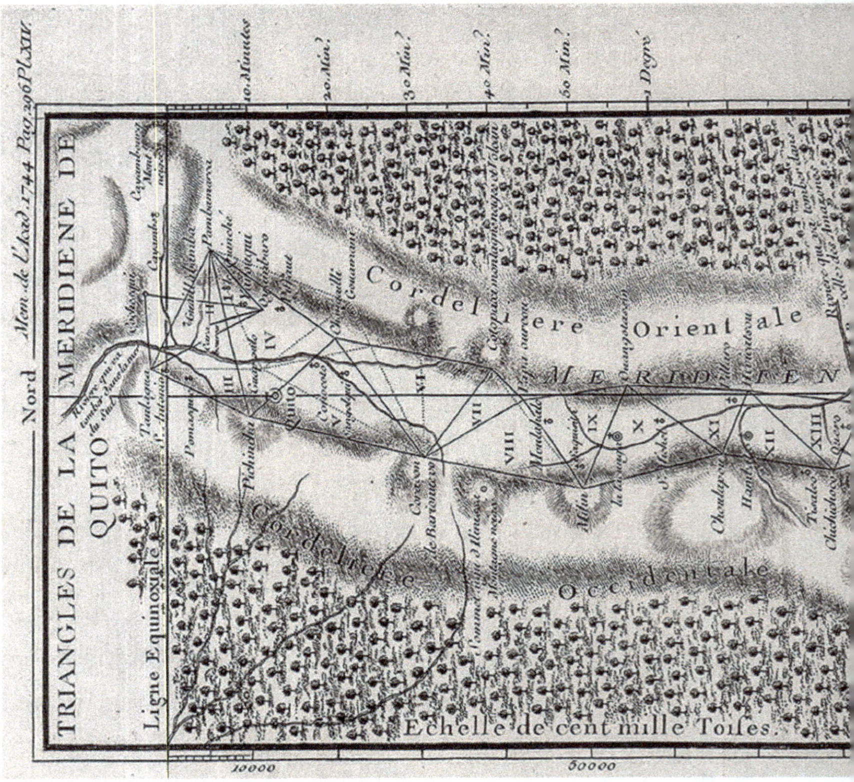

Figure II.12. A map of the Quito meridian with the triangles measured by Bouguer and La Condamine during their expedition to Spanish Peru (*Annals of the Academy of Sciences*, 1744).

This story, whose details I had forgotten, came back to me a few years ago during a visit to Ecuador. Claudine and I arrived in an ancient hacienda converted into a hotel at the foot of the Cotopaxi volcano, some seventy kilometers south of Quito. A plaque affixed by the Paris Academy of Sciences reminded us that La Condamine had stayed there in 1742. Performing his measurements on the meridian that descends from Quito to Cuenca

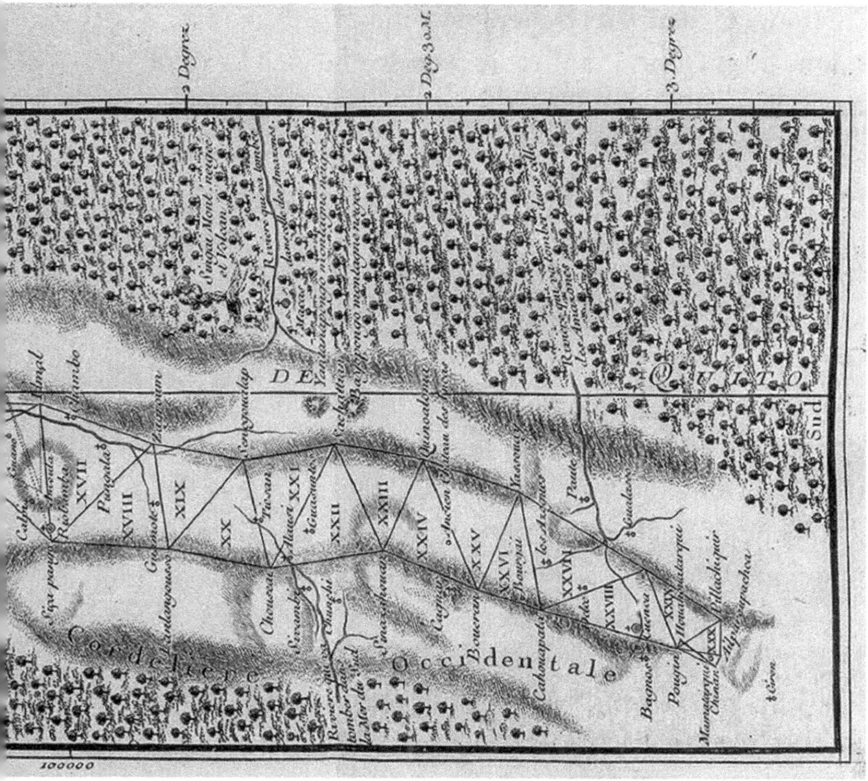

south of the equator, he observed an eruption of the Cotopaxi. Some sixty years after La Condamine's stay, this hacienda had also hosted the great naturalist Alexander von Humboldt, who during his trip to South America discovered the cold current that now bears his name along the coasts of Chile and Peru. Claudine and I had the privilege of spending two nights in the suite Humboldt had occupied in 1802.

La Condamine's journey lasted thirteen years. He came back from Peru along the course of the Amazon, before reaching Cayenne on the Atlantic coast.

As Bouguer had already reported the expedition's scientific results, his communication to the Academy of Sciences, which described the hitherto unknown lands he had crossed, was a report more concerned with geography and ethnography than with physics.

At the time, physics and geography were closely related sciences. One of the major problems of navigation was the precise determination of longitudes at sea: the natural way to achieve this was to compare local time, observable by the height of the sun or the position of the stars, with the time at the same moment under a reference longitude such as that of Paris or Greenwich. Another possible method was to observe an astronomical phenomenon such as an eclipse of Jupiter's moon Io, whose recurrences had been tabulated in local time by the Paris Observatory. Thus, one could in principle know the time in Paris anywhere in the world, provided that Jupiter was visible and one had a good, stable telescope on board. This was not really practical.

The other method was to keep on board a clock accurate enough to maintain the local time at the home port throughout the weeks or months of travel. Huygens had tried to make his pendulum clocks resistant to the pitching of ships, which were far from Galilean frames of reference.

The results were disappointing, so the British Admiralty sponsored a competition to reward the watchmaker who could make a watch more compact and more resistant to turbulence than a pendulum clock. This watch had to be based on the oscillations of a small spring rather than those of a bulky pendulum, and its imprecision had to be limited to a few seconds per month. An English watchmaker named John Harrison won this competition; his marine chronometer equipped English ships from the second half of the eighteenth century onward.

Precise measurement of the Earth continued to occupy scholars through the French Revolution and beyond. The universalism of the Enlightenment made it a noble scientific goal to define a unit of measurement acceptable to the whole world. In addition to idealistic scientists, realist merchants saw the advantage of replacing hundreds of disparate units—the feet, inches, and rods of different lengths used around the world—with a single standard. It was thought for a time that a natural unit could be the length of the pendulum with a period of two seconds whose measurement coincided approximately with the actual meter, but the variability by latitude of this standard cooled the ardor of its proponents. It would have had to be accompanied by a latitude correction, making this unit impractical.

During the French Revolution, the National Convention decided to base the length standard on the dimensions of the Earth, which belongs to all mankind, thus lending the chosen unit the desired universal character. The meter would be one ten-millionth of the distance from the pole to the equator. It was in order to define this universal meter that Delambre and Méchain undertook the triangulations I recalled at the beginning of this chapter. In this they followed in the footsteps of their planet-surveying predecessors: Cassini father and son, Maupertuis, La Condamine and Bouguer, and the many colleagues who accompanied them.

I have linked the beginnings of the history of light with that of the measurement of the Earth because these two adventures have many points in common. Both revolve around the same characters and the same discoveries in mathematics, astronomy, physics, and geography. It was by progressing in both adventures that modern science was formed, affirming its principles and values. The essential characteristics of the scientific approach as we know it today were already well established. The complementarity of fundamental discoveries and the development of instrumentation is particularly striking. This story also illustrates the unpredictability of research, which often leads scientists along surprising paths. Think of Galileo, who, after failing to measure the speed of light with his naive method, perfected the instruments that made it possible to do so some fifty years later. Think also of Jean Richer, who, having set out to measure the distance from the Earth to Mars, unexpectedly made the observation that revealed the shape of our planet. Modern science is full of such surprises as well.

A passion for precision

The history of science in the seventeenth and eighteenth centuries also shows the importance of measurement and of the passion for precision. It was by making increasingly precise measurements that the scientists of the time revealed sometimes unexpected effects with important consequences. This passion continues to drive science today. The question of a flattened Earth gives me a chance to illustrate, with the help of an analogy, where the quest for ever more precise measurements can lead. The physics of the Earth's rotation is that of an angular momentum, just like that of electrons and nuclei, discussed in the previous chapter. In this sense we can say that Huygens, Newton, and Maupertuis were interested in the spin of our planet. They showed that the existence of this spin, of the Earth's intrinsic rotational motion, caused a flattening effect and broke a fundamental symmetry considered obvious by the Ancients. According to them, the Earth must indeed have been a sphere, the perfect shape that God—or *Nature*, in Fermat's terms—had certainly chosen for it. It took a very delicate experiment, accompanied by subtle calculations, to show that this symmetry was only approximate, and that the shape of the Earth deviated from the ideal form that had been attributed to it.

A similar situation can be seen in atomic physics today. The electron has a spin, an intrinsic property of rotation, associated with a magnetic moment, as is classically the case of any distribution of rotating charges. Quantum physics considers the electron to be an elementary charge surrounded by a cloud of virtual particles, photons, electrons, and positrons, which are created and disappear at a rapid rate. I have already evoked, describing the *Lamb shift*, this image of the quantum vacuum as a swarming cloud of ephemeral particles fluctuating around the electron.

By virtue of a fundamental symmetry of the current theory of particle physics, which physicists call the Standard Model, the distribution of electricity in this cloud should not favor any direction in space. Thus the electron would produce around itself, in addition to its spin magnetic field, which has an orientation, a perfectly spherical and symmetrical electrical field. This field decreases in all directions according to Coulomb's law, inversely proportional to the square of the distance. This law is similar to that of gravitation for a mass uniformly distributed in a perfect sphere.

But some clues, coming in particular from cosmology, allow us to suppose that the Standard Model is not the final word, that the electron in fact has a very slightly non-spherical charge distribution due to the presence of hitherto unobserved particles in the quantum vacuum. I will briefly describe the cosmological arguments in support of the existence of these hypothetical particles, as they illustrate the fundamental unity of physics and the deep links between astrophysics and quantum physics.

The rotational motion of galaxies, as observed by measurements taken over ever greater distances, cannot be explained by the laws of gravitation exerted on these objects by known, optically observable matter. In order to understand the properties of these rotations and the stability of galaxies, cosmologists consider the existence of *dark matter*—so called because, not interacting with light, it is undetectable by optical instruments.

By the attractive gravitational force it exerts, this matter would ensure the cohesion of galaxies and prevent them from breaking up by counteracting the gigantic centrifugal force of their rotation. Cosmological estimates indicate that the total mass of this dark matter should be about five times greater than that of all the atoms in the universe that make up the known matter constituting us and all the objects around us. The search for evidence of the existence of this invisible matter is very active today; experiments on the distribution of charges in the electron are part of it.

If such invisible particles fleetingly populated the quantum vacuum, there could be a slight asymmetry of charge distribution in the electron, adding to a perfectly round charged sphere a small electric dipole moment made up of charges of opposite signs, slightly shifted relative to each other. A small excess of electricity in the direction the spin points would correspond to a small charge deficit in the opposite direction. The charge distribution would no longer be exactly round, but have a very slightly non-spherical shape.

This hypothetical electric dipole, aligned along the electron's spin, would modify the distribution of the electric field it produces around itself. The situation would be somewhat similar to that of the Earth, whose flattening changes the distribution of its gravitational field. But this effect would be fantastically weaker. It is difficult to assign a precise value to the size of the electron, but it can be represented qualitatively as a small ball of classical electricity with a radius of 10^{-15}

meters, surrounded by a spherical cloud of virtual particles a thousand times larger. The most precise current measurements have verified that the electron's defect of sphericity, expressed as the difference between its radii in the direction of its spin and in the transverse direction, does not exceed 10^{-31} meters. The difference of the radii of the electron charge in two orthogonal directions would thus not exceed a fraction on the order of $1/10^{16}$ of the electron's classical radius—an amount equal to one ten-millionth of one billionth! This is far from the 1/300 factor that measures the breaking of the Earth's spherical symmetry. If the electron had our planet's dimensions, the difference between its equatorial and polar radii would not exceed the size of a molecule!

As incredibly small as these orders of magnitude seem, physicists have been able to verify the perfect sphericity of the electron with this fantastic precision. These experiments, which seek to further lower the upper limit of the electron's electric dipole moment, could, if it exists, soon detect an asymmetry in the electronic charge distribution of the order of 10^{-32} meters or less. The devices that allow such measurements are much more sophisticated than Richer's clock and the telescopes used by the surveyors of the Pole and the Peruvian Andes. Lasers play an essential role in achieving such precision, which was totally unimaginable before these instruments appeared. The current precision of measurements has already made it possible to disprove the existence of certain hypothetical particles in the cloud surrounding the electron.

If a break in symmetry were finally found thanks to improved accuracy in measurements, we might discover new particles in the quantum vacuum that could shed light on the nature of the mysterious dark matter that makes up a large part of the material mass of the universe. The fundamental idea behind this type of research is the same one that led the Royal Academy of Sciences to measure the Earth in the eighteenth century. If we can answer a fundamental scientific question, we must try to do so—even if it is difficult and requires a great deal of effort. In this sense, scientists today are no different from those of the Enlightenment.

Basic science, business, power, and technology

The history of science during the reigns of Louis XIV and Louis XV thus sheds light on the links forged very early on between fundamental and applied research. The questions scientists sought to answer to satisfy their curiosity about the speed of light or its propagation in air and matter led to discoveries with practical applications for navigation and the measurement of time. They thus contributed to the development of trade in the era of expanding colonial empires in America and European trading posts in Asia. Conversely, voyages made for mercantile purposes made it possible to draw up maps and measure distances that advanced knowledge, for example by allowing the first determination of the speed of light, for which knowledge of the distance between Cayenne and Paris was fundamental.

Since this time, it has been clear that research is an expensive activity. When the Royal Academy of Sciences sent Maupertuis and La Condamine to measure the shape of the Earth, they had to finance two teams of astronomers, mathematicians, and craftsmen responsible for the rulers and other measuring devices, as well as a large supply staff. And the expeditions took months in one case, years in the other. The question of whether it was legitimate to spend so much money to satisfy the Academicians' curiosity was probably being asked even then, as the text by Maupertuis quoted above shows. He was careful to write that these expeditions were not simply about the speculations to which "the useless subtlety of philosophers sometimes attends," but that they could be of major importance for astronomy and especially navigation.

The implicit opposition between "useless" basic science and applied research useful to human activities was already there. When today's physicists write a funding proposal for their most fundamental experiments, they are bound to use the same type of argument. I suppose my colleagues seeking financial support for extremely delicate experiments measuring the electron's electric dipole moment should imitate Maupertuis and talk about the potential importance of their experiment for our understanding of cosmology.

Science as a means of ensuring a country's power and influence was also an important factor in the development of research in the seventeenth and eighteenth centuries. The almost simultaneous founding of the Royal

Society in England (1660) and the Royal Academy of Sciences in France (1666) testifies to the rivalry between the two great powers of the time. The story of Huygens, a member for many years of the Academy of Sciences and the Royal Observatory of Paris, who wrote his works in French and dedicated them to Louis XIV, also shows France's influence during the Grand Siècle. The arrival in Paris of the Italian Jean-Dominique Cassini, who founded a dynasty of scientists and astronomers, and that of the Dane Römer also demonstrate France's attractiveness in the seventeenth and eighteenth centuries.

The great scientific expeditions were undoubtedly financed to testify to the prestige of a king who could show he had the means to send his scientists around the world to answer profound questions. Today the construction of large instruments such as particle accelerators or giant telescopes, whose purpose is to answer fundamental questions, also serves to affirm the scientific and cultural influence of the governments that finance them.

Research on light and on the measurement of time and space, which intersected in the Grand Siècle and the Age of Enlightenment, continued to progress together into the present day. Visible light would soon be accompanied by invisible radiation such as radio waves, microwaves, and X-rays. The instruments that generated and exploited this radiation made it possible to measure time and survey the Earth with a precision that Huygens or Maupertuis could not have dreamed of—and, beyond that, make it possible for us to understand and to do much more.

CHAPTER 3

Daydreams in Faraday's laboratory

A few years ago, I was invited to give a lecture at the Royal Institution of Great Britain (RIGB), for this venerable academy's Friday Evening Discourses series. Founded in 1799, the RIGB is housed in a palace on the elegant Albemarle Street in London's Mayfair district, near Buckingham Palace. Friday Evening Discourses are organized according to an immutable rite, which was explained to me in detail before the event: I was to arrive, in tails, at the red-curtained amphitheater that has hosted the leading figures of English science for more than two centuries. I was to enter at 8 p.m. through the door to the left of the lectern while the session's chairman entered silently through the door on the right. I was to begin my talk on time, with no preliminaries, and finish it at the stroke of 9 p.m., marked inexorably by the clock counting down the minutes in front of me. All the more daunting, I was locked alone in the room next to the amphitheater while the audience arrived. This isolation was to ensure that even if I were overwhelmed by stage fright I would be unable to escape before the fateful hour, as one speaker had done in the nineteenth century, forcing Michael

Faraday, director of the institution, to replace him on short notice.

I accepted these constraints because one doesn't turn down the honor of speaking on light in the place where so many fundamental discoveries in the history of optics and electromagnetism have been made and shared. Thomas Young lectured there on light interference between 1801 and 1803, Faraday presented his discoveries on electromagnetic induction, Maxwell discussed color vision in 1861, and Lord Kelvin spoke in 1900 of the two famous clouds looming in the sky of classical science, heralding the revolutions of quantum physics and relativity. Such history—much more impressive than the bizarre rite in which I was to take part—convinced me to accept the challenge of speaking about quantum light in my strongly accented, Gallicism-sprinkled English.

RIGB is a unique institution, a kind of English version of the Collège de France, which also offers public lectures on major developing research topics and does not issue diplomas. It is to the Royal Society what the Collège is to the French Academy of Sciences. Members of the RIGB are generally *Fellows of the Royal Society*, just as professors at the Collège are mostly *Academicians*. RIGB and the Collège are places where laboratory research in various fields is carried out, while the two Academies are the institutions where this research is presented in the form of dissertations and recognized by prizes or awards. Young's papers to the Royal Society were answered in the nineteenth century by those Malus and Fresnel presented to the Academy of Sciences; Faraday's RIGB experiments echoed those Ampère and Biot were carrying out concurrently at the Collège.

When I visited Faraday's laboratory in RIGB's basement after my lecture, I felt with emotion just how much the history of science permeated this place. There were Faraday's devices: the first versions of electric motors, transformers, and dynamos, simple assemblies of coils and magnets that had led the self-taught scientist to introduce the concept of field in physics. There were the glass vessels and electrodes that had enabled him to establish the laws of electrolysis. In the adjoining library, I was able to look through the letters Faraday had exchanged with Ampere, each writing in his own language. Beneath their exquisite politeness, I could read the two men's intense scientific competition and the difference in their educations. Faraday's intuition made up for his lack of mathematical training. By his own admission, he was unable to follow the calculations of his French colleague—not that this prevented him from criticizing this or that aspect of his work, or from firmly maintaining his claim of precedence in the discovery of certain effects when his French counterpart contested it.

Such contact with the science of the past left me with a mixed feeling of otherness and familiarity. Otherness because I could feel the doubts and uncertainties of scholars asking themselves questions that today's students can easily answer, demonstrating the difficulty of putting ourselves in the place of the researchers of the past, of following their reasoning and conjuring their mental images when we know how the story ends. Familiarity because I glimpsed in this correspondence the same relationships that researchers today have with their colleagues around the world. Across the Channel, across the rivalries among nations just emerging from a long war, there was finally a common search for scientific truth, which didn't exclude competition or claims of precedence. On this point, nothing has changed in two hundred years.

That evening at the RIGB also left me with a certain sense of frustration. For all the treasures it offers to the initiated, the Faraday Museum is a relatively unknown and seldom visited place. And yet it was there that concepts and devices that have revolutionized our lives were born. The general public still knows little of these scientific revolutions, much less the emotion a modern-day scientist can feel faced with the traces of this past. That very night I conceived my plan to one day write a history of light—the one I began in the previous chapter, and which I continue here. In order to understand the scope of contemporary research on light and its interaction with matter, I realized, it was necessary to place it in context: to share with a non-scientific audience the lineage of ideas and the staying power of the scientific method, to make them understand in a word the profound meaning of the oft-cited adage, generally attributed to Newton, that every scientist sees the world by "standing on the shoulders of giants" who came before him.

So let us pick up the story of light where we left it, after the competing works of Huygens and Newton. Despite its name, the Enlightenment will bring few new ideas or experiments to our knowledge of radiation, and will not allow us to truly separate the wave and corpuscular theories. The decisive advances on this question will occur in the first twenty years of the nineteenth century, during the Napoleonic Wars and the Restoration. Our protagonists on both sides of the English Channel will be English and French scientists who, despite the conflicts and rivalries between their two countries, never stop collaborating scientifically.

Young v. Newton

The two primary scientists responsible for these advances were Thomas Young in England and Augustin Fresnel in France. Young, like many of his great predecessors of the seventeenth and eighteenth centuries, was a polymath: a linguist, physician, and physicist. He is known to the general public in England for having begun deciphering the hieroglyphics that French and English military expeditions brought back from Egypt. It was his French counterpart, Champollion, who would provide the complete key by analyzing the famous Rosetta Stone kept in the British Museum—but by then Young had already made some hypotheses that proved useful to the task. Contention regarding the authorship of this essential linguistic and historical discovery has long bristled between France and England, with arguments of more or less good faith, but the scientific rivalry between Young and Fresnel is a different matter, one that never abandons the objectivity that befits science.

Beginning in 1801, taking up Huygens's ideas and opposing Newton's—which took great courage in a country where the latter was revered—Young developed, in his Bakerian Lectures at the Royal Society, a wave theory of light based on a series of experiments that explained the phenomenon of light interference for the first time. In the inaugural lecture, he proposed a theory of color based on the idea that light is made of waves analogous to sound. Huygens had already made this argument, but Young added the fundamental concept of wavelength. Light vibration consisted of a succession of crests and troughs propagating in the aether—like a wave on the surface of a lake—at the speed of light. These waves overlap, add to, or subtract from each other. If they crest at the same time at the same point, their effects combine, resulting in greater illumination. If one wave's crest corresponds at one point to the other's trough, they counteract each other, resulting in darkness.

Young understood that the color of light is directly associated with its wavelength, which demystified Newton's rings. If we observe white light from the sun transmitted by a plane-spherical lens touching a transparent glass slide, we see iridescent circles centered on the point of contact. This is caused by the interference between the waves of various colors, transmitted directly, and those that bounce once between the glass plane and the spherical cap of the lens. These waves are reinforced when the difference of their paths, equal to twice the separation between the plane and spherical diopters, is an integer number of wavelengths. Conversely, when this difference is an odd multiple of half-wavelengths, the waves are destroyed. As the thickness of the air wedged between the two glass surfaces increases with the distance to the point of contact, constructive interference occurs on concentric circles whose radii depend on the wavelength. Newton's fuzzy concept of *fits* could now be replaced with a very clear model. Measuring the radii of the rings of various colors, or simply using Newton's values, Young was able to estimate the wavelengths corresponding to the different colors of the spectrum, between 0.0000266 and 0.0000174 inches, from red to violet.

These wavelengths are equal to the product of their periods and the speed of light, then known to a fairly high degree of accuracy thanks to Bradley's astronomical measurements. Young was thus able to measure the periods of light waves for the first time. Yellow light had a frequency equal to 561 million million undulations per second, a number close to 2^{49} (2 multiplied by itself 48 times!). Never before had a scientist been able to accurately estimate a quantity so remote from those of our daily experience. It was only two centuries later that lasers allowed us to design devices to directly count these gigantic frequencies and thus produce optical clocks of fantastic precision, billions of times more precise than Huygens's clocks or even Harrison's chronometers.

Thomas Young
(1773-1829)

The absolute length and frequency of each vibration is expressed in the table: supposing light to travel in $8\frac{1}{8}$ minutes 500 000 000 000 feet.

Colours.	Length of an Undulation in parts of an Inch, in Air.	Number of Undulations in an Inch.	Number of Undulations in a Second.
Extreme	.0000266	37640	463 millions of millions.
Red	.0000256	39180	482
Intermediate	.0000246	40720	501
Orange	.0000240	41610	512
Intermediate	.0000235	42510	523
Yellow	.0000227	44000	542
Intermediate	.0000219	45600	561 (= 2^{48} nearly)
Green	.0000211	47460	584
Intermediate	.0000203	49320	607
Blue	.0000196	51110	629
Intermediate	.0000189	52910	652
Indigo	.0000185	54070	665
Intermediate	.0000181	55240	680
Violet	.0000174	57490	707
Extreme	.0000167	59750	735

Figure III.1. Young's table of wavelengths (in inches) and light frequencies of different colors, from red *(top)* to violet *(bottom)*, via orange, yellow, green, blue, and indigo. The frequencies are based on the value, known since Bradley, of the speed of light (five hundred billion feet traveled from the Sun to the Earth in eight and a half minutes) (Young's Bakerian Lecture in November 1801).

Young also gave the first precise description of the phenomenon of diffraction. The shadow of a point light source illuminating a fine wire has dark and bright fringes parallel to the wire, outside and inside its geometric shadow. Young showed that these fringes correspond to the interference of light waves passing on either side of the wire. The crucial proof of this interference was the disappearance of the fringes when the light illuminating the wire was blocked on one side by a cardboard screen. The diffraction fringes observed in white light showed iridescent colors that were also explained by the notion of interference, the waves of various colors having different fringe positions.

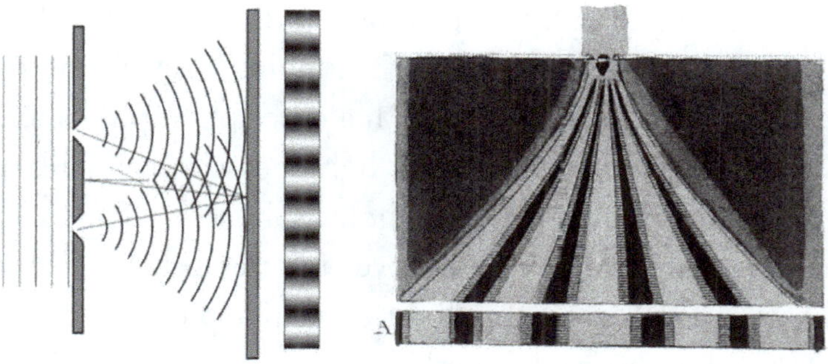

Figure III.2. Young's slits. (a) Principle of the experiment: two slits pierced in an opaque screen illuminated by an incident light beam behave like two point sources radiating waves that interfere and produce fringes on a screen (Wikimedia Commons). (b) Drawing of the interferences presented by Young in the *Philosophical Transactions of the Royal Society* in 1807.

Young's studies on diffraction led him to discover the dispersive properties of grooves traced on a reflective surface. He explained these by the interference of light reflected from these grooves' microscopic peaks and valleys. We can all observe these color interferences by looking at the light from a lamp or the sun reflected on a CD or DVD whose grooves are read by laser light in our modern audio or video players. The diffraction gratings made of parallel thin wires used by Fraunhofer to study the spectrum of sunlight exploited the same interference phenomenon (see Chapter 1). The gratings made by engraving regularly spaced grooves on glass or metal surfaces have since become essential components of light dispersion in spectrometers, replacing Newtonian prisms to advantageous effect.

The English physicist's most famous experiment is known to posterity as Young's double slit experiment, likely performed in 1801, with a description published in 1807. Young dropped a light beam on a screen pierced with two adjacent holes, obtaining two very close light sources whose waves interfered beyond the screen. Gathering this light on a flat piece of cardboard, he observed hyperbolic fringes that were alternately bright and dark. These fringes corresponded to the set of points where the path difference from the two holes was equal, respectively, to an even or an odd number of half-wavelengths. As with the wire diffraction images, the fringes observed in white sunlight were iridescent, as the interference phenomenon occurred for different path differences depending on the color.

This experiment, which Young compared to another he had performed with vibration waves produced by two divers on the surface of a body of water, gave striking evidence of the wave nature of light. Proponents of the corpuscular model tried to frame it as an interaction effect of the rays deflected by a force exerted by the holes' edges on the light particles; a similar explanation was proposed for diffraction by a wire, the rays being deflected from their rectilinear trajectory in the geometric shadow zone by their interaction with the outside of the wire. These explanations wouldn't last long in the face of the clarifications Augustin Fresnel soon offered to Young's wave theory.

Light is polarized

But before delving into Fresnel's work, we must return for a moment to the luminous glow on the windows of the Palais du Luxembourg, reflecting the rays of the sunset toward Malus's apartment (see Chapter 2). This engineer, a graduate of one of the first classes of the École Polytechnique and a veteran of Napoleon's Egyptian campaign, had the idea of observing the light reflected from the palace windows through a crystal of Icelandic spar. One face of this crystal was normal to light rays. When he rotated it around this normal, Malus observed that the light intensity transmitted by the ordinary ray passed through a maximum for a certain angle and through a minimum for an orientation increased by 90°, while the extraordinary ray evolved in the opposite way. For a certain incidence of the rays on the windows, the ordinary ray could be completely extinguished, in which case the light followed the path of the extraordinary ray. When the crystal was turned and the ordinary ray was transmitted perfectly, it was the extraordinary ray that was extinguished. From maximum transmission to extinction, the light intensity varied as the square of the cosine of the angle of rotation, a property now called *Malus's law*. (As a reminder, in a right triangle, the cosine of the angle formed by the hypotenuse and one side of the right angle is equal to the ratio of this side to the hypotenuse. The cosine decreases from 1 to 0 when this angle passes from 0 to 90°.)

These effects were analogous to those Huygens had observed by analyzing the light transmitted by two spar crystals traversed one after the other. The light reflected by a transparent surface, in this case glass (though water gives similar results), presents, like the light transmitted by a spar crystal, an anisotropy that breaks the rotational symmetry around the direction of the light beam. Malus described this property of light anisotropy as polarization, which he knew from experience was not merely a characteristic acquired by light through certain exotic crystals, but a general property of radiation manifested in all phenomena of reflection and refraction.

Though he made these observations at the same time Young was accumulating evidence across the Channel of the wavelike nature of light, Malus persisted until his death in 1812 in trying to explain the phenomenon of polarization within the corpuscular model, attributing it to an anisotropy of "light molecules." Most physicists of the time followed him on this path, notably Jean-Baptiste Biot in France and William Wollaston in England, demonstrating the resistance that still remained to wave theory.

Fresnel and the triumph of the waves

By picking up Young's experiments and extending them to the analysis of light polarization from 1815 onward, Fresnel ensured the ultimate triumph of wave theory. He showed that it explained, quantitatively and precisely, all the experiments of radiation propagation in ordinary transparent bodies as well as in birefringent crystalline media. His explanations were simpler and more precise than the increasingly complicated models proponents of the corpuscular theory were inventing in an attempt to defend Newton's point of view. Above all, Fresnel established clearly that light was a transverse wave, vibrating in a plane perpendicular to its direction of propagation. Polarized light, produced by the phenomenon of birefringence or by reflection on the surface of a transparent medium or a mirror, vibrates in a well-defined direction, whereas natural sunlight or lamplight vibrates in a random transverse direction, changing rapidly with time.

We must pause here to look more closely at Fresnel's work, so crucial a milestone does it mark in our knowledge of light and optics. While he was unaware of the precise nature of Young's work, published in a language he did not speak, Fresnel began by redoing most of the English physicist's interference and diffraction experiments, unwittingly walking in his footsteps a few years behind. He carried out his first experiments in the difficult conditions of a troubled time. A royalist, he had to flee Paris during the hundred days between Napoleon's return from Elba and the battle of Waterloo, taking refuge in his family home in Normandy without access to his optical instruments. He experimented there with makeshift means, an illustration of his great ingenuity: to create a point source of light, he isolated a ray of sunlight entering his room through a small hole drilled in a shutter, using the method Newton had already demonstrated. Since he didn't have a glass lens, he used a drop of honey to focus the ray, creating a quasi-point source of light that he used for his diffraction experiments.

During this forced exile, and in the years following his return to Paris after the Restoration, Fresnel established the quantitative laws of diffraction by obstacles by precisely generalizing Huygens's principle of secondary waves. He observed and explained with greater mathematical rigor the phenomena of fringes and interferences that Young had described a few years earlier. He showed that an obstacle of very large dimension compared to the wavelength of light

Augustin Fresnel
(1788 - 1828)

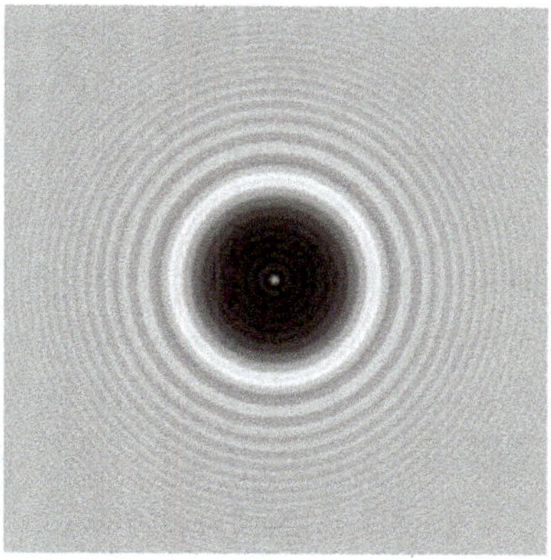

Figure III.3. The Poisson–Fresnel spot: a bright point appears in the center of the shadow of an opaque circular disc (Institut d'optique, Paris).

generally produced a sharp shadow, with a rapidly decreasing illumination within the geometric shadow. This explained the rectilinear propagation of the radiation, answering Newton's criticism that wave theory could lead only to the absurd appearance of light behind screens, even far from the edges of these obstacles.

In the discussions following the presentation of Fresnel's dissertation on diffraction at the Paris Academy of Sciences in 1819, the Academician Siméon Denis Poisson pointed out that the wave model led, in one particular case, to an impossible result—which, he argued, confirmed Newton's criticisms. The shadow of an opaque disc illuminated by a beam of parallel rays had to present a bright spot in the center of the circular shadow, since all the secondary waves coming from the wave plane around the obstacle must, due to symmetry, interfere constructively at that point. This result was manifestly ridiculous, Poisson said, so Fresnel's theory must be wrong. The experiment was performed by François Arago in front of a group of Academicians who were amazed to see a bright point actually appear in the center of the shadow. Thus did Fresnel's theory generally justify the existence of rectilinear light rays, and also explain non-intuitive phenomena that the corpuscular theory was unable to predict or interpret.

To carry out experiments demonstrating the laws of optics in the simplest possible way, Fresnel often used light of a single color, that of the sun filtered through a colored stained glass window: he called this *homogeneous light*, while today we call it *monochromatic light*. The diffraction fringes appeared clearly. Experimenting with white light, Fresnel also found all the chromatic and blurring effects of the diffraction fringes already observed by Young.

Fresnel also performed a variant of Young's double slit experiment by causing the light of a point source to interfere with itself, reflecting it by two mirrors arranged at an obtuse angle very close to 180°. The radiation from the two very close images produced the alternating bright and dark fringes Young had observed with his holes. In Fresnel's version, light did not graze any material obstacle, and the interpretation of an interaction between matter and luminous corpuscles could not be invoked.

Combining vectors and interfering waves

The generalization of the notion of wave superposition and of the phenomenon of interference is an essential aspect of Fresnel's work. Huygens's principle and Young's work laid the foundations, but it was Fresnel who gave a hitherto qualitative idea a precise form based on rigorous mathematics. To follow Fresnel's ideas, we must start by introducing the notion of the phase of a monochromatic light wave. Fresnel clarified this notion by associating an angle between 0 and 2π radians with the state of vibration of a monochromatic light at a given point and at a given time (the radian, the natural unit of angle measurement in mathematics, is defined in Chapter 1).

To visualize this phase, imagine that you've attached a small ball to the extremity of a spoke of a bicycle wheel rotating around a horizontal axis. If you look at the wheel "from the edge" by placing your eye above it in the plane in which it rotates, you see the ball oscillating in a linear sinusoidal motion with an amplitude equal to the wheel's radius. This oscillation, which is the projection on a straight line of a rotary motion, gives a solid representation of the transverse vibration of a linearly polarized light wave. The ball represents at each moment the extremity of the light vibration.

If you now look at the wheel sideways, you see the ball turning at the end of the spoke to which it is attached. The direction of this spoke makes an angle with the horizontal that increases by 2π radians with each turn. At any moment, you can associate the projected position of the ball *seen from above* with the angle that the spoke to which it's attached makes with the horizontal direction. This angle is called the *phase* of the oscillation at that instant. The choice of the phase's origin is arbitrary: we can decide that it is 0 and π when the radius is horizontal, at those moments when the ball seen from above reverses direction. The phase is then equal to $\pi/2$ and $3\pi/2$ at the two instants when its projected speed in the horizontal direction is maximal in one direction or the other.

Now imagine that you've attached another ball to another spoke of the wheel. The two balls seen from above perform oscillations of the same frequency and amplitude, but of different phases because the spokes to which they're attached are horizontal at different times. Now you should have a solid image of two waves of the same amplitude and different phases polarized in the same direction. You can vary the experiment by attaching three balls to three rays at an angle of 120° to each other, which will give you a good representation of three waves of the same amplitude out of phase, two by two, by $2\pi/3$ radians. If instead of attaching the balls to the ends of the spokes you place them at a smaller distance from the wheel's axis, you can represent waves of the same frequency but different amplitudes in the same way.

To describe the state of vibration of a wave at any point, then, Fresnel associated a vector in an abstract plane, now called a *Fresnel plane*. In the representation above, this is none other than that of the wheel to which the balls—the ends of the light vibrations—are attached. The Fresnel vector has a length proportional to the amplitude of the light vibration and rotates in its plane like the hand of a clock, making one revolution (2π radians) in one optical period. When we represent light as a vector at a given instant, we freeze it the way a snapshot "stops" the rotation of the bicycle wheel. Fresnel vectors are swift clock hands that rotate, as Young first discovered, at hundreds of thousands of billions of revolutions per second; they can be conceived of as the hands of the modern optical clocks that now count these frequencies. These vectors were already used in Fresnel's calculations two centuries ago, to virtually follow the course of time along the light rays.

This vectorial representation allows a pictorial description of interference phenomena. When several optical waves of the same frequency and polarization are superimposed at a point, the vector representing the resulting field is obtained by adding the vectors associated with these waves in the abstract space of the Fresnel plane. To add two vectors, we must translate the second one parallel to itself and make its origin coincide with the extremity of the first one. The sum vector is then the one that joins the origin of the first to the extremity of the second. If two vectors point along the same direction, they have the same phase; the corresponding waves reach their maximum amplitude of oscillation at the

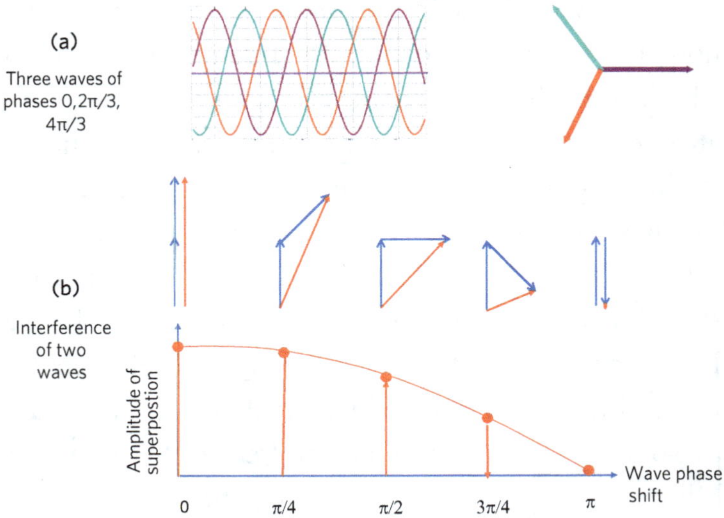

Figure III.4. Fresnel's representation of a monochromatic wave. (a) Evolution in time (*represented horizontally*) of three waves of the same amplitude, mutually phase-shifted by 120° (2π/3 radians). Their Fresnel vectors, equal in length, point toward the vertices of an equilateral triangle in the Fresnel plane. The sum of these three vectors is zero. (b) Interference of two waves whose phase difference varies between 0 and π radians: the amplitude of the resulting field (red arrows) decreases from a maximum equal to twice the amplitude of each wave to 0 when the two Fresnel vectors are opposed. The light intensity thus changes from a bright fringe to a dark one.

same time and combine their effects, giving greater illumination. In the experiment of Young's double slit or Fresnel's mirrors, the combined amplitude radiated by the two light sources is, at these points, twice that given by only one of the two sources; the intensity observed, proportional to the square of this amplitude, is quadrupled. The corresponding point is then on a bright fringe.

If, on the contrary, Fresnel vectors associated with two interfering waves point in opposite directions, they are said to be in *phase opposition* and their vector sum is canceled. The corresponding point is on a dark, unlit fringe. Between these two extremes, the Fresnel vector of the two superimposed waves is geometrically represented by the diagonal of the rhombus whose two vectors define adjacent sides. The length of this diagonal changes continuously as the path difference between the two waves changes according to a sinusoidal law, from a situation where its length is twice that of the sides of the rhombus (bright fringe) to one where it is zero (dark fringe).

These effects of superposition and interference can be generalized to the combination of any number of waves of the same frequency and polarization. The interference of three waves of the same amplitude whose Fresnel vectors point at 120° from each other, for example, gives a null resulting field. The addition of the Fresnel vectors forms an equilateral triangle, the end of the third vector coinciding with the origin of the first. Generally speaking, the sum of vectors of equal amplitude whose phases are distributed in a regular range between 0 and 2π radians forms a closed regular polygon. The superposition of the corresponding waves is thus canceled by destructive interference.

Representing the Fresnel vector makes it possible to retrieve Fermat's principle simply and generally. Let us evaluate the amplitude at point B of the light field emitted by a monochromatic source located at point A.

According to the Huygens–Fresnel principle, we can imagine a large number of successive surfaces of arbitrary shapes surrounding A, leaving B outside, and describe the field at B as the superposition of the fields emitted by the virtual sources distributed on the last of these surfaces. In turn, the field at any point on this surface can be seen as the superposition of the fields emitted by the sources distributed on the penultimate surface, and so on. The field at point B will thus result from the sum of an infinite number of Fresnel vectors associated with all the zigzag paths crossing all the surfaces from A to B, along which the virtual sources on the different surfaces will have accumulated their successive phases.

The direction of each vector on the path leading to B will indicate, like a clock hand, the time it took for the light to arrive from A on this path. The corresponding waves will finally overlap by constructive interference only for paths whose travel times are the same to within a small fraction of a period. All paths that deviate from this stationary time interfere in a globally destructive way because their Fresnel vectors fan out in the phase plane and do not contribute to the light amplitude observed in B. They can therefore be suppressed, for example by interposing diaphragms between A and B that eliminate them completely without changing the observed state of the radiation at point B. Again, then, we find the notion of light rays that follow the trajectories of stationary time.

A rotating vibration: circular polarization

The preceding reasoning applies only to waves oscillating in the *same direction* in real space, not to be confused with the abstract phase space. Fresnel found this out by conducting experiments with polarized light, showing that waves vibrating in different spatial directions could not give dark fringes. He showed in particular that ordinary and extraordinary rays produced by the transmission of a spar crystal never interfered destructively.

This is another consequence of the principle of superposition, seen in real space rather than the Fresnel plane. Two sinusoidal vibrations oscillating along different directions in space can never cancel each other out. This can be demonstrated by considering the combination of oscillations along two orthogonal directions, for example the Ox and Oy axes of an orthonormal coordinate system. If these are in phase and of the same amplitude, the resulting oscillation will be a vibration of 45° in the direction of the bisecting angle formed by the Ox and Oy axes. If they are in phase opposition, the resulting vibration will be aligned on the other bisector of the two axes, at 135°, but it will not be of zero amplitude.

To establish the rules of superposition of polarized light fields, Fresnel treated light as a transverse vibration of a hypothetical aether with elastic properties. The arrangement of the amplitudes of vibration from different directions can then be qualitatively understood with a mechanical analogy: consider the motion of a pendulum made of a small mass at the end of a wire. Move the mass away from its point of equilibrium in a given direction and let it go without giving it an initial speed: it will then start to oscillate along a straight line segment, producing a situation similar to that of a linear polarization. The pendulum can then oscillate either in the Ox direction or in the orthogonal Oy direction. If we launch the pendulum in the same way starting from a point with initial non-zero coordinates along Ox and Oy, we will obtain an oscillation, always linear, in a direction forming an adjustable angle with Ox. By combining the motions of two oscillators of the same phase, polarized in two different directions, we obtain a new linear polarization.

Now suppose that we move the pendulum away from its equilibrium position and at the same time give it an initial transverse kick. We predict intuitively that the mass will start to oscillate by describing an elliptical movement, which will be circular if the amplitude of the initial impulse is properly adjusted. Moving the mass in the Ox direction and giving it a kick in the Oy direction is like combining two oscillatory motions of different orthogonal directions and phases. It is thus understood that if we superimpose two linearly polarized vibrations at right angles, and give one of the components a phase advance or phase delay, we will create an elliptical polarization wave whose vector rotates at the frequency of the wave in the plane perpendicular to the light beam. If the phase advance of one vibration component relative to the other is 90° and their amplitudes are equal, the resulting polarization will be circular in one direction. If the phase shift changes its sign from +90° to -90°, we will obtain a circular polarization rotating in the opposite direction. For other values of the phase shift, the polarization of the light will be elliptical.

To polarize light circularly or elliptically, Fresnel passed a polarized beam through a thin birefringent plate with the property of propagating two orthogonal polarizations of light with different speeds, without spatial separation. Illuminating this plate with light polarized at 45° from the directions corresponding to the slow and fast propagations, he obtained a wave whose vibration components along these two axes had the same amplitude but an adjustable phase delay depending on the thickness of the plate.

For the so-called *quarter-wave* thickness, the two components had a phase shift of 90°, resulting in a circularly polarized transmitted wave. To study this new form of polarization, Fresnel passed the wave through a second plate with properly oriented axes, reconstructing a linear polarization that he could then analyze with a spar crystal. The circularly polarized light waves he discovered carry angular momentum along their direction of propagation, which they can transmit to matter when absorbed. As we saw in Chapter 1, these played an essential role in the optical pumping experiments at the beginning of my own scientific career.

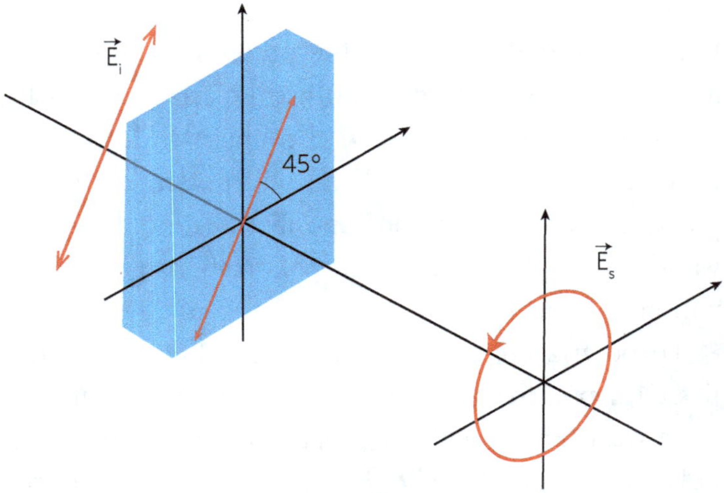

Figure III.5. A quarter-wave plate creates a phase shift of $\pi/2$ radians between linearly polarized waves along its fast (*horizontal*) and slow (*vertical*) axes. An incident wave linearly polarized along the bisector of these axes emerges circularly polarized.

To complete his analysis of interference in polarized light, Fresnel had to explain why interference was observable with *unpolarized* natural light, which he described as having a random, rapidly changing transverse polarization over time. The essential point is that the interfering waves come from a single source whose light has been separated into two components, then recombined in the plane of detection. The light polarization changes in the same way in both waves, and the interferences result from the combination of Fresnel vectors associated with light vibrations in the same direction. This direction changes rapidly with time, but the interfering waves follow these changes together and can therefore interfere constructively or destructively, depending on their path difference.

This remains true as long as the difference is not too great. Once it is, the two superimposed waves are emitted by the source at distant moments, for which the direction of the vibrations detected in the two waves and their phase are no longer the same. This loss of coherence of the two waves' vibrations results in an attenuation of the contrast of the fringes when moving away from the central fringe. The decrease in contrast is also due to the fact that light, even when filtered through colored glass, is not perfectly monochromatic. The fringes associated with components of similar wavelengths end up blurring as the path difference increases. The notion of spatial and temporal wave coherence, very important in modern optics, was already germinating in Fresnel's work.

These mechanical analogies, supplemented by simple hypotheses about the continuity constraints that light vibrations had to satisfy when crossing the surfaces separating two transparent media, allowed Fresnel to establish formulae giving the fractions of light reflected and refracted by a diopter, according to the angles the rays make with the normal to the surface. These formulae, known ever since as *Fresnel's laws*, gave an account of the polarizing property of the vitreous reflection that Malus had discovered. They showed that light falling on the interface between two media and polarized perpendicular to the plane of incidence was reflected more than the light polarized in this plane. Natural light, whose polarization changes rapidly over time, thus has its polarization component parallel to the surface favored by reflection.

For a certain angle of incidence, depending on the medium's index, the reflection changes unpolarized natural light into reflected radiation polarized parallel to the surface of the diopter. For this particular angle of incidence, called the *Brewster angle* after the English physicist who first identified it, the light reflected by the diopter is polarized perpendicular to the plane of incidence while the reflected and transmitted rays form a right angle.

Fresnel was also interested in the phenomenon of total reflection, which Descartes and Newton had already observed. When a ray propagates in a refractive medium and falls on a diopter of separation with the air, it is totally reflected if the sine of its angle of incidence is greater than the inverse $1/n$ of the medium's refractive index. No ray then crosses the diopter. Thus a diver swimming underwater and looking up at the surface sees it as a perfectly reflecting mirror if he looks in an oblique direction, whereas he can see what's happening in the air directly above his head. Fresnel's wave theory shows that in fact some of the incident light, even when totally reflected, actually penetrates the air, where it forms what is called an *evanescent wave*, appearing on the surface of the separation diopter as a thin layer of light whose thickness is on the order of the wavelength.

Light illuminated by mathematics

By representing a monochromatic polarized light wave with a vector in an abstract plane, Fresnel introduced complex numbers into physics. The vectors of the Fresnel plane—none other than what mathematicians call the complex plane—are in fact associated with a geometric representation of these numbers. While an ordinary number corresponds to a point on a so-called *real* axis, a complex number, the sum of real and imaginary parts, corresponds to the end of a vector in the complex plane, associated with the real axis and an orthogonal *imaginary* axis. The projections of this vector on these axes are equal, respectively, to the real and imaginary parts of the complex number considered. The rules of combination and superposition of Fresnel vectors that govern interference phenomena only translate the algebraic addition properties of the complex numbers they represent.

A light that consists of vibrations of *different* frequencies can be analyzed as a sum of monochromatic waves obeying the principle of superposition. In the 1820s, the physicist and mathematician Joseph Fourier, a contemporary of Fresnel, developed the mathematical tools—called *Fourier series* and *Fourier integrals*—to describe a time-dependent quantity as a superposition of components associated with amplitudes represented by vectors rotating with different frequencies in the complex plane. Fourier analysis found immediate application in the description of light waves.

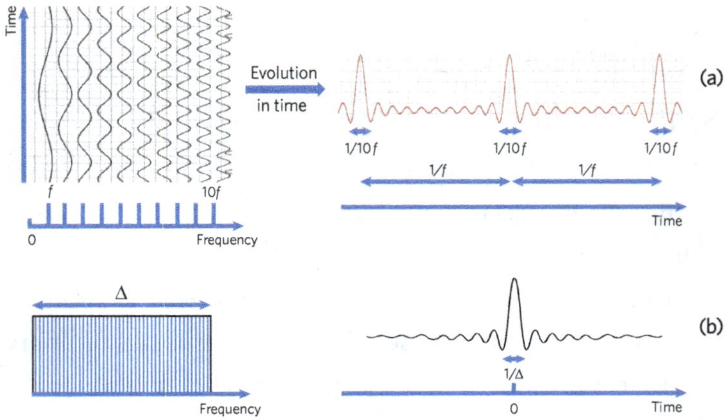

Figure III.6. Examples of Fourier series and Fourier integrals. (a) Sum of N waves of the same amplitude of frequencies Nf ($N = 1$ to 10), in phase at time $t = 0$. The discrete spectrum of this sum is a comb of equidistant frequencies. Its evolution in time is described by a periodic succession of pulses of duration $1/10f$, of period $1/f$. (b) When f tends toward 0 keeping $Nf = \Delta$ constant, the spectrum of the sum wave becomes quasi-continuous. Its temporal evolution is described by a Fourier integral, taking the form of a single light pulse of duration $1/\Delta$. In general, a superposition of monochromatic waves in phase at a given instant can be described in a complementary and equivalent way either by its frequency spectrum (*left*) or by its evolution in time (*right*). The mathematical operations which make it possible to pass from one description to the other are called *Fourier transforms*. As shown in this particular example, Fourier transforms cause the product of the widths of the spectral and temporal distributions to be on the order of 1. This mathematical result is of particular importance in quantum physics.

To start with a simple example, suppose that the polarized light emitted by a source is a superposition of monochromatic waves of the same amplitude whose frequencies are multiples of a fundamental frequency f, forming a comb of N equidistant frequencies.

Let's further assume that all these waves are in phase at an initial instant. From this moment, the Fresnel vectors associated with these waves start to rotate in the phase plane. At the instant $1/Nf$, the Fresnel vector of the fastest wave will have rotated one revolution more than that of the slowest wave. There follows destructive interference of all the waves whose Fresnel vectors span a range of 2π radians in the phase plane, and the resulting light intensity cancels itself out.

Out of phase with one another, the various components of light nonetheless continue to rotate. After a time $1/f$, they will all have completed an integer—albeit a different one—number of revolutions and return to phase. The sum wave will reappear. The optical frequency comb is thus composed of a succession of light pulses, of durations on the order of $1/Nf$, regularly repeating themselves every multiple of $1/f$. This comb constitutes a periodic Fourier series. In Fresnel's and Fourier's day, such waves dephasing and rephasing at regular intervals were only theoretical mathematical constructions. They can now be generated experimentally with lasers, and play an essential role in the extremely precise optical clocks I have already mentioned.

Let us now compress the frequency interval f toward 0 by increasing the number N of components so as to keep Nf constant, equal to a finite spectral width Δ. The Fourier series then becomes a continuous sum, called a *Fourier integral*. As the frequency interval between the components tends toward zero, the time period $1/f$ of the sum wave becomes infinitely long and the light field is made up of a single pulse of duration $1/\Delta$.

The Fourier integral thus describes how an isolated light pulse is analyzed as a sum of monochromatic waves dispersed in frequency over a spectrum whose width is inversely proportional to its duration. The functions representing the spectral distribution of the pulse and its temporal profile are called *Fourier transforms* of each other. The fact that these functions have widths inversely proportional in frequency and time is a fundamental property of wave superposition. The representation of the phase plane introduced by Fresnel allows us to give a simple explanation that generalizes to interference in time the one he gave for interference in space. We will see this fundamental mathematical property of reciprocity of the widths of Fourier transforms in quantum physics when we talk about Heisenberg's uncertainty relations—it will then take on a new and surprising physical meaning.

According to Fourier analysis, a perfectly monochromatic light wave, therefore one of zero spectral width, must have an infinite duration—in other words, must be a light of constant intensity for an infinite time. If the wave settles and then attenuates in a finite time interval, it automatically acquires a spectral width inversely proportional to its duration. By interrupting a continuous monochromatic wave with a shutter of aperture duration T, a pulse of spectral width on the order of $1/T$ is obtained. We can try to verify this with a spectroscopy experiment, dispersing the light of the pulse in a spectrograph. In Fresnel's time, the shortest shutter times were on the order of one microsecond, which corresponds to a spectral width on the order of 1 MHz, very small compared to the resolving power of prism or grating spectrometers of the time. Nowadays, thanks to lasers, we can achieve very short light pulses on the order of the picosecond (10^{-12} seconds), the femtosecond (10^{-15} seconds), or even less, and their spectral spread, which reaches hundreds of billions of hertz, can be easily measured, confirming the predictions of Fourier analysis.

The properties of the Fourier transform make it possible to establish an essential condition that a frequency measurement must satisfy in order to achieve a given accuracy. If we want to distinguish in a spectrum two neighboring frequencies f_1 and f_2, the experiment must last at least a time $1/(f_1 - f_2)$, the longer the duration the smaller the difference. This leads us to a principle not unlike a moral precept: the greater precision one seeks to obtain, the more time (and care) one must devote to the measurement.

It should be specified that a spread spectrum corresponds to a short light pulse only if the waves that constitute it are in phase with each other at a given moment and they all continue to oscillate at their natural frequency afterward, without phase perturbation over time. This is not the case with natural light, sunlight, or the lamps Fresnel used, whose different frequency components undergo rapid and random variations in amplitude and phase. For such sources, frequency spreading does not imply reciprocal spreading in time. This explains why sunlight, which covers a wide spectrum, produces continuous illumination at one point rather than a brief flash of light.

Physics has since taught us that natural light is in fact a superposition of short light pulses overlapping in space and time, emitted independently by the source's atoms. In keeping with Fourier's laws, each atom emits a wave train whose duration is limited by the inverse of its emission time.

Back to the speed of light

As early as the 1820s, the wave theory of light had made decisive advances in describing all observable optical phenomena with great precision, in a mathematical formalism that gave a simple and general account of the principle of wave superposition. It was becoming increasingly difficult to support the corpuscular theory, which had to multiply *ad hoc* hypotheses and which predicted phenomena contrary to what was observed for certain effects of light propagation in anisotropic crystalline media.

One decisive test remained to be done: measuring the speed of light in a material medium. The corpuscular theory predicted that this would be greater than the speed of light in a vacuum, while wave theory suggested the opposite. Proposed by Arago in 1838, the test was carried out independently by two young French physicists, Hippolyte Fizeau and Léon Foucault. Their experiments were the first terrestrial measurements of the speed of light, which until then had only been evaluated by astronomical observation.

In a first experiment in 1849, Fizeau measured the speed of light in the air using Galileo's original idea.

This consisted, as we have seen, in quickly interrupting a light beam making a long round trip and measuring the time the signal took to return to the light source. Fizeau carried out his experiment not between two Tuscan hills but between two Parisian locations: his house in Suresnes, a suburb of Paris, and his parents' apartment in Montmartre, a hill in the north of the city—a distance of 8.6 kilometers. He also improved on Galileo's method by replacing the human assistant who reflected the light after an inevitable latency with a mirror that reflected it instantly.

Fizeau used two optical devices, analogous to astronomical telescopes, one in Suresnes and the other in Montmartre, so that the light rays would make the 17.2-kilometer round trip without diverging too much. Instead of discovering and hiding the light by hand, he used a rapidly rotating gear wheel whose indentations let the light pass through in short pulses of a few tens of microseconds. The light source was an incandescent piece of chalk. Its light, focused by a condenser lens, was sent toward the mirror in Montmartre by reflection on a properly oriented semi-reflective blade. Just after this blade, the light ray crossed the edge of the rotating gear wheel at the focal point of the condenser lens, before being transformed into a parallel light beam by a telescope. The wave arriving in Montmartre was focused by the second telescope on the mirror and reflected before retracing its path in the opposite direction, returning to Suresnes through the gear wheel. The light finally fell back on the semi-reflecting blade, and the ray transmitted through this blade was focused by an eyepiece toward Fizeau's eye.

When the wheel turned slowly, the light reached Fizeau's eye in the form of intermittent light pulses, leaving a persistent impression on his retina of a continuous weak luminous flow. When the rotation speed reached a speed of 12.5 revolutions per second, the light would suddenly go out, reappearing when the speed increased beyond this value. Fizeau understood that for this speed of rotation, a notch in the gear wheel was replaced by the next tooth in the time it took the light to make its round trip, thus blocking the light's reflection. Repeating the observation several times, using a tachometer to measure the rotation speed corresponding to the cutoff, Fizeau determined the time light took to travel from his house to Montmartre and back, deducing a value of 315,000 km/s for the speed of light in air—a result close to the one deduced from the aberration of stars measured by Bradley the century before. It should be noted that the speed of light in air must have differed from that determined by astronomical observation in a vacuum by only 0.03%, the air index being close to 1.0003.

Further reducing the distance traveled by light, Foucault was the first to measure its speed in the confined space of a laboratory. By reducing the wave's path to a few meters, he created a device that allowed a direct comparison between light's propagation times in air and

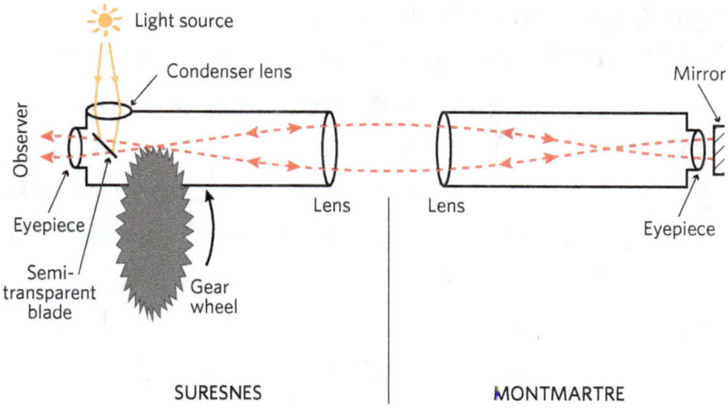

Figure III.7. Schematic diagram of the first terrestrial measurement of the speed of light by Fizeau (1849).

in a material medium. The path the light traveled was only fifteen meters, requiring a high rotational speed for the device filtering the light beam. Foucault replaced Fizeau's wheel with a small rotating mirror *m*, whose rotation frequency could reach 800 Hz. A light ray (coming from an image of the sun produced by a device called a heliostat) crossed a diaphragm at a point *a* before being focused on the mirror *m*. When the mirror *m* was stationary and properly oriented, the ray reached a concave reflecting mirror *M* placed 7.5 meters away and returned to *m to* be finally refocused in *a*. In order to distinguish between incident and reflected light rays, a semi-reflecting plate in front of the detection eyepiece separated the forward and return beams and formed the image of *a* in *α*, symmetrical to *a* with respect to the plate. When the mirror *m* was rotated, the reflected beam was spread over a wide range and periodically managed to return to the eye, forming a low-intensity residual image at a point almost coinciding with α. When the mirror's rotation frequency reached a few hundred hertz, Foucault observed the image no longer at *α* but at a point *α*' shifted by a fraction of a millimeter. This shift corresponded to an angle of a few minutes between the directions of the incident (*am*) and reflected (*ma*') rays. The ray coming back to the source was indeed deflected by twice the angle at which the mirror *m* had turned during the light's propagation time on the round trip *mM*. By measuring the distance between *α* and *α*' and the mirror's rotation frequency, Foucault deduced the speed of light in air, obtaining a result close to Fizeau's.

He then repeated the experiment by placing a tube filled with water between *m* and *M*, and found that the light arrived at a point *α*" farther away than *α*'. It was reflected by the rotating mirror *m* in a direction (*ma*") forming with (*am*) an angle greater than (*ma*'). The light thus had a lower speed in water than in air, in the ratio $n = 1.33$ of the indices of these two media. Finally, Foucault modified the device so that the light deflected by *m* passed along two symmetrical beams at once, a space in air and another in water, the two rays reflected by two identical mirrors *M*' and *M*. Comparing the positions *α*' and *α*" of the source point images obtained in the same experiment, he was then able to directly measure the ratio of light propagation speeds in the two media without even having to measure the rotation

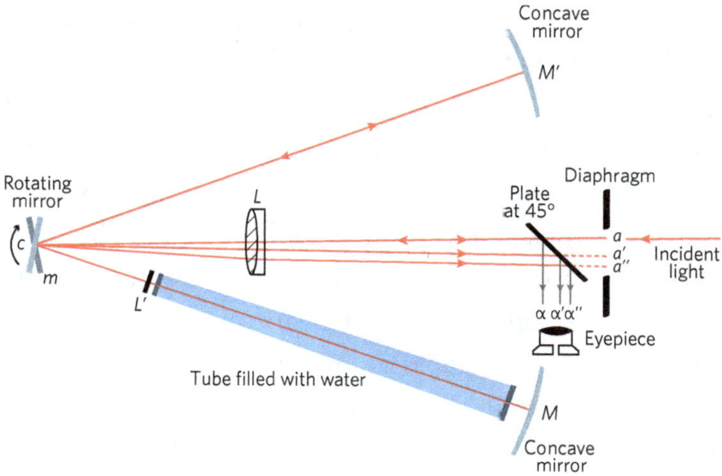

Figure III.8. Diagram of Foucault's experiment to measure the speed of light in air and water (1850). This was the first determination of this speed in a laboratory, over a distance of a few meters.

frequency of mirror *m*. This crucial experiment delivered what was thought to be the *coup de grâce* to the corpuscular interpretation of light.

It was also, incidentally, the nail in the coffin of the friendship of Foucault and Fizeau, who had collaborated closely in the 1840s in their research on the emerging fields of photography and interferometry. A few weeks after Foucault's experiment, Fizeau carried out a very similar one to measure the speed of light in water, differing only in a few technical details. He confirmed that light went slower in water than in air. Though he was the first to prepare the experiment, he was not the first to make this important discovery. When he learned that his friend was about to step on his toes, he suggested in vain that they avoid a "race to the bell tower" and instead partner to carry out the work together. Foucault declined the offer and went ahead to preempt Fizeau. This would not be the last time a friendship between researchers succumbed to competition to obtain an important scientific result.

Fizeau consoled himself the following year by measuring the speed of light in a stream of water moving in the direction of the light beam or in the opposite direction. He used a method radically different from his gear wheel, carrying out an interference experiment that measured the path difference between a light beam in a water current traveling in the direction of the current and a light beam traveling in the opposite direction. He showed that the current transmitted to the wave's speed in the fluid only at part of the speed of its flow. This property, which Fresnel had intuited, would not be truly explained until half a century later, in the framework of Einstein's special theory of relativity.

Foucault, meanwhile, refurbished an improved rotating mirror experiment in 1862, at the request of Urbain Le Verrier. The discoverer of Neptune had become director of the Paris Observatory and wanted as precise as possible a value of the speed of light in a vacuum in order to refine the estimate of the astronomical unit, the distance between the Earth and the Sun. Foucault increased the accuracy of his 1850 measurement by lengthening the path of the light, which he bounced several times across a zigzag of fixed mirrors before it returned to the rotating mirror; he obtained the value of 298,000 km/s, very close to the one accepted today.

This experiment was performed at the Paris Observatory, the same place where Römer first evaluated the speed of light. Thus we return, at a crucial moment in our journey through the history of classical light, to the very place where it began.

Indeed, by the early 1860s we had reached a turning point in this history, where discoveries about light would meet those that had progressed independently to elucidate electrical and magnetic phenomena. At the confluence of these two lines of research, Maxwell would establish between 1861 and 1865 a theory that unified electricity, magnetism, and optics and unraveled some of light's mysteries while heralding new ones. But before we turn to this last chapter in the history of classical light, let's quickly summarize where it had led science in the mid-nineteenth century.

It had been established that light was a superposition of transverse waves whose lengths were distributed between 0.4 and 0.7 microns from violet to red, vibrating at frequencies on the order of millions of billions of hertz and propagating in a vacuum at a speed c close to 300,000 km/s. Transparent material media were characterized by an index n, a real number greater than 1 depending on the frequency and, in anisotropic media, the direction and polarization of the light ray. The speed of light in a transparent medium, c/n, was smaller than in a vacuum and depended on the light's color, which explained the chromatic dispersion effects of prisms and thin plates. All the effects of interference, diffraction, and spectral dispersion of light were governed by a principle of superposition, expressed mathematically in a formalism that represented the amplitudes of the waves with complex numbers combining as vectors in an abstract plane. The attenuation of light in an absorbent medium was treated by introducing the notion of a complex *n-index*, the length over which the light extinguishes being inversely proportional to the imaginary part of this index.

A number of issues remained unresolved, most obviously the nature of the light vibration. All known vibratory phenomena propagated in a medium whose elements they moved, whether it was sound animating air molecules with oscillatory movements; waves on the surface of a liquid making a floating plug oscillate in the vertical direction; or vibrations of a solid surface, such as a drumhead, with standing waves that could be materialized by sprinkling it with a powder that accumulated along the lines where the surface was stationary.

The idea that naturally presented itself to physicists was thus that light was a vibratory manifestation of a hypothetical medium, the aether, whose contradictory properties had intrigued them since the time of Descartes and Huygens. This medium was thought to be constituted of very tenuous elements, since it penetrated all the transparent media. At the same time, it had to be extremely rigid to be able to transmit the luminous vibrations at such a great speed. Finally, it had to have very particular elastic properties to propagate only transverse vibrations of a very different nature from the longitudinal vibrations of sound. In the first half of the nineteenth century, several physicists took an interest in this aether and built variously convincing models intended to describe its paradoxical properties.

Beyond the problem of the aether, another mystery implicit in the scientific work of the time was the nature of the matter that light encountered in its propagation, whether the surface of the diopters on which it was reflected or transmitted, or the homogeneous media—glass, water, or crystals—through which it passed. What, for example, made certain media optically anisotropic? Was it a characteristic of the material that constituted them or a property that the aether that bathed these media acquired on contact with it? When an opaque body absorbed light, it was heated. What was the nature of this caloric element that passed from light to material bodies? What role did the aether play in this exchange?

It even seemed that this heating could occur when an opaque body was placed in the extension of the visible spectrum of sunlight spread by a prism. This observation, made by the astronomer William Herschel in 1800, was confirmed a few years later by Thomas Young, who thought this "caloric" radiation was an undulation of the same nature as light. Fizeau and Foucault, during their period of friendly collaboration, had carried out experiments to prove that there was indeed infrared radiation giving rise to interference, but their experiments were not very precise. On the other side of the spectrum, physicists had also suspected, since the observations of the German Johann Wilhelm Ritter, that there was ultraviolet "chemical radiation" with the property of blackening a paper soaked in silver chloride. What was the nature of this radiation?

To answer all these questions, it was not enough to explain what the aether was. Fundamental questions had to be answered about the nature of the matter with which light interacted. The mysteries of light could not be separated from the mysteries of the material environment—this at a time when the existence of atoms was still hypothetical. The second half of the nineteenth century, starting with the work of Maxwell, and in particular the beginning of the twentieth, with the advent of the theories of relativity and quantum physics, would bring decisive answers to all of these questions.

From the salons of the Enlightenment to Faraday's laboratory

We come now to the convergence of optics and the other great current that traversed physics in the first half of the nineteenth century: discoveries in electricity and magnetism. Until the end of the previous century, the properties of magnets and the manifestations of electricity had remained largely mysterious. Natural magnets, mainly iron oxides such as magnetite, had been known since antiquity, and various forms of compasses, consisting of small iron needles magnetized by rubbing against magnetite and pointing in a direction close to the Geographic North Pole, had been used for navigation by Chinese and European merchants since the Middle Ages.

The first scientific treatise on magnetism was written by William Gilbert, an English physician and astronomer of the Elizabethan period, contemporary of the young Galileo. Gilbert described the attractive and repulsive properties of magnetic poles: two poles of the same nature repel each other and two poles of a different nature attract each other. He also suggested that our planet was a gigantic magnet exerting a force on the needles of compasses, their north-seeking poles being attracted to the Earth's Magnetic Pole, close to the geographic pole indicated by the North Star. He carved a model of our planet into a piece of magnet, and on this replica described the network of lines of force that guided sailors' magnetized needles. It is striking to note what close links both magnetism and optics had to astronomy and navigation from their inception.

Gilbert also considered the magnetization produced by rubbing iron against magnetite in comparison to the electrification of substances such as amber, which acquire the property of attracting light objects, dust or small pieces of paper, after being rubbed by a piece of fabric, leather, or fur. He found that while the forces exerted by magnets were permanent, those produced by electrified substances vanished after a time. He coined the term *electricity* (from the Greek word for amber) to describe these fleeting effects, maintaining that electricity and magnetism had nothing in common even though the two phenomena could be induced in a similar way.

The eighteenth century saw the development of electrostatic machines that accumulated electric charges by friction on two metal parts insulated from one another. The friction could be obtained by rapidly rotating a glass wheel against leather pads, the charges collected by metal combs rubbing against the wheel. Scientists learned to store these charges in electrical capacitors such as the Leiden bottle, a construction of crumpled tinfoil inside a glass enclosure impervious to electrical charges, with a rolled sheet of metal on the outside. The study of these devices led to the understanding that there were two forms of electricity, positive and negative, coexisting in equal quantities in neutral matter. Electrostatic machines used friction to create an imbalance between these two forms, accumulating an excess of positive charges on one side of the bottle's insulating glass wall and negative charges on the other. When the inside and outside of the bottle were connected with a metal wire, the charges balanced each other as they flowed along the wire. This electric current could also pass through an experimenter's body if he bridged the two poles by placing his hands on it; it could also be transmitted along a chain of people holding hands. The resulting electric shock, a veritable lightning strike that caused the participants' muscles to contract violently, was a fashionable attraction among the spectacular experiments carried out in princely courts and bourgeois salons. Benjamin Franklin, ambassador in Paris of the American insurgency against the English and a regular visitor to these salons, was the first to distinguish between the two forms of electric charge. By inventing the lightning conductor, he demonstrated that lightning was of the same nature as the currents produced by the flow of charges between the poles of a Leiden bottle.

The first truly quantitative electricity experiment was carried out in 1784 by Charles Augustin de Coulomb, who used a torsion balance to measure the force between two electrically charged metallic balls. This balance consisted of a horizontal insulating rod suspended from a wire in the middle. Two identical small metal balls glued to the ends of the rod could rotate in the horizontal plane under the effect of the twisting of the wire when the assemblage was moved from its equilibrium position. One of the spheres, electrically charged by contact with one pole of a Leiden bottle, approached another sphere carrying the same charge and placed on the circumference traveled by the spheres attached to the torsion pendulum. The force between the two charged spheres made the mobile contraption rotate until it was exactly balanced by the wire's twisting force, proportional to the angle of rotation with respect to the equilibrium position. After calibration of the instrument, measurement of this angle gave the value of the electrical force between the two charges. Coulomb showed that charges of the same sign repel each other and charges of opposite sign attract each other, and that the force of attraction or repulsion is proportional to the product of the charges and inversely proportional to the square of their distance. He repeated the same experiment with magnets to show that the attractive force between a south pole and a north pole is also inversely proportional to the square of their separation. The laws of electric and magnetic forces discovered by Coulomb are mathematically similar to that of universal attraction—also inversely proportional to the square of the distances, except that gravitational masses always attract each other.

It was with the aid of a torsion balance similar to Coulomb's that the Englishman Henry Cavendish measured, in 1797, the absolute value of the gravitational force between two masses of lead. This experiment determined the gravitational constant G involved in the expression of Newton's law, giving the first estimate of the mass M of the Earth. The acceleration of the Earth's gravitation g is indeed equal to GM/R^2, where R is the Earth's radius. The measurement of G when g and R are known leads immediately to the value of M, equal to about 6×10^{24} kilos. Thus the Enlightenment, during which the shape of the Earth was determined, culminated scientifically in a fundamental experiment that established its mass, another important parameter of our planet.

Coulomb's experiments were concerned with the forces that stationary electrical or magnetic charges exerted on each other. Electricity and magnetism studies would make major advances when the experiments were extended to include moving charges. A first step in this direction was taken by the Italian physician Luigi

Galvani, who conducted the first electrophysiology experiment at the end of the eighteenth century when he showed that electrical fluid flowing along metal wires could produce muscle contractions in dissected frogs. He noticed that contractions also occurred when he connected two points on one of the frog's nerves to the ends of a loop formed by two wires of different metals placed end to end in a humid atmosphere. This experiment gave Alessandro Volta the idea for the battery that bears his name, the voltaic pile. In 1799, setting aside Galvani's frogs and biological considerations, he made a stack of zinc and silver discs separated by cardboard discs soaked in salt water, and obtained an electric current by using a metal wire to connect the zinc and silver discs at the base and the top of the stack.

Subsequent experiments at the beginning of the nineteenth century showed that this current was caused by chemical reactions that occurred on contact with metal surfaces soaked in salt water, zinc oxidizing on contact with water and hydrogen being released on contact with silver discs. We know now that the oxidation of zinc releases electrons, which flow into the wire and through the cell, reducing the hydrogen present in the water in the form of positive H+ ions. This hydrogen is transformed into a neutral H_2 molecular gas that escapes the cell as bubbles in the salt water.

Galvani's and Volta's experiments caused an immediate sensation in Europe, attracting the attention of scholars and even politicians. First Consul Bonaparte attended a demonstration by Volta in 1800 and founded an Academy prize for the author of the best paper on what were then called *galvanic currents*.

Gian Domenico Romagnosi, an Italian jurist from the city of Trento, took part in this competition and submitted a text to the Academy of Sciences in 1802 that rather obscurely described an experiment which tended to show that a galvanic current had an effect on a magnetic needle. This observation contrasted with the idea—entrenched since Gilbert and supported by Coulomb—that magnetism and electricity were totally different phenomena.

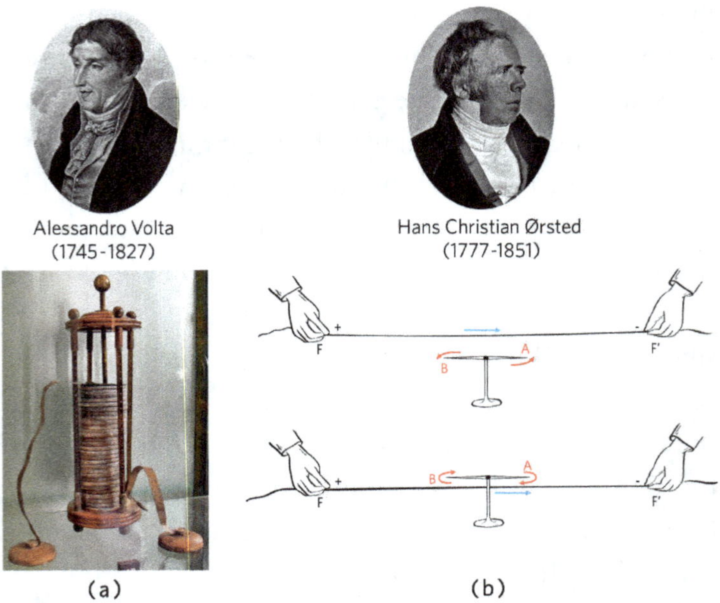

Figure III.9. (a) The voltaic pile. (b) A sketch of Ørsted's experiment showing the reversal of the direction of the force exerted by the current on the magnetic needle depending on whether it is above or below the thread.

What became of Romagnosi's paper is a mystery. No trace of it has survived besides the date of its receipt in the Academy's correspondence log: no report was written on it, and of course no prize was awarded to it. Was it lost, or perhaps rejected because it opposed the then-dominant theory of electricity and magnetism in enigmatic terms? Largely ignored by the scientific community, Romagnosi's work is known only from a succinct and imprecise account he gave a few years later in a Trento newspaper. It was not until eighteen years later, in 1820, that a fundamental experiment by the Danish physicist Hans Christian Ørsted unequivocally established a link between electricity and magnetism. He showed that a magnetized needle, oriented parallel to a horizontal metal wire, tended to orient itself perpendicular to the wire as soon as the circuit was connected to the terminals of a voltaic pile. He found that the transverse force acting on the needle changed sign depending on whether it was above or below the wire, thus showing that the magnetic forces exerted by the current wound in circles in planes normal to the conducting wire, the direction of these forces being reversed with that of the current. The phenomenon discovered by Ørsted was soon used in the galvanometer, an instrument that measures the intensity of the current in a circuit by its influence on the direction of a magnetic needle placed in its vicinity.

News of Ørsted's experiment, related with the precision lacking in Romagnosi's dissertation, quickly made its way around Europe. The Academician Arago, who had attended a demonstration of the experiment in Geneva, mentioned it to André-Marie Ampère, then professor at the École Polytechnique. Despite the bias he had inherited from his mentor, Coulomb, against any connection between electricity and magnetism, Ampère decided to repeat Ørsted's experiment and generalize it to other cases. He showed that a current flowing through a wire not only exerted a magnetic force on a magnet, but also attracted another parallel wire with a current flowing in the same direction, repelling that wire if the currents were in opposite directions. He experimented at his Parisian home on an apparatus combining electrical circuits with a scale to measure the forces acting on wire segments. This device, very simple in principle, is currently on display in the faculty room of the Collège de France, where Ampère was appointed to a physics chair in 1824.

Based on these experiments, Ampère worked with Biot, Savart, and Laplace to establish the formula giving the magnetic force exerted between two current elements according to their mutual distances and orientations, and that of the torque that a current element exerts on a small magnet. He also showed that a solenoid made of a wire wound around a cylinder behaves, when a current flows through it, like a magnet whose poles are located at both ends of the cylinder: the magnetic effects created by the solenoid are identical to those created by the magnet. Two solenoids carrying a current attract or repel each other depending on whether their opposite or identical poles are placed close together. This analogy

André-Marie Ampère
(1775-1836)

Figure III.10. Ampère's experimental setup for measuring forces between currents (Patrick Imbert, Collège de France).

led Ampère to hypothesize that magnetized matter is in fact made up of small loops of current on a molecular scale, a model heralding the modern atomic theory of magnetism that I sketched in Chapter 1.

Ørsted and Ampère focused on understanding the magnetic properties of permanent, time-invariant currents; the next step in understanding electromagnetism, as it was beginning to be called, was taken in the 1830s by Faraday, who showed that the motion of a magnet in the vicinity of a closed electrical circuit generated a transient current in that circuit. In some ways this was the reverse of Ørsted's experiment: while the Dane had shown that the establishment of a current in a circuit set a magnetic needle in motion, Faraday showed that the movement of a magnet in turn established a current in a circuit.

A simple variation of one of Faraday's experiments can be conducted by dropping a magnet through a vertically held solenoid. If the solenoid's circuit is open, the magnet will fall like any other mass in the Earth's gravitational field. If the circuit is closed, the magnet will pass through the solenoid much more slowly, as if held transiently by an invisible hand. The current it induces by its motion in the solenoid creates a force on the magnet that opposes its movement.

Faraday called this effect of a variable magnetic field acting on an electric current *induction*. He also showed that these experiments could be done without magnets, and instead with two independent electric circuits such as nested solenoids of different diameters. When he introduced the battery-powered solenoid into the external winding, the variation of the flux of the magnetic field created a transient current in this circuit, detected by the galvanometer connected to it.

Birth of the concept of field

Faraday had no mathematical background, just an intuitive, creative vision of the phenomena he discovered. By sprinkling iron filings on a sheet of paper stretched over a magnet or in a plane orthogonal to a wire with a current flowing through it, he observed what he called magnetic lines of force. These were wrapped around the wires of circuits carrying currents; they came out of one pole of a bar magnet and entered the other in elongated loops with their long axis parallel to the bar.

Michael Faraday
(1791-1867)

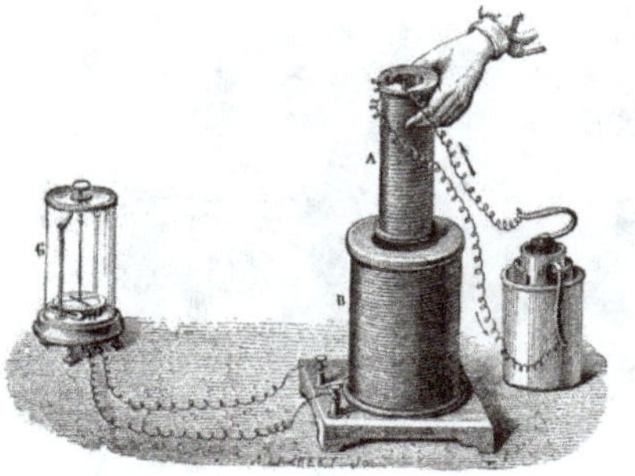

Figure III.11. A diagram of one of Faraday's induction experiments: a battery-powered solenoid A is introduced into winding B. The transient current induced by the variation of the magnetic flux in this circuit is detected by the deflection of the needle of the galvanometer C.

The vicinity of the magnets and circuits appeared to Faraday as a swarm of vectors indicating the direction of the magnetic forces that acted on the iron filings at each point in order to orient them.

Similar experiments make it possible to materialize the lines of force produced by static electric charges: they can be conducted by dispersing small grains of insulating material, in the form of diluted powder, in a bath of insulating liquid in which two electrodes connected to the poles of an electrostatic machine are immersed. The grains then acquire small dipoles and orient themselves along curves between the two poles, thus visualizing the electric field lines in the same way iron filings materialize magnetic field lines.

Through this kind of experiment, Faraday introduced the field concept in physics. Magnetic and electrical lines of force form a spatial interlacing, a sort of mesh

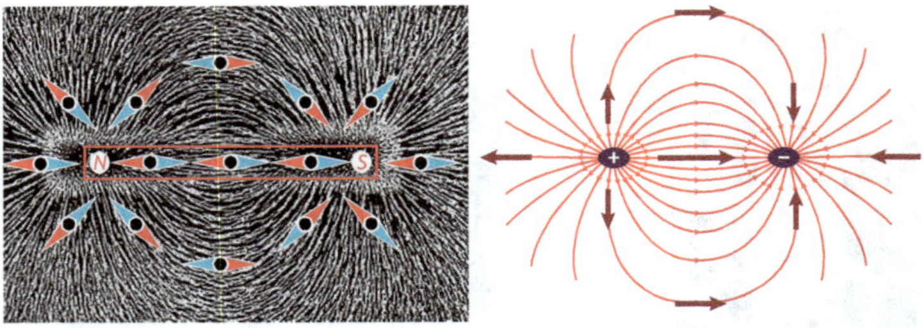

(a) (b)

Figure III.12. (a) Magnetic field lines materialized by iron filings sprinkled on a sheet of paper stretched over a magnet. Observing these lines inspired Faraday to introduce the concept of field to physics. Small compass needles (red north pole, blue south pole) superimposed on the photo mark the direction of the field created by the magnet (whose edges are marked by the red frame) at different points. On the inside of the bar, the magnetization is aligned in the opposite direction of the field lines looping on the outside. Around the poles of the magnet, the magnetic field flux through a closed surface is zero: it enters on one side of the surface (red tip of the needle facing the pole) and exits on the other (blue tip facing the pole). The nullity of this flux is due to the non-existence of magnetic monopoles. (b) Field lines of an electric dipole consisting of two opposite charges. The lines all exit from a surface surrounding the positive charge and all enter a surface containing the negative charge. The electric field flux around the charges is non-zero. Unlike the case of a magnet, the field on the line between the charges has the same sign as the one circulating along the loops that connect them.

whose structure is determined by the distribution of charges, currents, and magnets that give rise to them. For Faraday, these oriented lines had a concrete physical meaning. The vector tangent to these lines at each point gave the direction of the forces exerted on the charges or on the small compass needles placed there.

The intensity of these forces depended on the density of the field lines: the greater the density, the tighter the lines; the weaker the density, the farther apart they were from each other.

There is an important difference between networks of electric and magnetic field lines. The former form rays that diverge from positive charges and converge to negative ones, intersecting at the points where these charges are located. The latter, on the other hand, do not cross each other and always form closed loops. They resemble current lines in a flowing fluid that can form loops and vortices but never intersect.

This difference in the structure of electric and magnetic field line arrays can be analyzed quantitatively by defining the so-called *field divergence*, a mathematical function introduced by the German mathematician Carl Friedrich Gauss at the beginning of the nineteenth century. The value of the divergence at a point is a real number calculated by adding spatial derivatives of field vector components with respect to different directions in space. This number expresses the tendency of the field lines to converge toward or diverge from this point. In the vicinity of a positive charge, the electric field lines are divergent: they all point outward from a small area surrounding the charge. The field flux through this surface is said to be positive. The opposite occurs in the vicinity of a negative charge; the field lines all point toward the interior of the volume in which the charge is located, and the field flux through the surface bounding this volume is negative.

In general, the flux of any vector field through an infinitesimal element of a closed surface is a number equal to the product of the area of this surface and the projection of the field vector on the normal to the outward-facing surface. The global flux of the field through the total surface is obtained by adding the fluxes that pass through all of its infinitesimal elements. A theorem demonstrated by Gauss says that this flux is equal to the integral of the divergence of the field in the volume bounded by this surface. This theorem, combined with Coulomb's law, imposes on the electric field a divergence proportional at each point to the density of electric charges at that point.

On the other hand, the divergence of the magnetic field is zero everywhere, which expresses the absence of free magnetic charge, the non-existence of so-called *magnetic monopoles*. A magnet's positive pole is always associated with its negative pole. This duality is always reconstituted, even if we try to separate the magnetic charges by breaking the magnet. The zero divergence of the magnetic field implies, according to Gauss's theorem, that its flux through a closed surface is always zero and that there is thus a perfect balance between the contribution to the flux of the magnetic field lines leaving the surface and that of those entering it.

The magnetic field lines recall those of an incompressible fluid in stationary flow. The flow of particles exiting from one side of a small tubular element of the fluid must be equal to the flow entering from the other side so that the amount of flowing material is conserved in the volume. The magnetic field's properties of zero divergence and conservative flux were established by Gauss, who studied the laws of magnetism in collaboration with the German physicist Wilhelm Eduard Weber at the same time as Faraday.

The mathematical notions of flux and divergence were beyond what Faraday, a pragmatic and self-taught experimenter, could conceive: he did not venture more than a very qualitative description of the electromagnetic field whose existence he had glimpsed. Nonetheless, the physical intuition of the director of the Royal Institution was remarkable. The fields he introduced to physics were thus defined in a vacuum, without reference to a theoretical aether that he never discussed—in which sense we can say that, paradoxically, Faraday was more modern than the physicists who followed him in the second half of the nineteenth century. Despite his great intuition, however, Faraday did not realize the analogy between the vectors tangent to his field lines and the luminous transverse vibrations of Fresnel theory. This crucial step would be taken by a young Scottish scientist.

The confluence of light, electricity, and magnetism

Between 1861 and 1865, James Clerk Maxwell, then professor at King's College London, meditated on Ampère's and Faraday's experiments, trying to describe them by mathematical equations that clearly showed their properties.

These experiments established relationships between electric and magnetic fields distributed in space and the stationary or moving charges that were their sources. Faraday's induction effect showed that the variation of the magnetic field through a conductive coil produced an effect on the metal charges equivalent to that created in the metal wire by an electric field whose vectors rotated along the wire of the coil like the velocity of a vortex flow. Maxwell deduced that even in the absence of a hardware circuit, the temporal variation of a magnetic field at any point in space must create a swirling electric field around that point.

Having thus clarified the Faraday effect equation, Maxwell tackled Ampère's work. This showed that a permanent current created around itself a magnetic field described by equations whose mathematical structure was identical to that of the electric field vortices in the Faraday equation. In the absence of current, however, the Ampère equations did not generate any field, which broke the symmetry between electric and magnetic fields. To restore this symmetry, Maxwell supplemented Ampère's equations with the hypothesis that a magnetic field vortex could be generated not only by a charge current, but also by a temporal variation of the electric field. He simply added to the term representing current in the Ampère equation a complementary contribution proportional to the time derivative of the electric field.

To complete his theory, Maxwell finally added to the Faraday and modified Ampère equations those that Gauss had discovered. One expresses the absence of magnetic monopoles and the conservation of the flux of the magnetic field by imposing a zero divergence everywhere in the magnetic field; the other relates the flux of the electric field through a closed surface to the charge it contains, which translates Coulomb's law (Gauss-Coulomb equation).

In modern mathematical notation, the vortex properties of the fields described by Maxwell are expressed by introducing a vector mathematical function called the *curl* of the field, constructed from the spatial derivatives of its components on the three coordinate axes. Maxwell did not use this concept; the formulas he wrote took a more complicated form than the one we give his equations today through the field functions of curl and divergence.

In any case, Maxwell obtained a set of vector equations describing the dynamic evolution of the electric and magnetic field in space, in the presence of any distribution of variable charges and currents. These equations even described the evolution of fields in the absence of any material source—as if the charges and currents were a scaffolding whose removal revealed the beauty of a mathematical edifice describing the evolution of a field free of constraint, propagating like a wave in space at an arbitrary distance from any source. The variation of a magnetic field created an electric field whose variations in turn generated a magnetic field, the two fields feeding off one another during their propagation. These equations had an infinite number of solutions, depending on the initial conditions given. The simplest were transverse plane waves, whose electric and magnetic fields vibrated in a plane normal to the direction of propagation of the wave, in two directions orthogonal to this plane.

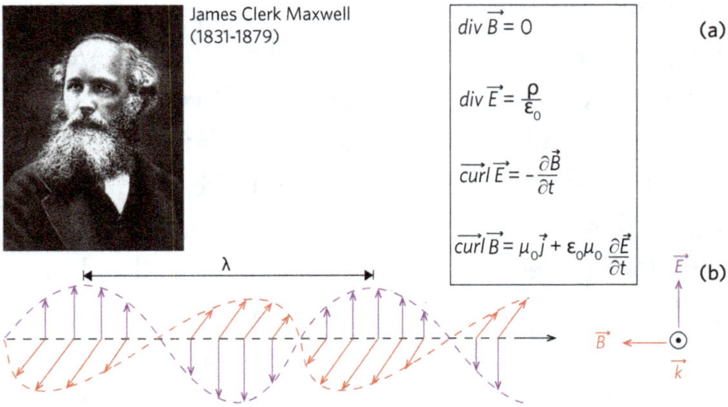

Figure III.13. Maxwell's electromagnetic theory. (a) The four equations linking electric $\vec{E}$ and magnetic $\vec{B}$ fields to charge (spatial density ρ) and current (density $\vec{j}$) distributions. The constants ε_0 and μ_0 come from Coulomb's and Ampère's laws. The notations *div* (divergence) and $\overrightarrow{curl}$ (curl) represent mathematical operations of spatial derivation of the coordinates of electric and magnetic fields, and the expressions $\partial\vec{E}/\partial t$ and $\partial\vec{B}/\partial t$ describe the temporal variations (derivatives) of these fields. The first equation says that the magnetic field cannot diverge from a point and expresses the absence of free magnetic charges. The second equation expresses Coulomb's law, which indicates that the lines of the electric field diverge from or converge toward the charges. The third equation expresses Faraday's law, which relates the variation of the magnetic field to the vortices of the electric field it induces. The last equation describes the reciprocal effect, resulting from Maxwell's modification of the equation describing Ampère's law. A vortexing magnetic field (expressed as $\overrightarrow{curl}\,\vec{B}$) is produced by the current to which the term of variation of electric field $\vec{E}$ must be added. (b) These equations admit among their solutions waves of the electromagnetic field propagating in space in the absence of charges and currents. The electric and magnetic fields of the wave are orthogonal to each other and transverse to the direction of propagation. The speed of the waves is $c = 1/\varepsilon_0\mu_0$. It is the coincidence of this value with that of the speed of light measured by Foucault that led Maxwell to identify light radiation with an electromagnetic wave.

The equations made it possible to calculate the speed of these electromagnetic waves. The square of this velocity had to be equal to the inverse of the product of two constants, ε_0 and μ_0, which Coulomb and Ampère had introduced to express, respectively, the electrostatic force between two electric charges and the magnetic attraction force between two wires through which a current flows. By calculating with the values experimentally determined for these constants at the time, Maxwell found for *c* a value of 308,000 km/s, very close to the 298,000 km/s value Foucault had just measured in Paris. This could not be a coincidence. The page of Maxwell's manuscript containing this finding is on display at the Royal Society in London. At the bottom of the page, one can read in Maxwell's beautiful, even handwriting:

> The agreement of the results seems to show that light and magnetism are affections of the same substance, and that light is an electromagnetic disturbance propagated through the field according to electromagnetic laws.

This sentence, in all its British understatement ("seems to indicate"), was surely the most important in all of physics at the time, perhaps even of all time. In a few words, it unifies three sciences that had been considered independent for centuries: electricity, magnetism, and optics. The realization that light and electrical and magnetic phenomena were one was a revelation as profound as Newton's realization that the force attracting an apple to the ground is of the same nature as the force keeping the Moon on its trajectory around the Earth.

It is fair to say that the relationship between the speed of light *c* and Coulomb's and Ampère's electrical and magnetic constants had already been established, some ten years before Maxwell, by the German Wilhelm Eduard Weber and his colleague Rudolf Kohlrausch. But they had given no explanation for this result.

Maxwell's theory opened fascinating possibilities. It made precise predictions that would be confirmed experimentally in the decades that followed. Light was only a small part of the spectrum of electromagnetic waves described by the equations; it had a special status only because the human eye, shaped by natural selection, was sensitive only to waves whose lengths were between 0.4 and 0.7 microns. The theory predicted the existence, in the domain of longer wavelengths, and therefore smaller frequencies, of infrared electromagnetic waves—and, on the other side of the spectrum, among waves of shorter length and therefore higher frequency, of ultraviolet radiation. These predictions confirmed what physicists had already glimpsed in the first half of the century, giving a much more precise meaning to what had until then been vaguely described as "caloric radiation" for the near infrared and "chemical radiation" for the near ultraviolet. These invisible waves would gradually be identified, first in frequency ranges close to that of the visible waves and then farther in the spectrum.

An important step came in 1887 with the generation, by the German Heinrich Hertz, of radiofrequency waves with lengths on the order of a meter. On the other side of the spectrum, another German, Wilhelm Roentgen, discovered in 1895 a mysterious radiation that he called *X-rays*, and that crystalline diffraction experiments later revealed to be an electromagnetic wave whose wavelengths, on the order of a few tens of nanometers, are one hundred to one thousand times shorter than those of visible light.

An essential property of Maxwell's equations is what is called their *linearity*—independent solutions of these equations combine to form another solution of the same equations. This property justifies the principle of superposition stated by Fresnel in 1820: the vibrations in his theory of light are none other than the electric fields of Maxwell's waves, which add up and interfere in the same way. Another important property of these equations is that the value of the fields on any closed surface uniquely determines the fields outside that surface. This property justifies the Huygens–Fresnel principle.

Some mysteries disappear but others remain

Maxwell's theory resolved some of light's mysteries. The interaction between light and the transparent media through which it passes or the opaque substances that absorb it could only be due to the coupling of the fields of the light wave with electric charges present in matter. The exact nature of these charges only began to become clearer at the turn of the century, with the discovery of the electron by the Englishman Joseph John Thomson in 1897. As early as the 1870s, however, the young Dutch physicist Hendrik Lorentz explained the reflection and refraction of light with a model in which electric charges, present in these media, were set in oscillatory motion by electromagnetic forces. His theory, based on Maxwell's equations, explains in particular, by an effect of superposition, that light moves slower in a material medium than it does in a vacuum.

To describe the electromagnetic field transmitted by a slab of homogenous transparent medium, we must add to the incident field the one radiated by the charges set in motion by this field. In the slab, these charges oscillate at the frequency of the incident field and radiate in the same direction an induced field that interferes with the incident light. If the wave has a frequency that does not coincide with an absorption frequency of the medium, the field thus induced in a forward direction with the same polarization as the incident field has a phase advance of 90°, or $\pi/2$ radians. According to the superposition principle, the total field is described in the Fresnel plane by the sum of a vector representing the incident field and a vector oriented at right angles, representing the field induced in the slab. The sum of the two describes a vector that has been rotated in the Fresnel plane. The total field has thus accumulated a larger phase as it passed through the slab than it would have had the slab not been in its path. As the field always oscillates at the same frequency, this means the wave propagated through the slab with a shorter wavelength, and therefore a lower speed.

The birefringence of an anisotropic slab can be explained with the same model by conceding that the charges in the medium vibrate with different amplitudes depending on their solicitation by an electric field oscillating in one of two orthogonal directions. This results in different phase shifts of the transmitted fields polarized at right angles to each other, and thus two different indices depending on the orientation of this polarization. An analogous model qualitatively explains the phenomenon of light attenuation in an absorbent medium. In this case, the charges of the medium are made to vibrate so as to emit in a forward direction a field in phase opposition with the incident wave. The result is a Fresnel vector, the sum of the incident field and the induced field, which is smaller than that of the incident field: the slab has absorbed the field and the transmitted intensity has decreased. The superpositions of waves and interferences are thus the explanatory keys to the phenomena of electromagnetic wave propagation in a vacuum and in material media.

Although they unveiled many of light's mysteries, Maxwell's equations did not solve the question of the aether, whose strange properties were simply extended to encompass the setting in which all electromagnetic phenomena took place. This hypothetical undetectable medium now had to propagate, in addition to light, the whole spectrum of invisible electromagnetic waves whose existence was predicted by the equations and which experimenters were gradually discovering. Since the theory gave a fixed value to the speed of electromagnetic waves, it was natural to conclude that the aether was the privileged medium in which these waves propagated at a velocity close to 300,000 km/s.

The Earth, in its path around the Sun, moved in the aether, meaning the electromagnetic waves had to feel the effects of the "aether wind" resulting from this motion, light being dragged by the aether in the same way it was dragged by a water current in Fizeau's experiment. The experiments intended to detect the effects of this drive give, as we will see in the next chapter, a negative result. The mystery would not be solved until the beginning of the twentieth century with the theory of relativity, which would finally banish the aether and its contradictory properties from physics.

The history of light in the nineteenth century illustrates even more vividly the features of scientific research that the discoveries of the seventeenth and eighteenth had already revealed. The unexpected nature of certain discoveries can be seen in Malus's observation of polarization by vitreous reflection. An essential aspect of the discovery lies in the scientist's ability to understand the importance of the clues luck has placed in his path. To wit, Ørsted systematically studied the properties of the mysterious force his initial observation had revealed, whereas Romagnosi did not—which undoubtedly explains the respective destinies of their work, one recognized and celebrated and the other controversial and largely unknown. The importance of accurate measurements was also affirmed: small differences in the predictions of the wave and corpuscular theories as to the polarization properties of birefringent media contributed to the triumph of the former. These differences had to be tested by increasingly precise optical experiments. The refinement of Foucault's and Fizeau's experiments to determine the speed of light also illustrates this race for precision.

The complementarity of fundamental and applied science became even more pronounced in the nineteenth century than it had been in previous eras. Fresnel, an engineer educated in one of the first classes of the École Polytechnique, was certainly motivated by pure curiosity in his fundamental work on the interference and polarization of light, but he also sought to apply his discoveries to the construction of useful devices, setting his fundamental research aside for long periods to devote himself to practical projects. He invented the lenses that bear his name and that long equipped lighthouses around the world, saving the lives of many sailors. These lenses, made of highly transparent glass prisms arranged in concentric crowns, have a very large aperture and focus much more powerful light beams than conventional lenses do, while being much less massive. Flat Fresnel lenses made of concentric circular engravings on plastic discs are still used today in various devices, such as cameras and video projectors; they are also used to concentrate sunlight on photovoltaic cells or in solar cookers.

Beyond optics, the byproducts of fundamental research have been crucial in the field of electromagnetism. Faraday's and Maxwell's discoveries led to developments in the electrical and communications industries; Faraday's induction phenomenon is used in dynamos and alternators that create electric current by rotating magnets in coils—this is how the energy produced by thermal or nuclear power plants is ultimately transformed into electricity. Engines that inversely transform electrical energy into mechanical work are also a direct result of Faraday's discoveries. His induction phenomenon also operates in transformers that couple currents in two coils that influence each other, raising or lowering the electrical voltage in industrial circuits and power lines. Faraday certainly didn't imagine all the applications of his discoveries, even though he saw their potential. It is said that when the finance minister Gladstone asked him what his work could be used for, he replied: "One day, sir, you may tax it."

As for the applications of electromagnetic waves, no need to insist on their practical importance. It is the radiofrequency waves discovered by Hertz that today carry most information from radios and cellular phones. On the other side of the spectrum, Roentgen's X-rays play a fundamental role in medicine and in many devices that probe the properties of matter. Once twentieth-century quantum physics has made it possible to understand the profound nature of the interactions between matter and radiation, other applications of electromagnetic waves will appear, largely dependent on the invention of the laser.

The link between basic science and technology is a two-way street. If the former is essential to the development of the latter, the reverse is also true, and the history of light in the nineteenth century offers many examples. The measurement of the speed of light on Earth, based on an ancient idea by Galileo, only became possible because technology had made enormous progress in generating and detecting light pulses a few microseconds in duration—unimaginable in the seventeenth century. The clock and turbine mechanisms that set the mirrors in rapid rotation and the revolution counters used by Fizeau and Foucault could not have been designed two centuries earlier. The interferometric devices invented to accurately measure wavelengths were used by Fizeau to demonstrate the dragging effect of a water current on the speed of light; at the end of the century, they would play an essential role in probing the mysterious properties of the hypothetical aether. It was also advances in visible light detectors, as well as infrared and ultraviolet detectors, that made it possible at the turn of the century to make precise measurements of thermal radiation, paving the way for quantum physics.

Another, subtler aspect of the relationship between technology and basic science deserves to be emphasized in Maxwell's decisive approach. His familiarity with the mechanical devices of the Industrial Revolution gave his intuition a concrete basis that greatly helped him elaborate his theory of electromagnetism. The description

I have given above is a modern vision, stripped of the mechanistic scaffolding Maxwell imagined to arrive at his famous equations. He in fact constructed a model of the electromagnetic aether, made of hexagonal cells separated by freely rotating wheels and dragging the cells in a rotational motion by friction; the translation speed of the wheels circulating between the hexagons represented the current and the rotation of the cells played the role of the magnetic field. As for the electric field, it was proportional to the force that the cells exerted on the wheels, causing them to spin. The torque exerted by this force on the hexagons was, according to the laws of mechanics, equal to the variation per unit of time of the cells' rotation speed. The expression of this relationship linking the electric field in a circuit to the variation of a magnetic field translated Faraday's law into mechanistic language. Analogies of the same nature led Maxwell to add to the Ampère equations the term proportional to the derivative of the electric field, restoring the symmetry of electric and magnetic fields in his equations. The Science Museum at the Cavendish Laboratory in Cambridge, which Maxwell directed until the end of his life, displays the complex device that he had the mechanics in his laboratory build to give concrete form to his theory of electromagnetism. The aether he sought to represent by this ingenious construction has now disappeared from physics and Maxwell's equations remain, like jewels released from the gangue in which they were discovered.

The history of light in the nineteenth century confirms at last what the previous centuries had already taught us: science is an indivisible whole, and discoveries often result from the conjunction of information acquired in various fields of knowledge, truth appearing at the confluence of apparently disparate streams of research. Optics, astronomy, and the measurement of the Earth progressed together until the Enlightenment, before being joined by electricity and magnetism 150 years ago. Often the same scientists were involved in these different fields even before their convergence was recognized: this was the case of Arago, who encouraged Fresnel, Foucault, and Fizeau in their work on light, and Ampère in his research on electromagnetism; or of Biot, who distinguished himself with his studies on the polarization of light and the forces between electric currents. Faraday is another example. In addition to his fundamental research on electrical induction, he discovered the effect of rotation of the plane of polarization of light propagating in a medium subjected to a magnetic field, which we now call the Faraday effect. All of these discoveries also benefited from advances in mathematics, which made it possible to calculate and model increasingly complex effects. The work of Fourier was essential, as was that of many other mathematicians, such as Gauss, Hamilton, and Green, to name but a few.

The conjunction of different sciences went even beyond physics. Galvani's work on the physiological effect of currents is the forerunner of current research in neurobiology, which uses physical techniques—certainly more sophisticated than those of the Italian physician of the late eighteenth century—to explore the properties of the brain. It can also be said that the voltaic pile, which paved the way for the study of electrolysis, was the precursor of modern chemistry. It was using one of these first batteries that Humphry Davy, patron of the young Faraday, discovered several chemical elements at the Royal Institution, where we began this journey through the history of light in the nineteenth century.

Physics, biology, and chemistry had another point in common, historical rather than scientific: it was in the decade between 1859 and 1869 that major discoveries revolutionizing these fields were announced. Darwin published his treatise on the evolution of species in 1859, Maxwell his article on electromagnetic theory in 1864, and Mendeleev his note on the periodic classification of elements in 1869. It is no exaggeration to say that it was in this marvelous decade, 150 years ago, that the essence of all modern science was born. As far as light is concerned, as we shall see, this birth would have unforeseen consequences.

CHAPTER 4

The two clouds of Lord Kelvin

One of the most famous lectures ever to grace the red velvet amphitheater of the Royal Institution in London was the one given in April 1900 by the eminent English physicist Lord Kelvin, on the two clouds darkening the sky of physics in the late nineteenth century. He began his talk with this sentence:

> The beauty and clearness of the dynamical theory, which asserts heat and light to be modes of motion, is at present obscured by two clouds.

The first of these clouds concerned the aether, the medium in which light was thought to propagate, whose paradoxical properties we have discussed. The second had to do with the distribution of energy between the different degrees of freedom in a system in thermodynamic equilibrium. The first cloud had been revealed by the negative result of Albert Michelson's experiment to measure the velocity of the Earth relative to the aether; the second was indicated by the absurd result physicists obtained when they tried to calculate, from the principle of *equipartition of energy*, the newly measured spectral properties of the radiation of heated bodies.

The year of Lord Kelvin's lecture was also the year of the great Paris World's Fair, a vivid manifestation of faith in the successes of science and the technological advances they foretold. Visitors—gentlemen in bowlers or top hats, ladies in crinoline—traveled between pavilions on the moving sidewalk that Georges Courteline immortalized in one of his short plays. All were amazed at the innovations presented by the great powers of the era, vying with one another to showcase the most spectacular advances in their industries. The postcards published in France and Germany for the occasion described a vision of the year 2000: one, for instance, showed how radioactivity would serve a century later to heat us. With a simple piece of radium radiating its heat into the hearth, there would be no more need for wood or coal. Other cards extrapolated from new technologies, such as the telephone or the cinematograph, but were nonetheless far from anticipating what the twentieth century would really bring us: electronics, computers, lasers, the GPS, and MRI scanners were unimaginable in 1900.

The carefree era then coming to an end, which Stefan Zweig nostalgically called "the world of yesterday," could not conceive of the technologies of the twentieth century, which immensely increased our means of communication, information, and action on the world around us. These technologies are the product of two fundamental theories of physics that would emerge in the first quarter of the new century, once Lord Kelvin's two clouds had finally dissipated. The first heralded the theory of relativity, the second quantum physics. These two revolutionary theories, born of questions about the paradoxical properties of light, changed our view of the world as profoundly as the Copernican revolution had in the sixteenth century.

Let us recall the context in which these revolutions emerged. The last years of the nineteenth century had been particularly fruitful for science and technology. This period of relative peace before the great conflicts of the twentieth century saw spectacular economic development in Europe and America that benefited from great scientific advances. Exploitation of the laws of thermodynamics had allowed the development of steam ships and locomotives, greatly reducing the length of land and sea voyages and bolstering commerce. The combustion engine was born with the first automobiles, which soon replaced horses for urban transportation. The Wright brothers were secretly working on building their first airplane. The discoveries of the laws of electromagnetism revolutionized the industrial world with the introduction of electric motors and lighting. Communications made tremendous progress with the invention of the telephone and the first tests of radio transmission using Hertzian waves.

Roentgen's 1896 discovery of X-rays and that of radioactivity by Henri Becquerel and Pierre and Marie Curie in 1898 were widely announced in the popular press, which soon led to speculation as to their possible practical use (the postcard of radium in the fireplace, for example). The first Nobel Prizes in Physics were awarded to Roentgen in 1901, to Hendrik Lorentz and his compatriot Pieter Zeeman in 1902, and to the Curies and Becquerel in 1903, all to widespread commentary, demonstrating a public enthusiasm for science beyond the limited circle of the scientific community of the time.

That community, consisting of a few dozen scholars, largely shared the layman's optimism and enthusiasm. The progress of fundamental science gave the impression that the great mysteries of nature had been solved. Maxwell had done for electromagnetism and light what Newton had done two centuries earlier for mechanics and gravitation, explaining seemingly disparate phenomena in a unified theory. Mechanics and electromagnetism were complemented by thermodynamics, culminating in the work of Josiah Willard Gibbs and Ludwig Boltzmann clarifying the concepts of energy and entropy. This fundamental knowledge had led to an industrial civilization from which everyone stood to benefit. Progress in experimentation had unveiled new phenomena with the discovery of radio waves, the electron, the photoelectric effect, X-rays, and radioactivity. Most scientists believed, however, that these discoveries could be explained within the framework of established classical theories. They would soon be proven wrong.

Although Lord Kelvin had indicated the two directions from which the great changes would come, he himself was far from grasping their importance. He must have thought that these two clouds were merely passing through a serene sky, that the storm they portended would be only temporary. As the sentence at the beginning of this chapter shows, he linked the problems of heat and light by considering both to be part of a "dynamic theory of motion." Indeed, heat would be described as associated with the motion of the particles that make up matter, transmitted to the motion of electromagnetic waves propagating in the aether that bathes it.

What Lord Kelvin was really pointing out, beyond his atmospheric metaphor, was the appearance of internal contradictions among the theories that had brought classical physics its successes at the end of the nineteenth century. Michelson's experiment showed that Maxwell's theory of light did not satisfy the principle of relativity of Galilean motion. Worse, Maxwell's equations, combined with the equation of the equipartition of energy—one of the dogmas of classical thermodynamics—led to an absurd model that failed to describe the most obvious properties of the spectrum of light radiated by heated bodies. In the twentieth century, the resolution of these two problems would replace the Newtonian postulates with new principles allowing physics to explore the worlds of the infinitely small and the infinitely large. This would require overturning the concepts firmly rooted in the minds of physicists concerning time and space, mass and energy, determinism and chance.

Light played an essential role in this revolution in physics. The resolution of the mysteries hidden behind Lord Kelvin's clouds would not only illuminate light in a new way, but also apply to all phenomena in physics, radically altering the stage on which physicists observed the world. The symbiosis between theory and experimentation, between blue sky research based on pure curiosity and advances in technology, were also essential in this revolution, as it was in previous centuries. Progress was decisive in instrumentation, from radiation detectors such as bolometers, which measure the heat exchange between light and matter, to those based on the newly discovered photoelectric effect. Advances in interferometers for measuring very small displacements of interference fringes played an important role as well. And, as always, so did luck.

Luck was certainly behind the arrival in our story of a young employee of the Swiss Patent Office in Bern, who had a brilliant intuition and in a few years found himself propelled into the firmament of physics. At the beginning of the twentieth century, Albert Einstein orchestrated the great changes that the theories of relativity and of quanta would bring about. His imprint on modern physics is so strong that for over a century there would be hardly a discovery or an invention that did not, in one aspect or another, lead us back to him. Proof of his genius also lies in the great questions about our universe, still unresolved, that he helped formulate. Of course, Einstein was not alone in this adventure, which unites a great many remarkable scientists. But to give unity to our story I will place Einstein at its center, and describe this physics through the prism of his thought.

Michelson and the puzzle of the aether

Let's start with Lord Kelvin's first cloud. The American physicist Albert Michelson, with the aid of his compatriot Edward Morley, had attempted, in a series of experiments between 1881 and 1887, to measure the velocity of the Earth relative to the luminiferous aether, as the medium in which light supposedly propagated at the speed c of 300,000 km/s was known. The experiment, similar to the one Fizeau had done in 1852 to measure the velocity of light in a flowing liquid, was performed using a special interferometer that Michelson developed and which has since been named for him.

A light beam was divided into two branches by a semi-reflective beam splitter, creating two beams that propagated at right angles to each other toward two mirrors that reflected the light back to the beam splitter. The light beams which recombined on the beam splitter interfered with each other. When the interferometer was adjusted so that the wavefronts from the two beams coincided exactly, one would ideally see behind the beam splitter, in the output path orthogonal to the input, a flat tint, bright when the path difference between the two beams was an integer number of wavelengths and dark when this path difference was an odd number of half wavelengths.

Slightly misaligning one of the beams by rotating one of the mirrors at a very small angle replaced the flat tint with a figure of parallel fringes that moved when one of the mirrors was translated along the direction of the beam it reflected. The shift of the fringes corresponded to an interfringe distance when the path difference varied by one wavelength. This displacement made it possible to measure with great precision the path difference of the beams propagating in orthogonal directions or the difference in travel times of the light in these two directions—which amounts to the same thing. The device was placed on an optical table floating on a mercury bath, which could thus be oriented so that the incident light beam made an adjustable angle with the direction of propagation of the Earth in its orbit around the Sun.

The velocity v of our planet in this orbit, on the order of 30 km/s, is equal to roughly one ten-thousandth of the speed c of light. Assuming that the Earth was moving in the aether, and applying the Galilean velocity-addition law, Michelson expected light to propagate along the branch parallel to the direction traveled by the Earth in its orbit with the velocity $c - v$ from the splitter to the mirror, and $c + v$ back toward the splitter after reflection. The relative velocity with respect to the interferometer reference frame, first decreased by the velocity v of the Earth in its orbit, then increased by the same velocity after reflection, was roughly, on average, the same as if v were equal to 0. This is a first-order approximation in v/c.

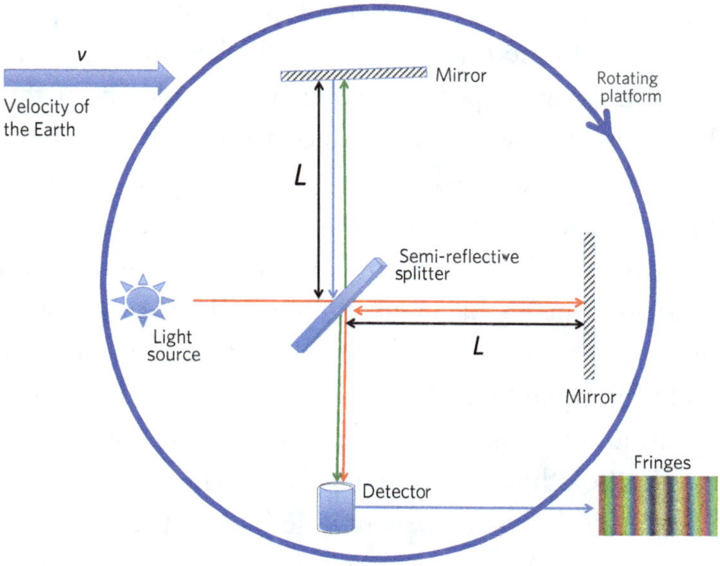

Figure IV.1. Simplified diagram of the Michelson–Morley experiment. The difference in the round-trip times of light between the two branches of the interferometer should vary between $+(L/c)(v^2/c^2)$ and $-(L/c)(v^2/c^2)$ when the base of the apparatus rotates, exchanging the branches parallel and perpendicular to the direction of the Earth's velocity in its orbit. In the actual experiment, additional mirrors reflected the light back and forth along the interferometer's arms, increasing L to a value of 11 meters. The displacement induced by the expected "aether wind," equal to 0.4 of an interfringe distance, was not observed.

A more precise calculation shows that the light was in fact slightly slower on its round trip than if the Earth were motionless, resulting in a propagation time $(2L/c)/[1 - (v/c)^2]$, where L was the length of the interferometer branches. This propagation time was slightly longer than $2L/c$, the one that would be observed for a stationary interferometer in the aether, the delay being, to a second-order approximation, equal to $(2L/c)(v^2/c^2)$.

The interferometer made it possible to compare this delay to the one the light underwent in the other branch, oriented perpendicular to the velocity of the Earth. As the light wave traveled through this branch, the mirror and the splitter moved at velocity v in a direction normal to the motion of light, lengthening the path it had to travel by an amount easily calculable using the Pythagorean theorem. The result was a propagation time from the beam splitter to the mirror and back equal to $(2L/c)/\sqrt{[1-(v/c)^2]}$, differing from $2L/c$ by an amount approximately equal to half of what it was in the other branch. Ultimately, when one branch of the interferometer was oriented along the direction of the Earth's motion, Michelson expected a difference in light propagation time between the two branches equal to $(L/c)(v^2/c^2)$. In order to increase the effect, sets of additional mirrors reflected the light back and forth along the interferometer arms, resulting in an effective L of 11 meters. The difference of expected light paths was then on the order of 10^{-7} meters, *i.e.* two-tenths of a wavelength. The interferometer's fringes should have moved by 0.4 of an interfringe distance when Michelson turned the device by 90° to exchange the direction of the two branches with respect to the direction of the Earth's velocity in the aether.

But the experiment did not detect any fringe displacement. The measurement was repeated in different seasons over several years, corresponding to different directions of the Earth's motion in the putative aether, always with a negative result that plunged the scientific community into perplexity. It seemed the aether was always pulled in a perfectly fixed way, accompanying the Earth in its orbit without allowing any perception of the "aether wind" the experiment was meant to measure.

An ingenious explanation for this negative result, which Lord Kelvin mentioned in his lecture, was proposed independently by the Irish physicist George FitzGerald and by Hendrik Lorentz, the young Dutchman we met in the previous chapter. They assumed that the interaction of the interferometer's matter with the aether led to a contraction of the material lengths in the direction of motion by a quantity that compensated exactly for the increase of the expected light phase shift. This effect, introduced arbitrarily in an attempt to preserve the aether, was to be found and reinterpreted a few years later within the framework of Einstein's theory of relativity—but then the aether disappeared from the picture.

Einstein comes on the scene: thought experiments

The young German physicist, it seems, posed the question of the speed of light in a purely theoretical way, Michelson's experiment having made little impression on him. It is said that when he was still a teenager, having learned Maxwell's theory, he wondered what would happen if he could fly at a speed close to that of light: not at 30 km/s, like the Earth around the Sun, but one thousand or ten thousand times faster. Would the light beam then appear to him to go much slower? Would it be conceivable to catch it, to surf on the light wave and see it as motionless? Clearly something was wrong with Maxwell's equations when approached in this way. Seeing the electric or magnetic field of light as static led to an absurd result. Indeed, it is these fields' temporal variations that allow them to maintain each other. Without variation in the electric field, we no longer understand how a magnetic field can appear; without variation in the magnetic field, the very existence of the electric field is no longer intelligible.

This daydream about racing with light showed clearly that Maxwell's equations cease to be valid if we change our reference perspective while preserving the classical conceptions of space and time, *i.e.* using the coordinate transformation rules of classical mechanics, which postulate that time intervals and spatial distances are the same for all observers in translation at uniform speed with respect to each other.

Einstein's solution to this enigma, presented in his famous 1905 paper "On the Electrodynamics of Moving Bodies," is—no pun intended—luminous. To the question of whether one could catch up to light, his answer was a simple no. This postulate of impossibility is the starting point of the theory of special relativity. Einstein had only to extend to the laws of electromagnetism the principle of relativity of motion stated in the seventeenth century by Galileo, who posited that the laws of mechanics should be expressed in the same way for all reference frame moving at constant speed relative to each other.

These frames of reference are called *Galilean* (or *inertial*, as Newton's law of inertia applies to them). In the absence of an applied force, any body in such a frame propagates in a straight line at a constant speed. As we have seen, the Galilean principle is verified daily on trains, ships, and airplanes in uniform motion. Without looking outside, a passenger cannot tell whether she is moving or not. All the mechanical experiments she can perform, such as throwing a ball in the air or pouring a liquid into a cup, will produce the same result as if she were motionless relative to the Earth.

Einstein generalized this principle by applying it to electromagnetism, then to all of physics. This has been called the postulate of "special" relativity, to specify that it was restricted to the description of the laws of physics in Galilean (or inertial) frames of reference. General relativity, as opposed to its special version, will later be freed from this restriction. For the time being, for as long as there is no risk of confusion, we will omit the qualifier *special* and speak simply of the postulate of relativity.

This postulate, then, demands that physics be expressed by identical equations in all Galilean frames of reference. Maxwell's equations give light the speed c = 300,000 km/s. This speed must be the same in all reference frames, so we cannot know by measuring it whether or not we are moving in relation to a preferred observer. Put another way, the constancy of this speed is a law of physics and must therefore be observed in all inertial reference frames. Just as there is no absolute motion according to Galileo, we cannot have some privileged inertial frames of reference in which Maxwell's equations are valid and others in which they take another form.

This postulate has fundamental consequences: the Newtonian velocity-addition law, based on the existence of a universal time and of identical spatial distances for all observers, can no longer apply, as the speed of light is not added to or subtracted from any other speed. To renounce this law is to admit that the time measured by an observer depends on the reference frame at which she finds herself, and to recognize that the variables of time and space are intimately linked.

To arrive at these revolutionary ideas, Einstein did not conduct real experiments, as Michelson and Morley had. At his little desk in the Bern patent office, the twenty-six-year-old physicist conducted virtual "thought experiments" that allowed him to explain the postulate of relativity simply and to play out all of its logical consequences. He considered the same luminous phenomenon as described either by an observer on the platform of a virtual train station or by a passenger in a train car passing the platform at a constant speed. For the effects he was studying to be truly observable, this virtual train had to have a very high speed, close to that of light, but this detail didn't matter to Einstein. What mattered was that these thought experiments allowed him to establish fundamental principles, even if they were still difficult to test in realistic experiments.

The first result to appear in such an imaginary experiment is that the notion of simultaneity, considered a given in classical physics, is no longer so obvious in the theory of relativity. The question of the simultaneity of two events certainly does not arise if they occur in the same place. An observer doing an experiment aboard the train has only to observe the coincidence of the two events with a precise position of the second hand of his watch (there were no digital watches at the time). Another observer, this one on the platform, could only confirm this local simultaneity. Similarly, if an event observed at the same point on the train occurs before or after another (coinciding with two different positions of the second hand of the watch of the observer on the train), the same temporal order must be observed by the observer on the platform, as a reversal of this order would create a serious problem of causality. So far, there was nothing particularly surprising in this.

The surprise came when Einstein considered the notion of the simultaneity of two events occurring at *different points* in space. To verify this simultaneity, he proposed linking the two events by means of light signals observed at an equal distance from the two locations. If the signals triggered by the events reach an observer located at this equidistant point at the same time, he will say they are simultaneous. This is because, the speed of light being constant for him (as for any observer in a Galilean frame of reference), the light takes an equal amount of time to travel the distance separating him from the two equidistant events.

Einstein then imagined a simple, albeit highly unlikely, coincidence of events. As the train passes through the station, and the back and front of the train pass points A and B on the platform, lightning strikes these points. An observer, whom we will call Alice, stands motionless in the station at an equal distance from points A and B. Seeing the two flashes of lightning at the same time, she concludes that the two lightning strikes are simultaneous. A passenger, Bob, sitting on the train at an equal distance from the front and the back, observes the same events. For Alice on the platform, the lightning emitted at point B reaches Bob before the lightning coming from A, since Bob is moving from A to B while the light propagates at the same speed in both directions.

Bob, detecting the two flashes at the same point, obviously comes to the same conclusion. Since the speed of light, whatever its direction of propagation, is equal to c in his reference frame, Bob can conclude only that the lightning strike at B occurred before the one at A. And his point of view is just as valid as Alice's, even though for her both flashes were emitted at the same time.

A similar thought experiment, described by Einstein in a later paper, allows us to deduce the relativity of the time intervals measured by two observers moving in relation to each other. He imagined a clock made up of two parallel horizontal mirrors between which a very brief light pulse bounces vertically. One of the mirrors, slightly transparent, lets through a small fraction of light, measured by a detector placed outside the cavity

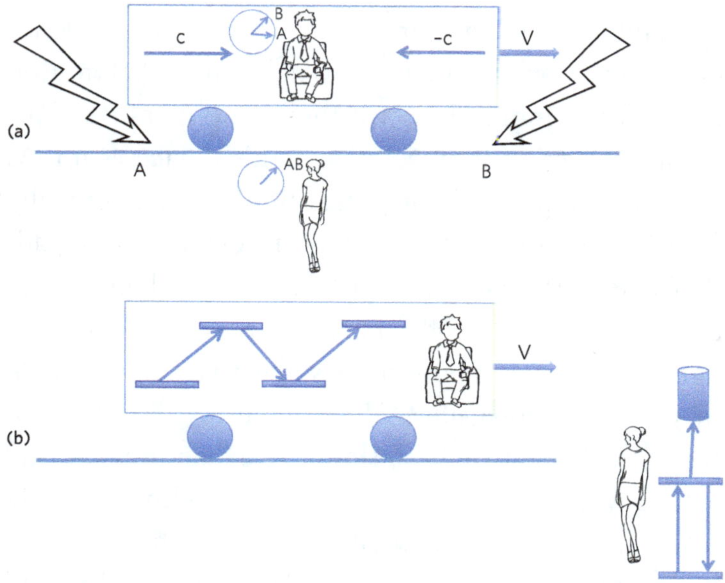

Figure IV.2. Relativist train thought experiments. (a) Lightning strikes the front and the back of the train simultaneously, according to Alice, who is observing from the platform. For Bob, the passenger, it strikes at the front of the train before striking the rear. (b) As observed by Alice from the platform, the rhythm of light pulses from Bob's clock is slower than that of her own clock. She concludes that Bob's time is passing more slowly than hers.

formed by the mirrors. Each time the light travels back and forth between the mirrors, the detector clicks, thus materializing the ticking of the clock. If we know the distance L between the mirrors, and that the speed of light is constant whatever the direction of its propagation and whatever the Galilean frame of reference, the interval between two clicks allows a precise measurement of time.

Suppose Alice, on the platform, and Bob, aboard the uniformly moving train, have two of these identical optical clocks, which they synchronized before Bob boarded the train. How will their time measurements compare? Let's put ourselves in Alice's place again. As she sees them, the light pulses that bounce vertically between the mirrors on the train travel an oblique path that is longer than the path taken by the light in the clock next to her on the platform. Since the speed of light is the same in both frames, regardless of the clock in which it propagates, Alice perceives the clicks of Bob's clock occurring at a slower rate than the ones she counts with her clock on the platform. Thus, not only is the notion of simultaneity relative to the observer; the measurement of time intervals is as well. The time measured by the clock moving at a constant speed relative to Alice runs slower than the time measured by the clock attached to her reference frame.

One could object that Einstein chose a particular type of clock based on an optical phenomenon; this time dilation may apply only to this type of clock. In other words, the relativity of time measured in this way may reflect only the properties of a particular phenomenon, with no reason to apply to other physical events, for example a biological rhythm such as a traveler's heartbeat, the succession of his movements on the train, or even his lifespan. Einstein responded to this argument with irrefutable logic, simply relying on the principle of relativity as he had stated it. He completed the thought experiment by imagining that Alice and

Bob, in addition to their optical clocks, have also synchronized two clocks of a totally different nature, for instance two mechanical chronometers counting the oscillations of a spring. Bob is now in possession of two clocks, one optical and one mechanical. If he finds that the two instruments are ticking at different rates, he will conclude without having to look out the train's window that he is moving relative to the platform. This would contradict the principle of relativity. It is therefore necessary, for this principle to be verified, that the optical and mechanical clocks indicate the same time. Seen from the platform, then, the two clocks are slowed down identically, and the same is true of any time-dependent phenomenon, optical or mechanical or even biological. In other words, it's not the clocks whose evolution is modified when they move—it's time itself that changes!

In a lecture on relativity in Bologna in 1911, the French physicist Paul Langevin, a friend of Einstein's, added a spectacular twist to this thought experiment. He imagined two twins separating, one climbing into a rocket to make a long journey through space at a speed close to that of light, and the other remaining on Earth. On the return trip, the traveling twin will have counted a number of ticks on all of his optical, physical, and biological clocks much smaller than the number counted by his twin who remained on our planet—and will therefore be much younger! Relativity permits this kind of time travel, allowing one of the twins to travel into the other's future. This seems quite paradoxical, but the conclusion is inescapable when we admit the sole postulate of relativity.

To be convinced of this result, we must raise an objection. Why couldn't we also say that it is the twin who remained on Earth who has moved in relation to the rocket, and that it is therefore his clock that has slowed down in relation to that of his traveling brother? Then it would be the latter who ages faster. To avoid this contradiction, should we not admit, via a kind of average, that time could only have unfolded in an identical way for both twins? This reassuring argument for classical physics does not hold water because the twins are not in symmetrical positions. After their instruments were synchronized, the first twin stayed in the Galilean frame of reference of the Earth while the second set off in an accelerated craft and traveled for a long time at cruising speed with a constant velocity before returning to Earth. The latter twin therefore underwent strong accelerations at departure and arrival, changing the Galilean frame of reference along the way—which he could not fail to notice. Thus he cannot apply the principle of relativity and say it was for his sedentary twin that time slowed down!

In order to fully analyze the two-twins experiment, it is necessary to calculate what happens to the traveler's clock during the acceleration phases of his journey, a matter of general relativity that we will discuss later. We will see that the flow of time is quite affected by the accelerations of the clock's reference frame and by the variations of the gravitational field in which it is located. These effects of general relativity must be taken into account rigorously to estimate the exact age of the twin returning from space. However, the general conclusion remains the same: the traveler ages less quickly than his sedentary brother. Without going into detail, we can be convinced that this result is obvious if the rocket's acceleration phases are much shorter—for both twins—than the periods of uniform motion. Whatever happens at the beginning, in the middle, and at the end of the trip, the time "gained" by the traveling twin during those long phases where he moves at a constant speed in relation to the Earth is, so to speak, a given for him. He must definitely age slower than his brother.

Analyzing the motion of light pulses allows us to evaluate the exact time dilation factor related to the motion by simply applying the Pythagorean theorem to estimate the lengthening of the path followed in the vertical clock by the light propagating in a zigzag on the train. The calculation is the same as the one we made in analyzing Michelson's experiment, even if its interpretation is now much more general. The time interval between two clicks of the clock aboard the train is dilated with respect to that of the same clock on the platform by the factor $\gamma = 1/\sqrt{[1-(v/c)^2]}$. At the limit where v tends toward c, the time dilation tends toward infinity and the time elapsed in the frame moving at speed c stops completely: for his twin remaining on Earth, the traveler does not age at all!

The clock thought experiment also leads to the conclusion that spatial distances are relative to the observer as well. Indeed, suppose now that Bob, our passenger, tilts his optical clock by arranging the mirrors—without changing the distance between them—in two parallel vertical planes perpendicular to the train's direction of motion. The light must now propagate along this direction, either in the same direction as the train's motion or in the opposite direction. Returning to the analysis of this clock's functioning as perceived from the platform, let us assume that, for Alice as for Bob, the two mirrors are always separated by a distance *L*. The light, at constant speed *c*, starting from the rear mirror, must now reach the front one, which flees before the light pulse at speed *v*, thus traveling a longer path than it would if the train were stationary. On the way back, the returning light pulse propagates toward a mirror coming toward it. Thus it has a shorter path to travel than if the train were immobile. The addition of the two times t_1 and t_2 of the outbound and return pulses is calculated the same way as the delay of the light in the branch of Michelson's interferometer parallel to the motion of the Earth. We find $t_1 + t_2 = (2L/c)[(1/(1 - (v/c)^{2)}]$. Comparing this result with that of the vertical clock's period, equal to $(2L/c)(1/\sqrt{[1-(v/c)^2]})$, we obtain different results, in the ratio $\gamma = 1/\sqrt{[1-(v/c)^2]}$! Will the clock work at a different rate if it is oriented vertically or horizontally? This would contradict the principle of relativity, as the difference would allow Bob to realize he is in motion without having to look at the platform.

To solve this paradox we must admit that, contrary to what we assumed in our calculation, the distance of the mirrors in Bob's clock measured from the platform differs depending on whether its axis is oriented perpendicular or parallel to the motion of the train. In the first case, it is not modified. In the second, it is contracted by the factor $1/\gamma = \sqrt{[1-(v/c)^2]}$. This contraction shortens the path that light travels in an optical clock with an axis parallel to the motion of the train, and makes the periods of clocks oriented along different axes identical, as required by the principle of relativity.

In fact, Einstein's thought experiment takes up all the calculations applied to Michelson's, explaining in an obvious way why the latter could only give a negative result. The reason is to be found not in any effect of length contraction due to the interaction of the interferometer's matter with a hypothetical aether, but simply in the postulate of relativity, which says that light has the same speed with respect to all inertial frames of reference. And this principle entails the relativity of times and lengths.

A relativistic change of perspective

The formulas for time dilation and length contraction had appeared in the scientific literature prior to Einstein's work. Lorentz, as we have seen, had already hypothesized, with George FitzGerald, the length contraction in the direction of motion to reconcile the negative result of Michelson's experiment with the existence of the aether. Lorentz—and, independently, the French mathematician Henri Poincaré—had also established the formulas for transforming space and time coordinates which, in a change of Galilean frame of reference, leave the form of Maxwell's equations invariant. Lorentz's formulas are equations that describe how the same event, defined by position x,y,z and time t in a Galilean frame R, is described by new coordinates x',y',z' and t' in another frame R' translating at velocity v with respect to R. The factor γ that appears in these equations, describing time expansion and length contraction, is called the *Lorentz factor*.

Until Einstein's 1905 article, however, Lorentz's equations were seen as esoteric properties of Maxwell's, established mathematically but whose physical meaning remained obscure. Lorentz described the time appearing in these formulas as "local" time, theoretically useful for calculations but different from real physical time, which continued immutably in the fixed reference frame with respect to the aether. Was the failure of Michelson's experiment to determine the aether wind caused by the peculiar nature of this aether, the properties of whose interaction with matter "conspired" to make it undetectable?

Einstein's theory did away with this question altogether. Instead of viewing Lorentz's equations as *ad hoc* adjustments to explain a negative experimental result, or as formulas describing the abstract mathematical properties of Maxwell's equations, Einstein simply deduced them from the postulate of relativity using simple thought experiments, without any other hypothesis. In his theory, light propagated in a vacuum, be it empty space free of

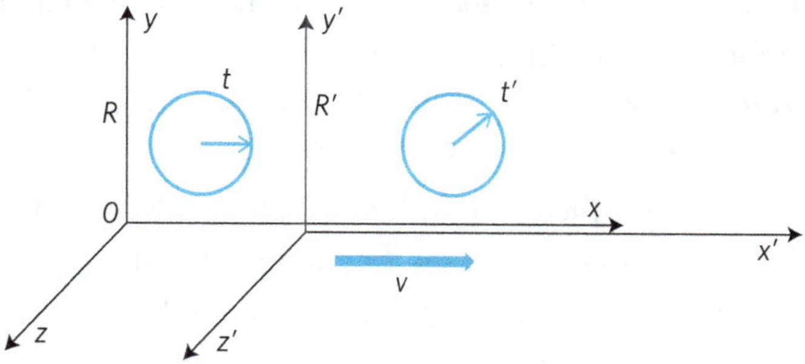

(a)

$$t' = \gamma\left(t - \frac{vx}{c^2}\right) \qquad \gamma = \frac{1}{\sqrt{1 - \frac{v^2}{c^2}}}$$

$$x' = \gamma\,(x - vt)$$

$$y' = y$$

$$z' = z$$

(b)

$$u' = \frac{u - v}{1 - uv/c^2}$$

(c)

$$c^2\Delta t'^2 - \Delta x'^2 - \Delta y'^2 - \Delta z'^2 = c^2\Delta t^2 - \Delta x^2 - \Delta y^2 - \Delta z^2$$

(d)

Figure IV.3. The Lorentz transformations. (a) In the inertial reference frames R and R', the coordinates of an event are respectively t,x,y,z and t',x',y',z'. Reference frame R' moves at velocity v along the Ox axis relative to reference frame R. (b) Formulas linking the coordinates of an event in the two reference frames, expressed using the Lorentz factor γ. At the limit where v is very small compared to c ($\gamma = 1$), the Lorentz transformations are reduced to the classical change of Galilean frame of reference. (c) Relativistic velocity transformation. This equation expresses how the velocity u' of a particle propagating along Ox in R' is deduced from its value u in R. The numerator corresponds to the classical velocity-addition formula. The denominator ensures the constancy of the speed of light. If $u = c$ in R, this equation gives $u' = c$ in R', regardless of v. (d) The Lorentz transformation conserves the four dimension distance between two events.

all matter or the empty space between the electric charges contained in the atoms of a material medium. No mysterious aether-bathing matter had to be invoked. The principle of relativity certainly had counterintuitive consequences, but these could be accepted as long as they were logically deduced from the theory's initial postulate, implying the constancy of the speed of light in all reference frames moving uniformly in relation to each other.

From the perspective of relativity, it is not correct to say that "everything happens as if the length of a ruler is shorter when it's moving relative to the measurer." It is in fact "really" shorter, when we compare measurements made in the two reference frames, per the only procedure allowed by the theory of relativity. This procedure consists, in any inertial frame of reference, in evaluating the distance between the two extremities of the ruler *at the same instant*, whether the observer is on the platform or on the train. Since the simultaneity is not the same for both, it stands to reason that the lengths should not be the same either. Both points of view are equally valid, and each expresses physical reality as it should be described.

Lorentz's formulas show that "adding" the speed of light to the speed (lower than c) of any reference frame always gives the value c, as is correct to satisfy the principle of relativity. The theory also shows that the speed of light in vacuum cannot be exceeded in any Galilean reference frame. If a particle in a reference frame had a speed greater than c, a light beam from a source placed behind that particle would never catch up to it in that frame. On the other hand, if we placed ourselves in a reference frame accompanying this particle, the light, always going at speed c with respect to this new frame, would end up overtaking it and interacting with it. The contradiction between these two results would invalidate the principle of relativity, as physics would not be the same from both perspectives. Mathematically, the impossibility of a "superluminal" reference frame results from the fact that the Lorentz factor γ would become the square root of a negative number. The coordinates in this frame would then be imaginary numbers, losing all physical meaning.

Note that only the speed c of light *in a vacuum* is an absolute limit. In a transparent medium of refractive index n, the radiation propagates at the speed c/n, lower than c. If the medium is itself moving in a Galilean reference frame with a velocity v in the direction of light, the speed of light waves seen by an observer in this frame is greater than c/n while remaining lower than c. Lorentz's velocity-addition formula shows that if v is small compared to c, the wave moves in this frame with velocity $v' = c/n + v\ (1 - 1/n^2)$. This velocity would be equal to $c/n + v$ if the classical velocity-addition law applied. The principle of relativity thus shows that a flowing transparent liquid only partially carries light with it. The higher the index of light in the medium, the stronger this entrainment is. Fizeau observed this effect in his 1852 interferometric experiment. At the time, he had measured the speed of light in a tube through which water was flowing. He interpreted his result as a property of the liquid's *partial* entrainment of the luminiferous aether. Einstein's relativistic explanation evidently eliminates the aether. Fizeau's experiment, performed half a century before the birth of relativity, must have reassured Einstein that his theory was correct.

Space mixes with time

The Lorentz transformations mix space and time coordinates in a way fundamentally different from the Galilean reference frame changes of classical mechanics. In the latter, time is invariant while the space coordinates x', y', z' of an event in a frame R' appear as combinations of the space coordinates x, y, z and time t of the same event in another frame R. These combinations simply show that R is deduced from R' by a translation in space at the relative velocity v of the two reference frames, possibly accompanied by a change of perspective linked to a rotation of the coordinate axes. These operations of translation and rotation maintain the spatial distance ΔL between two events, and thus its square, which can be expressed with the Pythagorean formula using the identity $\Delta L^2 = \Delta x^2 + \Delta y^2 + \Delta z^2 = \Delta x'^2 + \Delta y'^2 + \Delta z'^2$. In these expressions, the quantities Δx, Δy, Δz and $\Delta x'$, $\Delta y'$, $\Delta z'$ are the differences between the Cartesian coordinates of the events in the two Galilean reference frames. As for the time interval between these events, it is unchanged in classical physics and the same is evidently true of its square, which translates to the trivial identity $\Delta t^2 = \Delta t'^2$.

When we replace Galileo's transformations with Lorentz's, there is no *separate* conservation of time and distance, only that of the difference D^2 of the square of the path light travels in the time between the two events reduced by the square of their distance. The conserved quantity is written $D^2 = c^2\Delta t^2 - \Delta L^2 = c^2\Delta t'^2 - \Delta L'^2$. This conservation law, derived directly from Lorentz's equations, was established by the mathematician Hermann Minkowski, one of Einstein's professors in Zurich. It expresses the constancy of the "quadridistance" of the two events in space-time by extending the Pythagorean theorem of Euclidean geometry from three dimensions to four. Events of the theory of relativity are thus described in a pseudo-Euclidean space-time called *Minkowski space*, in which the "metric," the means of calculating distances, affects the square of the time coordinate of a sign opposite that of the squares of the spatial coordinates.

The relativistic distance D between two events is thus the same for all Galilean observers. Its square is positive when the distance traveled by light in the time interval between the events is greater than their spatial distance in all frames of reference. On the other hand, this square is negative if the events are so close in time that the light cannot pass from one to the other in any reference frame within the time interval between them. In this case, the "distance" D, the square root of a negative number, is purely imaginary.

In the first case, the distance is said to be *timelike*. Information has thus been able to propagate from one event to another at a speed less than that of light between the two instants separating them. The earlier event may have had a causal influence on the subsequent one, and the temporal order of events is the same for all observers. There is then a reference frame R_0 in which the two events take place at the same point, for which $\Delta L = 0$. In this frame the distance D is simply equal to ct and D/c is what is called the *proper time interval* associated with the two events.

The notion of trajectory in classical Newtonian physics can be extended to Minkowski space. In relativity, the history of a particle moving in space is described by a continuous sequence of events corresponding to its passage through different points at successive instants. This sequence of points constitutes the particle's *world line* in Minkowski space. A length along this line is defined by dividing it into small intervals separating very close events and adding the distances between them, in the sense of Minkowski's metric. On each of these elementary segments, we can consider the particle's velocity to be constant and associate with it a proper Galilean reference frame in which it is locally stationary. The quadridistance that it travels is then equal to cdt, where dt is the time interval given by a clock in this proper reference frame. By summing these elementary time intervals, we calculate the length of the particle's world line, which is none other than c multiplied by the proper time accumulated by the clock attached to it.

In the case where $D^2 = c^2\Delta t^2 - \Delta L^2$ is negative, the distance D, an imaginary number, is said to be *spacelike* and no information can have been exchanged between the two events, which remain, in all reference frames, independent of each other, with no possible influence between them. The temporal order may even change from one frame to another. To take a dramatic example, the first event could be associated with you, the reader of this book, reading these lines at this moment, and the second with the possible—if highly improbable—disappearance of the Sun in a cataclysmic explosion that, by your watch, occurred five minutes ago. Since light from the Sun takes about eight minutes to reach us, nothing in your present life would herald what may happen in three minutes. In the situation I have just evoked, there may even be observers in the universe, moving at a high speed relative to the Earth and the Sun, according to whom you are reading these lines *before* the Sun goes out, not after. When one such observer receives signals from our solar system, they will be justified in saying the Sun disappeared five minutes after you read these lines, not five minutes before!

Finally, what can we say, in the borderline case between the two previous ones, about events that are, in a given frame *R*, at a zero Minkowski distance from an "event" at the origin of the space-time coordinates ("event 0")? All of these events have their coordinates satisfying the condition $ct = \pm\sqrt{x^2 + y^2 + z^2}$. This relation defines a three-dimensional conical surface in Minkowski space-time, known as the *light cone* of event 0. This cone has two sheets, one in the half-space $t > 0$—the sheet of the future of 0—and the other for $t < 0$, the sheet of its past. All light rays passing through the origin at the instant $t = 0$ have reached in the past, or will reach in the future, the space-time points located on this cone.

Any event located inside the cone is separated from the origin by a timelike distance. It is in the future or in the past of *0* depending on whether it is inside the upper or lower sheet, and this chronological order is the same for all Galilean observers, whatever their frame of reference.

The two zones inside the cone thus define the set of events in the *absolute* future and past of the original event. The events located outside the cone are at a spacelike distance of 0. In another reference frame *R'*, they occurred before or after *0*, or simultaneously with this event, depending on this reference frame's velocity, direction, and value in relation to *R*. They cannot be causally related to event *0*.

Lorentz's formulas thus mix space and time coordinates, stripping these two concepts of the absolute meaning they had in classical physics. This consequence of the principle of relativity, born of a deep analysis of the physical content of Maxwell's equations, extends to the physical quantities that are electric and magnetic fields and the distributions of charge and current that create them. An electric charge that is stationary for Alice appears as a moving charge to Bob, who is in motion relative to Alice. There is therefore a current in Bob's frame that does not exist for Alice. For Alice, the stationary charge simply creates a static electric field described by Coulomb's law. For Bob, the moving charge generates a time-dependent electric field, but also a magnetic field, which is zero for Alice. What is a "pure" electric field for one observer thus acquires a magnetic field component for another.

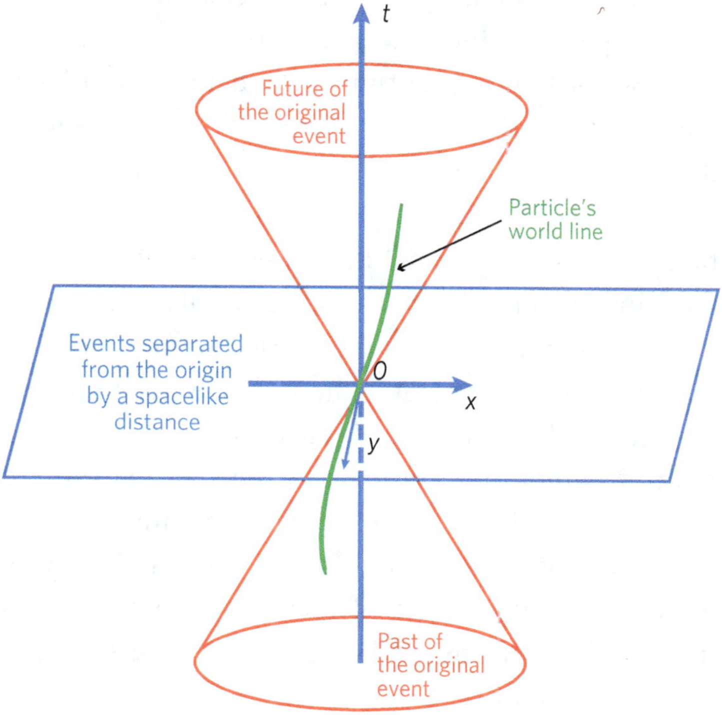

Figure IV.4. Light cone in Minkowski space. To visualize four-dimensional space-time in a drawing that can represent only three dimensions in perspective, we limit the space to two dimensions, marked by the coordinates x and y, and reserve the vertical dimension for time. All events separated from the origin event 0 ($x = y = t = 0$) by a timelike interval are represented by points inside the upper (absolute future of 0) or lower (absolute past of 0) sheet of the cone. A stationary particle in the frame at point $x = y = 0$ has the time axis for its world line. If it moves through $x = y = 0$ at time $t = 0$, its world line is a curve inside the cone. The points outside the cone, separated from the origin by a spacelike distance, correspond to events that occur before or after event 0 depending on the observer's reference frame. They cannot have any causal relationship with the origin event.

Thus, a relativistic change of reference not only mixes space and time; it also mixes, profoundly, the laws of Coulomb, Ampère, and Faraday. Fields, charges, and currents lose their absolute character, as do time intervals and distances. In these transformations, however, an invariant remains. Maxwell's equations must, in all inertial frames of reference, express the same relationships between the spatial and temporal derivatives of the fields and the distributions of charge and current that generated them. Likewise, the forces exerted on the charges and currents must depend on the distribution of electric and magnetic fields in the same way in all reference frames. This fundamental invariance of the equations' form expresses the very essence of the principle of relativity, which requires that the laws of physics be identical in all Galilean reference frames.

Mass and energy combine: $E = mc^2$

Having unified the concepts of space and time, Einstein completed the theory of special relativity by unifying two other concepts that classical physics considered distinct: mass and energy. This led him to establish the equation $E = mc^2$, undoubtedly the best-known formula in all of science—but also, for non-scientists, one of the most mysterious. This famous formula appeared in a short note Einstein published, still in 1905, a few months after "On the Electrodynamics of Moving Bodies." Let's try to show its origin simply.

Let us first recall how classical mechanics describes, in a Galilean reference frame, the dynamics of a very simple system, a point particle of mass m. The quantity of motion or momentum $\boldsymbol{p} = m\boldsymbol{u}$ of this particle is a vectorial quantity (having three space coordinates: p_x, p_y, and p_z) equal to the product of the mass m by its velocity $\boldsymbol{u}$ (a vector of components u_x, u_y, u_z, of modulus $u = \sqrt{u_x^2 + u_y^2 + u_z^2}$). In the absence of an applied force, the momentum remains constant, the particle continuing to move in a straight line with uniform velocity $\boldsymbol{u}$, whose value naturally depends on the Galilean reference frame in which it is observed. When the particle is subjected to an external force, its momentum changes by an amount equal to the product of this force and the time over which it is applied. The mass m, a constant parameter in classical physics, thus measures the particle's inertia, its resistance to a change in velocity. The latter is weaker, for a given force, the greater the particle's mass. For a set of particles isolated from the outside world, the total momentum, the vector sum of the various components' quantities of motion, is preserved in the interactions that the various parts of the system exert on each other.

Classical physics associates another important concept with a system of interacting bodies: its energy, a quantity that measures the system's capacity to provide work or heat to its environment. An example is the kinetic energy a moving body acquires when it passes from stasis to a final velocity $\boldsymbol{u}$. This energy, $mu^2/2$, is equal to the work of the force that accelerated the object. It increases proportionally to its mass m and as the square of its velocity. It can be restored in the form of heat, *i.e.* of disordered kinetic energy in the atoms or molecules of the mass m or its environment, when the moving object is slowed by frictional forces. This kinetic energy can also change into potential energy when the object's position changes in a force field. Thus, a mass launched upward in the Earth's gravitational field sees its kinetic energy transformed into potential energy as its altitude increases, with the inverse exchange between potential and kinetic energy occurring when it falls. The first principle of thermodynamics expresses the conservation of the total energy of an isolated system, the sum of its kinetic and potential energies and its calorific energy stored in the thermal agitation of the atoms that constitute it.

These classical properties of mass, momentum, and energy needed to be reconsidered within the framework of the theory of special relativity. It was clear that Newton's physics had to change to satisfy the principle of relativity. Not only did the velocity-addition law, based on the existence of absolute time and space, need to be modified to account for the constancy of the speed of light in all Galilean reference frames, but the fundamental law of Newtonian dynamics itself needed to be revised in order to prevent a material object from reaching, in any frame of reference, a velocity greater than c. If the momentum of a uniformly accelerated particle increases proportionally to time, its velocity, proportional to momentum in classical physics, would finally, after a finite time, reach and exceed the speed of light c—which is forbidden by the principle of relativity.

To make this infinite acceleration impossible, Einstein simply posited that the mass of an object in a Galilean reference frame system should not be constant but rather increase with its velocity. If the mass at rest is m_0, it becomes $m_u = m_0/\sqrt{[1 - (u/c)^2]}$ when the particle is moving at velocity u. The mass increase factor has the same expression as the Lorentz factor introduced in the formulas that describe time dilation and length contraction in a change of Galilean reference frame. The particle's velocity u simply replaces the velocity v of the relative motion of the two reference frames. This factor, which we will call γ_u, tends toward infinity when the velocity u of the particle tends toward c. The momentum thus becomes $\boldsymbol{p} = \gamma_u m_0 \boldsymbol{u}$. The introduction of the factor γ_u is the only modification necessary to make the law of Newtonian dynamics compatible with the theory of special relativity. This simple change makes the form of the dynamic equations invariant in the Lorentz transformations and thus reconciles the laws of mechanics and electromagnetism, both of which respect the principle of special relativity according to which physics must be the same in all inertial frames of reference.

Making the mass depend on the velocity by the factor γ_u prevents any material particle from reaching the limit velocity c. The application of a constant force always increases the momentum proportionally to time, but this increase is divided between mass change and velocity change. At the beginning of the motion, when the velocity u is small compared to c, the factor γ_u is barely different from 1, the difference between m_0 and m_u is undetectable, and Newtonian physics applies: velocity and momentum increase proportionally with time and the motion is uniformly accelerated. When u becomes an appreciable fraction of c, however, the particle's mass begins to increase and, gradually, most of the momentum increase is translated into a mass increase, at almost constant velocity. The velocity tends toward c without ever reaching it, while the particle's mass increases more and more.

The energy stored by the particle due to its motion also depends on its relativistic mass $m_u = \gamma_u m_0$. An elementary calculation shows that in order to bring the initially stationary particle to velocity u, the accelerating force must provide work equal to $(\gamma_u - 1)m_0c^2 = (m_u - m_0)c^2$. This expression suggests the equivalence between the kinetic energy and inertial mass of a material particle. The particle's kinetic energy is none other than the excess inertial mass it has acquired from its motion, multiplied by the square of the speed of light. As long as the velocity u is small compared to c, the relativistic expression is reduced, with an excellent approximation, to the classical expression of kinetic energy. For u/c very small compared to one, $\gamma_u - 1$ is indeed practically equal to $u^2/2c^2$ and the particle's kinetic energy takes the classical value $m_0u^2/2$. Relativistic effects become important only when u reaches an appreciable fraction of c.

As the variation of the particle's kinetic energy generally appears as the difference between the two terms $m_u c^2$ and $m_0 c^2$, Einstein deduced that the former, $m_u c^2$, represented the particle's energy when it has velocity u, and the latter, $m_0 c^2$, its energy when at rest. Generalizing this result of relativistic dynamics to all energetic phenomena, Einstein postulated the equivalence of mass and energy, stating that any transformation of a physical system accompanied by a change in energy ΔE resulted in a change in the system's mass Δm, such that $\Delta E = \Delta mc^2$. This formula had to be valid whether it described a mechanical process (such as the velocity change of a moving body), a thermal phenomenon (such as the heating or cooling of a solid, liquid, or gas), or an interaction involving electromagnetic forces (such as a chemical reaction rearranging atoms between molecules).

The conservation of energy and that of mass, which Antoine Lavoisier expressed with the famous maxim *Nothing is lost, nothing is created*, are two distinct laws in classical physics. In relativity, they are replaced by a law of global conservation of mass-energy. If energy is produced by a physical system (in the form of radiation or heat or work), this production is accompanied by a corresponding loss of mass according to the formula $\Delta E = \Delta mc^2$. Conversely, if a system receives energy, its mass increases in the same proportion.

The theory of special relativity shows that mass and energy are physical entities of the same nature and can be measured either in the standard unit of energy (the joule, which corresponds approximately to the work of a hundred-gram weight descending one meter in the Earth's gravitational field) or in the standard unit of mass (the kilogram). Because of the enormity of the square of the speed of light expressed in meters per second, one kilogram of matter is equivalent to a huge amount of energy—9×10^{16} joules, equivalent to the energy dissipated by the explosion of twenty megatons of TNT, or the energy it would take to free 1.5 million tons of matter from the Earth's gravitational field! The gigantic value of the energetic equivalent of mass implies that the usual energy processes observable in everyday life correspond to extraordinarily small relative changes in mass, which were undetectable in Einstein's time. For example, bringing matter from 0 to 300 degrees kelvin (ambient temperature) imparts to the atoms and molecules constituting it a thermal motion velocity v_t on the order of one millionth (10^{-6}) the speed of light. The corresponding change in mass is then on the order of $(v_t/c)^2 = 10^{-12}$, *i.e.* only one thousandth of one billionth. This mass gain corresponds to only one microgram per ton of matter.

A typical chemical reaction changes the masses of the reacting products by an amount on the order of 10^{-10}, also undetectable with the measuring instruments available at the beginning of the twentieth century. As for the much more energetic nuclear reactions, which change the mass of the nuclei involved by an amount on the order of 10^{-3} to 10^{-4}, their nature was not yet known in 1905. These nuclear reactions, which transform roughly a thousandth of the matter of stars into heat and light, are at the origin of the energy they radiate. The formula $E = mc^2$ thus unravels the mystery of solar energy, indispensable for the development of life on Earth. It also heralds the possibility of exploiting the enormous amount of energy stored in atomic nuclei for military or peacetime purposes.

As for the total transformation of mass into energy, it occurs only under extreme conditions, in accelerators where the annihilation of matter and antimatter particles can be observed with the production of very short-wavelength electromagnetic gamma radiation. We can also observe conversions of large fractions of mass into energy in the universe in the coalescence of black holes, which generate enormous energy in the form of gravitational waves that reach our planet after propagating through space for billions of years. These phenomena were unimaginable at the time Einstein established his theory of special relativity, but an inkling of their possibility was already there. It is remarkable that Einstein was already able to announce—in such diverse fields of physics, from the infinitely small to the infinitely large—effects that would become observable thanks to technological advances only fifty to one hundred years later.

Let us complete this panorama of special relativity with a remark that forges a link to the lifting of Lord Kelvin's other mysterious cloud. The 1905 theory of relativity established some essential properties of the photon, the light particle Einstein introduced in another essential contribution that year, his landmark article on quantum theory. To make the connection, let's begin with a simple question. Is the equivalence of mass and energy deficient in the case of the production of light? By definition, light transports energy at... the speed of light. The particles of light known as photons must have zero mass to fly at this speed. And yet they do transport energy through space: solar energy, to give an obvious example. How, then, can we reconcile this apparent contradiction with the famous formula $E = mc^2$?

This paradox finds its answer in an essential relationship between relativistic energy and momentum, which appears when we consider how these quantities are transformed in a change of Galilean reference frame. What is mass-energy in the frame of reference where a particle is at rest becomes a mixture of energy and momentum in a frame where the particle is in motion. By changing point of view, the energy divided by c and the three coordinates of momentum p_x, p_y, and p_z appear as the components of a four-vector which mix together according to the same Lorentz transformations as those of an event's space-time coordinates. Neither the energy nor the momentum has an absolute reality. On the other hand, the square of the energy divided by c^2, minus the sum of the squares of the momentum components—a quantity analogous to the square of the relativistic distance of two events—is preserved and takes the same value in all reference frames. In the frame where the particle is at rest, the value of the square of the energy is equal to $m_0^2c^4$. We then deduce the identity, valid in all reference frames: $E^2 - p^2c^2 = m_0{}^2c^4$, where we have condensed the sum $p_x{}^2 + p_y{}^2 + p_z{}^2$ into p^2, the square of the modulus of the momentum vector.

From this we can immediately deduce a universal expression of the relativistic energy of a particle, $E = \sqrt{[p^2c^2 + m_0{}^2c^4]}$, more general than the simple formula $E = m_0c^2$. In the reference frame, if it exists, of the particle at rest ($p = 0$), the general expression is of course reduced to this simple formula, all the energy now being concentrated in the particle's mass. If, on the other hand, its mass is null, the general expression becomes simply $E = pc$. The particle must move at the speed of light (in all reference frames) because there can be no Galilean observer to whom it would be motionless. If that were the case, the particle would have zero energy and zero momentum in that frame of reference, resulting, by Lorentz transformation, in $E = p = 0$ whatever the inertial frame of reference. Thus, energy can indeed be transported without mass by particles, provided they propagate at speed c. This is the case for the particles of light we call photons.

Relativistic theory thus requires that the photons of a plane wave have a momentum equal to their energy divided by the speed of light. When radiation is absorbed by matter, it transfers to it not only its energy E, but also its momentum p, in the ratio $p/E = 1/c$. The momentum received by the matter corresponds to a force pushing it in the direction of the incident light. This is the phenomenon of radiation pressure. The opposite occurs when matter emits light. An atom, by losing energy, can emit a photon which also carries away the momentum, causing a reaction in which the atom recoils in the opposite direction from that of the photon. We will return later to the interaction between atoms and photons and to the consequences of the conservation of energy and the total momentum of matter and light in these processes.

The analogy between space-time and energy-momentum four-vectors leads us, finally, to another important property conferred to a particle by special relativity. The Lorentz transformations not only keep the square of these vectors (with the rule associating opposite signs to the contributions of time and energy coordinates on one hand, and position and momentum on the other); they also preserve the *scalar product* of these four-vectors between them, a quantity equal to the product of the time and energy coordinates minus the sum of the products of the position and momentum coordinates. Thus, to a particle of energy E and momentum (p_x, p_y, p_z) passing at time t at the point (x,y,z), the theory of relativity associates the quadruple product $Et - xp_x - yp_y - zp_z$, a quantity intrinsic to the particle, which takes the same value in all Galilean reference frames. We will see that this relativistic invariant is proportional to the phase of the matter wave that describes the particle in quantum theory.

Special relativity gave an answer to the enigma posed by the speed c of light; as we have seen, this answer changed our conception of space and time. It showed that this speed intervened in the description of physical phenomena whose links to light had until then been unsuspected. The speed c was not only the velocity of electromagnetic waves in a vacuum; it was also the limit velocity for the propagation of all forms of information or influence at a distance. Through the famous formula $E = mc^2$, the value of c also revealed the enormous energy hidden in all possible forms of matter. And this revolution in our concepts of time and space, mass and energy, germinated in Einstein's mind based on a very simple idea borrowed from Galileo: *Physics must be the same in all inertial frames of reference.*

Einstein's "happiest" idea comes from Galileo again

Reflecting on what the theory of special relativity had accomplished, Einstein realized as early as 1905 that a great enigma remained to be solved. What was the true nature of gravity? It is the first force a baby perceives when it becomes aware of the world, acting on all bodies by pulling them toward the Earth. Newton gave it a mathematical expression around the same time he discovered the law of inertia linking an object's acceleration to the force it experiences. The theory of relativity had led Einstein to modify this Newtonian law of inertia by showing that a body's inertial mass depended on its velocity, but describing the law of gravity still posed a problem. Newton had indeed postulated that this force acted at a distance, instantaneously. According to his theory, if the Sun suddenly disappeared, the Earth and all the planets would simultaneously slip from their orbits and fly off, moving away at uniform speed from the point at which they were located at the time of the cataclysm. This moment was the same for all observers since, according to Newton, time was an absolute concept. Of course, this was in profound contradiction with Einstein's theory. Since light takes eight and a half minutes to reach us from the Sun, nothing could change the Earth's orbit during this interval.

As Newtonian gravity was in tension with special relativity, one or both had to be modified if they were to agree—and these modifications had to preserve the achievements of both theories. Newton's laws of gravitation had made it possible to calculate with extraordinary precision the positions of the planets (recall Le Verrier and the discovery of Neptune). The new theory therefore had to be at least as precise, and to coincide practically with Newton's theory in its description of falling objects and the motion of known celestial bodies. Special relativity had allowed a deep understanding of the laws of electromagnetism and inertia. The new theory of gravitation had to preserve these achievements as well. Einstein took eight years, from 1907 to 1915, to solve this problem and establish a theory of general relativity that satisfied all of its constraints.

In the meantime, his 1905 work had given him a reputation in the physics community that allowed him to leave the obscurity of the patent office in Bern. After a brief stint as a professor in Zurich and Prague, he became the youngest member of the Prussian Academy of Sciences in Berlin in 1912, at the age of thirty-three. It was to this academy that he presented, in November 1915, during World War I, the theory that would make him not just a star in physics circles, but a worldwide celebrity.

General relativity is an extension of special relativity. It simply generalizes the principle of relativity of motion to all frames of reference, not just inertial frames moving at uniform velocity with respect to one another. When I say "simply," I am glossing over the mathematical difficulties Einstein took eight years to overcome, between his initial idea in 1907 and the publication of his complete theory of gravitation in 1915. I will not discuss these mathematical problems, which would take us far beyond the scope of our subject. I will simply sketch the broad strokes of this theory so that even a non-scientific reader may understand the extent of Einstein's intuition and genius.

The spark—the happiest idea Einstein ever had, as he would often recall afterwards—came, once again, from Galileo. While studying the fall of bodies, the Italian scientist had realized that they all fall in the same way in the Earth's gravitational field. Whether you throw a heavy stone or a pea, you will see them follow identical trajectories (provided, of course, that you can disregard the frictional forces caused by air resistance, which act more strongly the lighter the object). According to legend, Galileo observed this universality of falling bodies by dropping two different weights simultaneously from the top of the Leaning Tower of Pisa and observing that they reached the ground at the same time. He thus disproved Aristotle's notion that a heavy body should fall faster than a light object.

Friction in the air ensures that a feather will float longer before reaching the ground than a billiard ball. But this is an artifact, a non-essential effect that can be eliminated by, say, observing the fall in a vacuum tube: there, feather and billiard ball fall at the same speed. Without a vacuum pump, Galileo understood this not by experimenting from the Tower of Pisa but likely through his famous experiments with balls rolling on an inclined plane that minimized friction. Let us recall that he had also shown that a pendulum's period does not depend on its mass, which expresses the same principle. All massive bodies follow the same law of motion in the Earth's gravitational field.

Newton, establishing his law of universal gravitation, had given a mathematical expression to Galileo's empirical observation. If a body is attracted to the Earth, it is because our planet exerts a force on it. This force is proportional to the product of the Earth's gravitational mass M and the gravitational mass m_g of the body. This gravitational mass is a characteristic parameter of the body in question, which makes it possible to calculate the intensity of the force it receives from the Earth. If this body were electrically charged as well, it would also have a charge q, which is the parameter that defines the force an electric field would exert on it. In this sense, gravitational mass is, like the quantity q, a parameter of "charge."

According to Newton's law of inertia, gravitational force imparts to the body an acceleration inversely proportional to its inertial mass m_i. This mass is a property of the body that expresses how it reacts to any type of force, gravitational or electrical or other. Gravitational acceleration is therefore proportional to the ratio m_g/m_i of the gravitational and inertial masses of any body attracted by the Earth. The observation that all bodies have the same acceleration in the Earth's gravitational field led Newton to postulate that the ratio m_g/m_i must be a constant, independent of the body in question. The value of this constant depends only on the chosen units and can simply be equated to 1. This equivalence, known as the principle of equivalence of gravitational and inertial masses, exists only for gravitation. The ratio q/m_i of the electric charge to the inertial mass is not constant. It is about two thousand times greater for an electron than for a proton, for example, and these two particles do not follow the same trajectory at all in an electric field.

The principle of equivalence of the two forms of mass had been well known since the seventeenth century, and astronomers took it into account daily in calculating the motion of the planets. Nonetheless, it was considered a curiosity, a strange coincidence of unknown origin. Einstein, who used it as a starting point for his theory of general relativity, viewed it differently.

He started from the observation that gravitational acceleration is not the only acceleration to have the same effect on all bodies. The same is also true of inertial accelerations describing the motion of bodies in non-uniform moving frames of reference. We encountered this type of force in Chapter 2, when we discussed the rotating merry-go-round. The centrifugal force is proportional to the inertial mass of the objects that experience it, leading to the same acceleration for all objects located near the same point in the reference frame rotating with the carousel. Let's take another example, that of a rocket far from any massive body, star, or planet, impelled by a uniformly accelerated motion. If you were on that rocket, you would feel the motion by being "pulled" in the direction opposite the acceleration, one you would naturally call "down." To stand upright, you would have to use muscular tension to exert a reaction force on the wall you called "the floor." If you released two objects of different mass, you would see them fall together toward "the bottom" with the same acceleration.

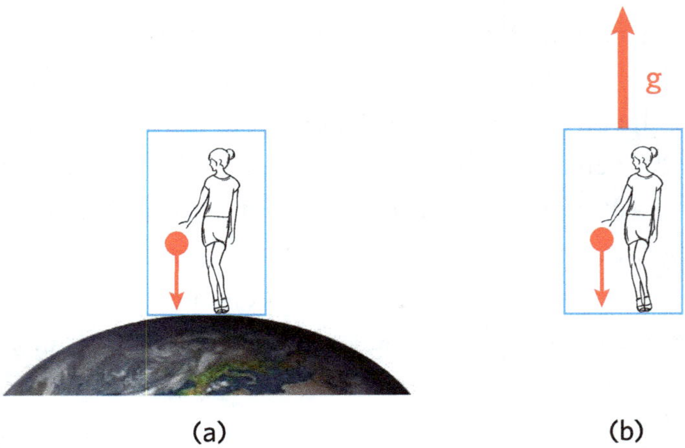

Figure IV.5. Local equivalence between gravitational force in (a) an inertial frame in the presence of a large mass and (b) inertial force in an accelerated frame in space, far from any mass.

In short, no mechanical experiment you could perform aboard the rocket would allow you to decide between two possibilities: you would be either in an accelerated rocket, far from any gravitational mass, or in a stationary enclosure in the gravitational field of a planet pulling you "down." Much more generally, you wouldn't know that you were not in fact in an accelerated reference frame *and* subject to a gravitational field at the same time, as the acceleration you observed on all the bodies around you results from the "real" acceleration of gravity, combined with that of inertial forces.

Another example will help visualize this. Suppose you are on a plane in freefall. The pilot has put the plane in an upward trajectory and cut the engines. The airplane and everything in it—including you—follow a parabolic trajectory in the Earth's gravitational field. If you let go of an object it will fall with you, meaning it will remain motionless relative to your hand, which is no longer holding it. You will feel no pressure from your seat; you will feel as if you no longer weigh anything. If you had been sleeping before the plane engines were shut down and woke up during the parabolic flight, you would have no way of telling whether you were falling into a gravitational field along with everything around you or floating away from any stars or planets. In the first case, you would be experiencing the Earth's gravitational force along with everything around you in the plane, but that acceleration would be exactly compensated for by the inertial force associated with the movement of all bodies in the reference frame of the cabin falling with you. In the second case, neither you nor any of the objects around you would be subject to any force.

You cannot decide between these two hypotheses unless you look through a window at the landscape outside. This experiment is performed on astronaut training flights to reproduce "weightlessness." It is also what happens on the International Space Station orbiting the Earth. Inhabitants of the Station know they are in orbit around a massive body to the force of whose attraction they and all objects around them are subject—but they could just as well be floating somewhere in outer space, far from any star or planet.

Imagining this kind of situation in thought experiments that replaced the trains of special relativity with rockets, elevators, and revolving carousels, Einstein gave the principle of equivalence another definition:

> There is equivalence between an accelerated reference frame in the absence of gravity and a stationary reference frame with respect to a gravitational source producing the same acceleration.

In both situations, all of physics should be describable in the same way, and an observer should be unable to distinguish them. More generally, in the presence of any gravitational field, physics should be described by the same equations in any reference frame, whatever its acceleration. What appears to be a "pure" gravitational force in one frame of reference can be seen in a second as an inertial force, or in a third as a mixed force combining a gravitational contribution and an inertial force. Playing out all the consequences of this principle, Einstein sought to establish the law describing the space-time variations of the gravitational field created by any distribution of mass and energy in the universe. This law had to respect the condition of relativistic covariance—*i.e.*, had to be expressed in the same form in all reference frames, accelerated or not—and coincide, for low speeds and for not-too-massive gravitational sources, with Newton's laws.

Gravitation and curvature of space-time

This course of study, which took eight years to complete, resulted in the theory outlined in Einstein's 1915 lectures at the Prussian Academy of Sciences. It led to a profound upheaval in our understanding of gravity and the structure of space and time. Einstein understood that a gravitational field is in fact only the physical expression of a curvature of space-time, meaning distances can no longer be expressed based on a four-dimensional extension of the Pythagorean theorem, as they are in special relativity.

In this curved space, called *non-Euclidean* because the Euclidean geometry we learned in grade school no longer applies, the position of events must be defined using a non-Cartesian, "curvilinear" coordinate system, and the distance between events must be expressed by more complex formulae than the one based on the Pythagorean theorem.

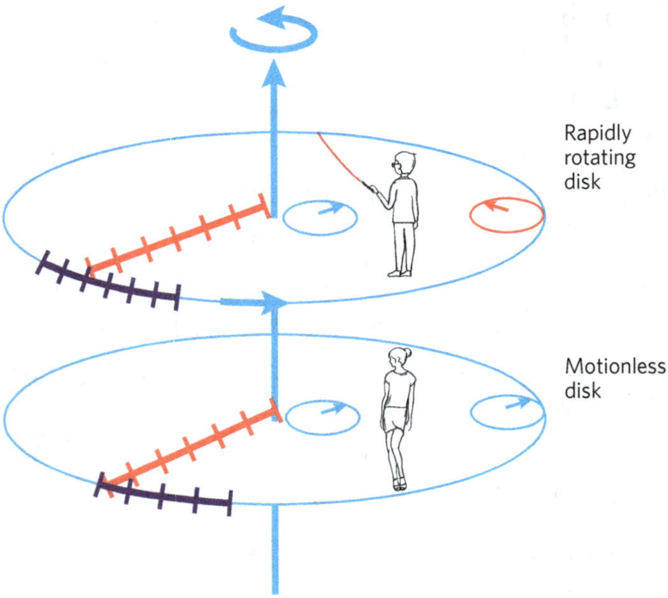

Figure IV.6. Revolving carousel thought experiment. Alice verifies that the ratio between the circumference and the diameter of her disk is equal to π and that the time of her clock is the same everywhere. Bob finds that the ratio between the circumference and the diameter of his rotating disk is greater than π and that his clock beats slower at the edge of his disk than in the center. By pointing a laser beam at the periphery of his disk, he finds that the light follows a curved path in his frame of reference.

To arrive at the essential notion of space-time curvature, Einstein started, as he had done in 1905, with a thought experiment involving rulers and clocks, using the concepts he had developed for the theory of special relativity to understand intuitively that gravitation and the deformation of space-time were in fact one and the same. He imagined two identical disks floating in space, one above the other, free to rotate around the same axis in two parallel planes, far from any source of gravitation. These disks carry two experimenters: we will call them Alice and Bob, as in the train experiments, even though Einstein did not give them these names (nobody troubled themselves with gender parity at the time).

Alice and Bob have identical standard rulers and clocks synchronized in advance. While Alice's disk *R* remains "stationary," Bob's disk *R'* is set in rapid rotation around the common axis of the two disks. Using her ruler, Alice measures the circumference and diameter of her disk by translating the ruler many times in the tangential and radial directions, and finds that the ratio of the number of these shifts is equal to $\pi = 3.14...$ as it should be in Euclidean geometry. She moves one of her clocks toward the edge of her disk, leaving another in the center, and finds that the two clocks remain synchronous and always indicate the same time, as is to be expected in her inertial reference frame.

She then turns her attention to Bob, spinning above her in the *R'* reference frame. Bob is doing the same experiments as Alice. He measures the circumference of his disk with his ruler. For Alice, this ruler, oriented along the tangential velocity of the edge of the disk *R'*, undergoes Lorentz's contraction. When Bob then starts to measure the diameter of his disk, he places his ruler along a radius. The ruler is then oriented perpendicular to the velocity and does not undergo any contraction according to Alice. She thus notes that the ratio between the number of Bob's transfers in the tangential and radial directions is greater than π. Clearly, Bob can only make the same observation. He is therefore in a *non-Euclidean* space in which the ratio between the circumference and radius of a circle is greater than π. If he repeats the measurement on a smaller circle still centered on the same axis, his gauge will undergo less contraction during the measurement of the circumference and he will find that, for this second circle, the value of the ratio of circumference to diameter is still greater than π, but getting closer to it. How does Bob interpret his result? In his revolving reference frame, he and all the objects around him, gauges and clocks included, are subject to a centrifugal force that he can interpret as the effect of a gravitational field. This field is manifested by the fact that the ratio between circumference and diameter of the circles is no longer equal to π and depends on the circles' position in the frame, which is a property of non-Euclidean curved spaces.

Bob then brings one of his clocks to the edge of his disk. Alice notices that this clock, which is moving in relation to her, is beating slower than the identical one Bob left in the center of his disk. Bob can only make the same observation. Thus he will notice that, in a gravitational field, not only is space non-Euclidean, but time flows unevenly at different points. Gravitational distortion affects both time and space. This is what it means to say that space-time acquires a curvature.

To convince himself at last that the geometry in his world has changed, Bob directs a beam of light toward the periphery of his disk. For Alice, the light travels in a straight line at speed *c*, but as the *R'* frame rotates the beam reaches the edge of the disk at a point not in the extension of the direction along which Bob pointed it. Bob sees this as well. For him, the light beam has been deflected by the gravitational field it has encountered and therefore follows, in a curved space, a path different from a straight line.

How can these effects be described quantitatively? Since we lack the intuition to easily conceive what curvature represents in a three- or four-dimensional space, let's get an idea by placing ourselves in a two-dimensional space and assuming we are confined to a flat surface, like a beetle moving on a tablecloth. According to ordinary "Euclidean" geometry—the kind we learn in grade school and which I recalled earlier in the context of special relativity—the plane's different points can be located using their Cartesian coordinates, here two-dimensional: a pair of two numbers, x and y, indicating its position in a system of orthogonal axes of arbitrary direction. The distances between two points are expressed using the Pythagorean theorem, as the square root of the sum of the squares of the differences of these two points' Cartesian coordinates. This theorem is based on the fact that the sum of the angles of any triangle is 180°, or π radians.

If instead of moving on a flat surface our beetle moves on a spherical cap, any triangle drawn on it will be formed of arcs of circles crossing at angles whose sum is greater than π. The geometry will be *non-Euclidean* and the Pythagorean theorem no longer valid. Cartesian coordinates will be replaced by the curvilinear coordinates u and v (*e.g.*, the latitude and longitude of a point on the Earth's sphere). The grid of parallel and orthogonal lines allowing a point to be located in a plane is then replaced by a grid of meridian and latitude lines, two sets of circles drawn on the sphere and intersecting at right angles. The distance between two points then becomes the length of the arc of the large circle passing through them. When the two points are adjacent, the distance depends, as in Euclidean geometry, only on the small difference du and dv of the two points' curvilinear coordinates, but is no longer expressed by the Pythagorean theorem.

The expression of the square of the distance $ds^2 = dx^2 + dy^2$ from ordinary geometry is replaced by a more general form, of the second order in du and dv, which can be written $ds^2 = g_{uu}du^2 + g_{vv}dv^2$. The coefficients g_{uu} and g_{vv} define the local *metric* of the surface. It is local because in general the coefficients g_{uu} and g_{vv} depend on the place on the sphere where it is defined. Near the equator the degrees of longitude and latitude are roughly equal in length, and g_{uu} is little different from g_{vv}, as in Euclidean geometry. Near the poles, however, the longitude coordinate is almost irrelevant, as all meridians converge at the same point and g_{vv} is much smaller than g_{uu}.

The case of a spherical surface with constant curvature is particularly simple. For any two-dimensional curved surface, one can, as the mathematician Bernhard Riemann did in the nineteenth century, define curvilinear coordinates u and v in an infinite number of ways, generalizing the longitudes and latitudes of the sphere, and a metric for each coordinate system defining the square ds^2 of the distance between two neighboring points by a second-order formula in du and dv, depending on three coefficients: g_{uu}, g_{uv}, and g_{vv}. The square of the distance is then written: $ds^2 = g_{uu}du^2 + g_{uv}dudv + g_{vv}dv^2$.

The curvature of four-dimensional space-time, impossible to represent intuitively, will be similarly defined by a metric depending on ten coefficients, four of which multiply the four squares of the curvilinear space-time coordinates and the remaining six the mixed products of these coordinates with each other. This set of coefficients defines the so-called *local metric tensor* of gravitational space-time.

Einstein thus arrived gradually at the idea that the gravitational field is nothing but the curvature of space-time, defined by the metric tensor at each point. The expression of this tensor depends on the distribution of masses present in space. In the absence of mass, this tensor is reduced to that described by the quasi-Euclidean geometry of special relativity, which has only four non-zero elements: $g_{xx} = g_{yy} = g_{zz} = -1$ and $g_{tt} = 1$ (if the time coordinate is defined as ct). A mass changes the curvature of space-time around it, similar to the way a weight placed on a stretched canvas deforms the surface by creating a trough. The matter and light present in this deformed space follow curved lines inflected by the presence of masses, calculable from the expression of the metric tensor at each point. These lines, called the *geodesics* of curved space-time, generalize straight lines followed by particles that are not subject to any force in the Euclidean spaces of inertial reference frames. As the American physicist Archibald Wheeler said in a pithy formula, *Mass tells space-time how to curve, and space-time tells mass how to move*. The equations of general relativity that describe this interaction between masses and the geometrical structure of space-time are complex and their solutions difficult to calculate in the general case of intense gravitational fields. At the limit of weak gravitational fields they coincide, give or take very minor corrections, with the solutions of Newton's equations.

The new predictions of the theories of relativity, special and general, were quite difficult to verify at the beginning of the twentieth century. The expected relativistic effects were extremely weak in a world where objects' velocities were very small compared to the speed of light. Even the highest velocities accessible to measurement, those of the Earth and the planets around the Sun, did not exceed one ten-thousandth the speed of electromagnetic waves. The effects of time dilation and length contraction and those of the gravitational curvature of space-time were extraordinarily small. It is remarkable that Einstein predicted the existence of these effects, solely by logical deduction from the simple principles of relativity, when their manifestations were so far from being observable.

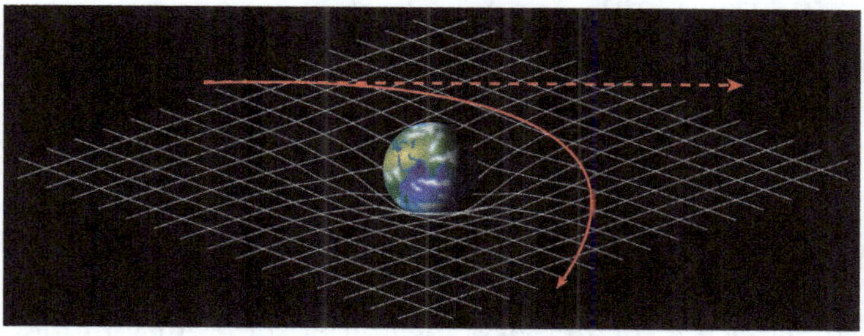

Figure IV.7. General relativity identifies gravitation with the curvature of space-time. Mass tells space-time how to curve, and in turn space-time tells mass-energy how to move in this curved space by following its geodesics.

Relativistic predictions and postdictions

Today, after a century of enormous technological progress, we know how to accelerate material particles—electrons, protons, or muons—to speeds very close to that of light. We have found that the inertial mass of these particles is indeed millions of times greater than their mass at rest, which requires colossal accelerating magnetic fields that can be produced only by gigantic machines such as the one at CERN in Geneva, whose circumference is twenty-seven kilometers. Unstable particles such as muons, which live only a few microseconds at rest, survive a thousand times longer in these accelerators, vividly illustrating the paradox of Langevin's twins. We are also able to measure the mass of atoms and nuclei before and after a chemical or nuclear reaction, and to verify with precision the validity of the formula $E = mc^2$.

Time measurements have made huge progress as well thanks to the development of atomic clocks. Commercial clocks on airplanes or satellites check special relativity's time dilation effect with an accuracy on the order of 10^{-14}. They have also made it possible to observe the effect of gravitational curvature on the flow of time by showing that two clocks, located at different altitudes, see the period of their ticking differ.

These gravitational and relativistic effects, which create clock delays of a few tens of microseconds per day, may seem negligibly small. They are important, however, and correcting them has become a routine matter in the GPS, the navigation system that allows us to position ourselves on Earth with great precision using triangulation measurements made by receiving electromagnetic pulses from different satellites. If the computers that calculate our position from these signals did not take relativistic effects into account, their readings would be off by several kilometers! By using the GPS every day, we unknowingly perform extremely precise tests of special and general relativity, the kind Einstein and his contemporaries were unable to perform or even design. We are far from the navigation methods of the eighteenth century, based as they were on observation of the periodic returns of Jupiter's satellites or the ticking of Harrison's stopwatch, but the spirit is the same. We use the knowledge gained from discoveries inspired by pure curiosity—blue sky research—for our daily navigation, and countless other practical uses.

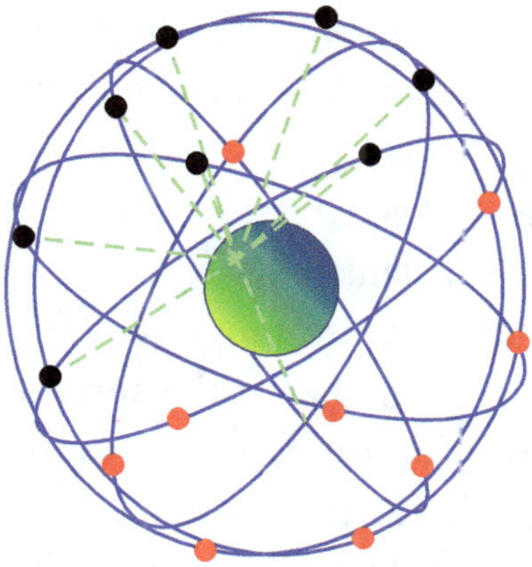

Figure IV.8. The GPS system uses the signals of atomic clocks embedded in a network of satellites orbiting the Earth. The corrections of special and general relativity are essential in signal processing. Without them, GPS readings would be off by several kilometers.

Advances in astronomy and astrophysics have also confirmed the predictions of general relativity in the area of strong gravitational fields. The existence of black holes, predicted by this theory, has been confirmed by numerous observations. These are the final state of large stars that have exhausted their nuclear fuel and collapse on themselves to produce a singularity of the gravitational field, a gravitational "well" from which not even light can escape. At the edge of these holes, the space-time curvature is so strong that time stops. Einstein encountered this situation in his rotating disk experiment. If the edge of the disk approaches the speed of light, Lorentz's dilation causes time to stretch to infinity. The disk's gigantic centrifugal force is then equal to the gravitational force experienced by matter on the horizon of a black hole, the limit beyond which matter is forever sucked into the gravitational well.

The equations of general relativity also show that gravity does not act instantaneously at a distance, but that the space-time curvature propagates from one region of the universe to another at the speed of light, like a wave of deformation of the cosmic web. Gravitational waves were predicted by Einstein's equations, though their existence was debated for a long time—including by Einstein himself—due to the great mathematical difficulty of solving these equations for strong gravitational fields. Their discovery, using gravitational antennas that are gigantic versions of Michelson's interferometer, made headlines in February 2016.

In the first event detected, the source of the waves was the coalescence of two black holes located more than one billion light-years from our planet, which orbited around each other and eventually merged into a single object, emitting in a fraction of a second a gravitational energy equal to three times the mass of the Sun! This cataclysmic phenomenon produced a tiny shift of the mirrors of the interferometers, on the order of a billionth of the size of an atom.

All of these observations were inconceivable in 1915. There were no clocks precise enough to measure the time dilation effects induced by motions of very low velocity compared to the speed of light, or by the weak gravitational effects produced by the planets or the Sun. The interferometers in which classical light circulated were, at best, sensitive only to displacements on the order of a tenth or a hundredth of a fringe distance. The sensitivity required to detect gravitational waves—a ten-billionth of a fringe distance—would have to wait for the invention of the laser and the implementation of multiple technological feats that could not be imagined at the beginning of the last century.

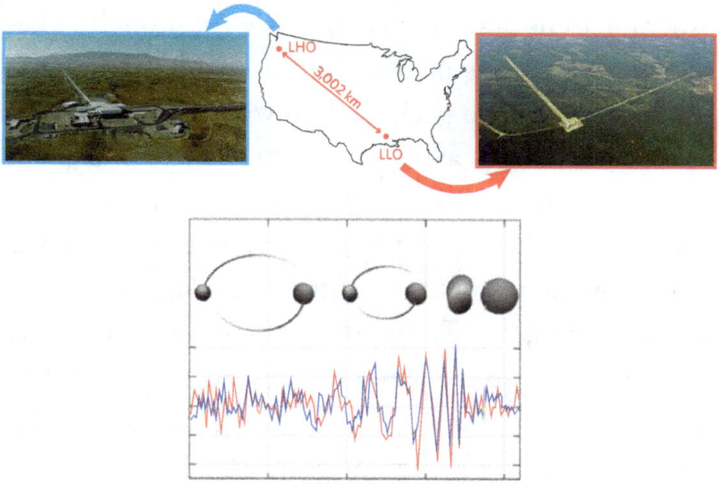

Figure IV.9. Detection of gravitational waves by the two LIGO (Laser Interferometer Gravitational-Wave Observatory) gravitational antennas located three thousand kilometers from each other in the U.S. states of Washington and Louisiana. These are giant Michelson interferometers sensitive to mirror shifts on the order of 10^{-18} meters, or a thousandth of a billionth of an interfringe distance! The oscillating signal detected in correlation by the two detectors (*red and blue signals*) indicates the passage of a gravitational wave emitted in less than a second more than a billion light-years away from the Earth by two black holes orbiting each other ever faster and eventually merging (adapted from *Physical Review Letters* (2016), 116, 061102).

Nonetheless, Einstein mentioned in his 1915 article two general relativity effects observable at the time by astronomical measurements, which would make a small correction to the predictions of classical Newtonian physics and thus help validate his theory. The first was in fact a "postdiction." Astronomers in the nineteenth century had observed with precision the orbit of Mercury, the planet closest to the Sun and therefore the one that experiences its gravitational effects most strongly. They had noted that this planet's trajectory around the Sun did not exactly loop back on itself with each revolution, but that the orbit's axis rotated very slowly in its plane of approximately 0.159 degrees (or 574 arcseconds) per century. Much of this tiny precession frequency was caused by the disturbance of the other known planets, mainly Venus, the Earth, and Jupiter. However, there was still a tiny difference of 43 arcseconds per century that was not understood. Le Verrier, the man who discovered Neptune, had tried in the 1850s to determine the orbit of a hypothetical planet close to the Sun that could explain the anomaly. No planet was found, and the mystery of the excessive precession of Mercury's perihelion remained unsolved until Einstein, using the solution at the limit of the weak gravitational fields of his general relativity equations, gave an exact account of the measured precession rate. This time, Le Verrier's observations had not given birth to a planet, but they had helped confirm the birth of a new theory of the universe.

The second effect predicted by Einstein and not exactly foreseen by Newton was the deflection of a light beam in a star's gravitational field. To say that the effect had not been foreseen by Newton requires clarification: as we have seen, Newton thought light was made of hypothetical particles of unknown mass, and suggested that these particles, whatever their inertial mass, should be deflected by the Sun's gravitational field. The effect is easy to calculate with classical mechanics, and we find, for a light beam tangent to the Sun, an extremely small deflection angle of 0.8 arcseconds. This angle was calculated by a German astronomer, Johann von Soldner, a century before Einstein. The theory of general relativity, which assimilates mass and energy, found the same angle again before Einstein took into account the additional effect of the space-time curvature in the vicinity of the Sun. By adding the two effects, he finally found a deviation of 1.7 arcseconds, about twice Newton's prediction. Einstein proposed verifying this prediction during the solar eclipse expected in May 1919, a few months after the end of World War I.

The Sun was to be obscured by the Moon against a background of several stars located very far from the solar system and thus fixed in the sky. A star seen at the edge of the solar disk would become visible at the moment of the eclipse, and its apparent position would have to be shifted by these 1.7 arcseconds. To measure this shift, it would suffice to compare the position of this star relative to the others during the eclipse and the following night.

The English astronomer Arthur Stanley Eddington led two expeditions—one to northeastern Brazil, the other to the island of Principe in the Gulf of Guinea—to make these measurements at two points on the trajectory of the shadow of the Moon. After developing the photographic plates and conducting complex statistical analysis of the resulting photos, he announced in September 1919 that the verdict was clear: the deflection was equal to Einstein's 1.7 arcseconds, not Newton's 0.8. The finding, announced by every newspaper on the planet, made Einstein an overnight global celebrity—for a difference of 0.9 arcseconds. He was the scientist who had just shown that we live in curved space, thus changing our vision of the world as profoundly as Copernicus had done four centuries earlier. Indeed, this has been borne out throughout the rest of history. But Eddington's evidence was flawed, to say the least, his measurements marred by errors and inaccuracies that his statistical analysis could not really correct. On some plates, the star's position appeared to be closer to Newton's 0.8 arcseconds than to Einstein's value. The problem was that Eddington knew this value and sought to confirm it. It so happened that the theory he wanted to verify was correct, and the small voluntary or unconscious allowances he made to confirm it, by rejecting certain clichés too favorable to Newton, would be forgiven by posterity.

This experiment is often presented as the decisive historical confirmation of Einstein's theory. It makes for a beautiful story, given the backdrop of an expedition to the tropics and scientific reconciliation between English and German scientists after a murderous world war. In truth, had this been the only proof of general relativity, it could have been widely questioned. The real evidence, as we know, came later, and overwhelmingly so. Eddington was lucky to be right even though his observations were dubious. There are many other examples throughout the history of science of observations tainted by errors because the experimenter knew too much about what he wanted to find. When intuition is right, experimental errors, whether intentional or unintentional, are eventually forgiven. In any case, Einstein needed no help. It is said that he never doubted the result of the eclipse, and that he went to bed calmly that night without waiting to hear from Eddington. The rest of the story proved him right beyond his hopes.

For the general public, Einstein's name has remained synonymous with relativity. His theory, which Eddington said only two people in the world really understood—guess whom he had in mind!—still plunges ordinary people into a feeling of wonder mixed with a vaguely unsettling strangeness. A century after Eddington, the detection of gravitational waves echoed the media frenzy of 1919. Once again, Einstein's name resonated with the evocation of these waves, tiny distortions of space-time carrying rich information about the universe.

But the theory of relativity is only one facet of Einstein's scientific work. His role in lifting Lord Kelvin's second cloud, and the veil on the laws of the microscopic world, was also essential. In 1905, the same year as his papers on special relativity, he wrote the first paper on the quantization of light. In the following years, he contributed to the enrichment of quantum theory, whose technological byproducts are innumerable today. It is reasonable, then, to say that Einstein had a greater influence on our daily lives through his contribution to the birth of quantum physics than he did through his discovery of relativity.

There is, however, a great difference between these two sides of Einstein's monumental body of work. Relativity came out of his head, from a single idea imposing itself by the sheer force of logic. Experimental observation had practically no influence on that theory's development. Even Michelson's experiment played no role at all. Einstein never mentioned it in his articles on relativity. For him, the constancy of the speed of light in an inertial frame of reference was a theoretical necessity, and had Michelson found a fringe shift in his experiment Einstein would likely have attributed it to an experimental error.

In fact, it is said that when he met Michelson face to face in 1931, during his first visit to California, Einstein greeted him politely, like the other professors to whom he was introduced by the president of the California Institute of Technology, without any indication that he may have been influenced by the famous experiment's negative outcome. His host was visibly disappointed by Einstein's lack of attention to the old scientist who had been trying in vain since 1887, by refining his experiment, to prove the existence of an aether that the theory of relativity had eradicated. Michelson's experiments, as well as all those that later sought to verify with increasing precision the constancy of the speed of light, had not become useless, but the burden of proof had been reversed. It was not for the theory to explain the observations, but rather for the observations to verify the validity of the principle of relativity.

For Einstein, Lord Kelvin's first cloud was dissipated by the simple force of rational argument, without the need for any form of observation. The second cloud, the one that mysteriously obscured the properties of thermal radiation, was different. To lift it, guided by new observations from the microscopic world, Einstein and other physicists developed quantum theory in a complex and fertile relay between theory and experimentation. Einstein still expressed his genius and his extraordinary physical intuition, but he never felt as comfortable with quanta as he had been with relativity. While the principle of the latter theory came to him very quickly, the postulates of quantum physics, which he helped develop sometimes reluctantly, left him for the rest of his life with a feeling of unease and incompleteness. The story of the birth of quantum physics, which we will discuss in the next chapter, is thus more complex, and in some respects more dramatic, than that of relativity.

CHAPTER 5

Light reveals the strange world of quanta

In a display case at the Niels Bohr Institute in Copenhagen, a strange little metal box hangs by a spring from a wooden frame attached to a graduated ruler. An arrow glued to the box and pointing to the ruler makes it possible, one surmises, to weigh the box when it's suspended from the spring. On the small black panel that closes the front of the box are a few formulas written in white, accompanied by the double signature A.E. ↔ N.B. The panel is slightly raised to reveal some instruments inside the box. This object, a small placard beneath the window explains, was a Christmas gift given to Albert Einstein and Niels Bohr in 1930 by a waggish young visitor to the Institute named George Gamow. It humorously reproduces the *photon box* from a famous thought experiment that Einstein and Bohr conceived during their debates on the principles of the nascent quantum physics.

Created ten years earlier, the Institute had become the mecca of this new physics and Bohr its undisputed leader, the man who had announced its laws and decreed how to understand them. The *Copenhagen interpretation* would become the doxa followed by the vast majority of physicists in the following decades. Like Gamow, most of the contributors to the theory's emergence passed through Copenhagen during this period. The famous uncertainty relations were established by Werner Heisenberg after one of his visits to the Institute. Bohr stated and clarified the meaning of the complementarity principle of quantum theory by analyzing the deeper implications of these relations. Not all of the Danish scholar's interlocutors were convinced, but he remained insistent that they accept his view of the quantum world as the only one that completely and coherently described nature's behavior at the scale of atoms and photons.

Figure V.1. Gamow's photon box, given to Einstein and Bohr in 1930 in memory of their famous debate on uncertainty relations at that year's Solvay Conference (see Figure V.10) (Haakon Bergest, Niels Bohr Archives, Copenhagen).

These debates found Bohr particularly opposed to Einstein, the man who had introduced quanta into physics a quarter of a century earlier but who refused to admit that Bohr's interpretation thereof was the final word. The photon box, a crucial thought experiment on which Einstein and Bohr had locked horns, symbolized a critical turn in this debate at the 1930 Solvay Conference. By affixing to it a stamp with Einstein's profile and the English word *patent*, Gamow sought to point out, perhaps ironically, that N.B. had hoped his debate with A.E. would give his theory a stamp of authenticity and completeness. The stamp may have also meant, paradoxically, that the debate was a purely theoretical one, from which the protagonists certainly could not hope to derive a patentable, practical invention.

Before we decipher the mysterious symbols written on the photon box's black panel, this chapter will follow the birth and evolution of quantum ideas. As in the last chapter, Einstein will serve as our guide. There are many reasons for this. He was the first, in 1905, to lay the foundations for this new physics by introducing the concept of the photon. Over the twenty years that followed, he made essential contributions to the development of quantum theory. He constantly encouraged the young physicists of the next generation to deepen their thinking and sharpen their experiments, until at last, in 1925, the veil hiding the mathematical laws of the microscopic world was lifted. It is remarkable that Einstein accomplished all of this while in parallel developing the theories of special and general relativity.

Einstein would later criticize the way these laws were interpreted in Copenhagen, seeking a truth beyond what they described. His critiques were beneficial in that they spurred Bohr and Heisenberg to clarify and refine the meaning of uncertainty relations and the principle of complementarity. Einstein's reflections finally led him, in the 1930s, to analyze the phenomenon of quantum entanglement, likely the most counterintuitive concept of this new physics. The ideas he put forward at the time continue to spur an extremely active and promising research stream today, that of quantum information.

Let's return once more to the miraculous year of 1905. Just two months before his seminal paper on relativity, Einstein published in a scientific journal an article with a surprisingly philosophical title: "On a Heuristic Viewpoint Concerning the Production and Transformation of Light." In this paper, he kills two birds with one stone. He gives an explanation of the theretofore misunderstood properties of thermal radiation and interprets the surprising characteristics of the newly observed photoelectric effect. And, in doing so, he introduces the revolutionary concepts of the photon and of *wave–particle duality*.

The ultraviolet catastrophe

Recall that the second cloud evoked by Lord Kelvin in his 1900 Royal Institution lecture concerned the spectral distribution of light emitted by heated bodies, known as *thermal radiation*, or *black-body radiation*.

All objects radiate light—visible, infrared, and ultraviolet—the spectrum of which depends on temperature. This radiation can be observed by maintaining at temperature T a cavity whose partition absorbs the light it receives (hence the name *black-body*) and measuring the radiation that escapes through a small hole pierced in the partition. The wavelength distribution of the light emitted at a fixed temperature had been measured by precise experiments at the end of the nineteenth century. The curve of radiated intensity as a function of frequency is bell-shaped, with a maximum that shifts toward high frequencies, *i.e.* shorter wavelengths, as the temperature rises. This experimental fact is easily observed: at ordinary temperatures, the maximum radiation is around 10 microns, a wavelength in the infrared range. When heated, the enclosure begins to emit red light. At higher temperatures the light turns orange, then yellow. At any temperature, the curve decreases to 0 for long and short wavelengths.

Surprisingly, classical physics cannot explain these simple characteristics. According to Maxwell, each frequency component of the electromagnetic field in a cavity is analogous to a small vibrating oscillator. The number of these elementary oscillators in a small spectral interval around a frequency ν grows as the square of this frequency. This growth can be understood qualitatively: the higher the frequency, the shorter the wavelength and the more field modes that can oscillate between the partitions of a cavity of a given size.

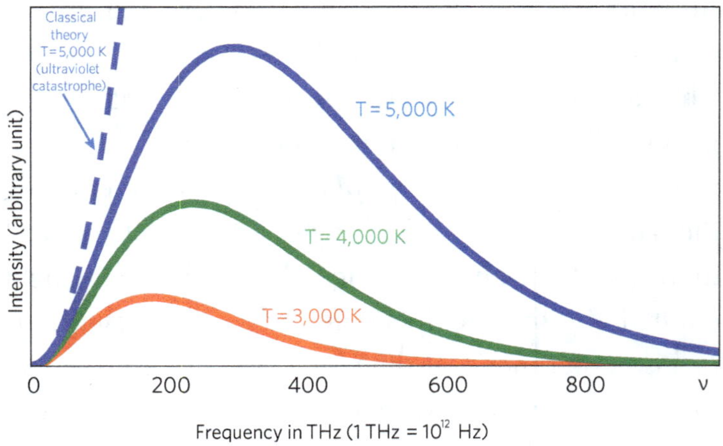

Figure V.2. Thermal radiation spectrum at different temperatures *T*. The maximum of the bell-shaped curves shifts toward the higher frequencies, and thus the shorter wavelengths, when *T* increases. The classical theory (shown in dashed lines for the highest temperature T = 5,000 K) predicts an intensity tending toward infinity at short wavelengths (the "ultraviolet catastrophe").

Conventional thermodynamics also says that the energy of the field must be shared equally among all its modes of oscillation, each mode having an average energy proportional to the absolute temperature *T*. This is equal to the ordinary Celsius temperature increased by 273.16°C. It is called *absolute temperature* because the zero of its scale corresponds to the cancellation of all thermal agitation, where all of a system's degrees of freedom are frozen in their lowest fundamental state of energy. Absolute temperature is measured in kelvins, in honor of the herald of the two dark clouds. Kelvin temperature, abbreviated with a capital K, is always positive, as there can be no thermal agitation below absolute rest. This state can never be strictly reached, but we now know how to get within a few billionths of a kelvin.

The proportionality constant k_B between the mean energy of an electromagnetic field mode and the absolute temperature T was introduced to physics by Ludwig Boltzmann. The equal distribution of energy among all modes expresses a "democratic" principle whereby nature tends, in thermal equilibrium, to equitably distribute the energy of a system at a given temperature among all its degrees of freedom. If at a given moment the energy is unequally distributed, the thermal exchanges will tend to suppress the differences in excitation between the oscillators. As a result, the intensity of thermal radiation escaping from the cavity at a given frequency ν should be proportional to k_BT and ν^2. This law, established by the physicists Lord Rayleigh and James Jeans, bears out the spectral distribution of thermal radiation at low frequencies, *i.e.* at long wavelengths. But at short wavelengths it moves away from the observed spectrum and leads to an absurdity: the radiation diverges and tends toward infinity along with the frequency, which is physically meaningless. This is called the *ultraviolet catastrophe*.

In 1900, Planck found a mathematical formula that was well suited to the experiment and made it possible to avoid this catastrophe by postulating that the exchanges of energy between light and matter could be done only by discrete elements or quanta of energy $h\nu$, proportional to the frequency. He then introduced the famous proportionality constant h between energy and frequency, which now bears his name. But this introduction of quanta seemed to be a mathematical trick, an *ad hoc* measure to try to save classical physics and dissipate Lord Kelvin's second cloud. In this sense, it played a role analogous to that of the length-contraction formula Lorentz had devised to solve the mystery of the first cloud. Einstein was unsatisfied with such an explanation.

Light between waves and particles

For the young physicist from Bern, quanta had a deeper meaning. It was not only the light–matter exchanges that were quantized, but also the energies of the field oscillators themselves. He hypothesized that each mode of the field has an energy scale the interval between whose values, $h\nu$, is proportional to the frequency. None of the intermediate energies between these values is allowed. To jump from one value to the next is to emit or absorb a particle of light, later called a *photon*. The greater the frequency, the greater the interval between the values and the more energy-expensive it becomes to climb the ladder. The thermal field in each mode is thus caught between opposing thermodynamic constraints.

Let's clarify this antagonism by recalling some simple notions from thermodynamics. This science, established in its completed form at the end of the nineteenth century by Josiah Willard Gibbs in the United States and Ludwig Boltzmann in Austria, introduced, in order to study the equilibrium of a physical system in contact with an environment at temperature *T*, a function *F*, called its *free energy*. This is equal to *E - TS* where *E* is the system's energy and *S* its entropy, the quantity that measures its degree of disorder. The second principle of thermodynamics states that any system maintained at temperature *T* tends to *minimize* its free energy.

To visualize this somewhat anthropomorphically, a thermodynamic system in equilibrium at temperature *T* is both "lazy" (it "likes" to lower its energy at constant entropy, which decreases *F*) and "messy" (it "wants" to increase its entropy and disorder at constant energy, which also decreases *F*). Depending on the temperature, one or the other of these two tendencies will prevail. At very low temperatures, it will prefer to be lazy and seek the lowest energy state. At high temperatures, it will prefer to be messy to increase *S* as much as possible. At an intermediate temperature, it will compromise by maintaining an optimal balance between *E* and *S* so as to minimize *F*.

The equipartition theorem that gives each oscillator an average energy k_BT is the result of this compromise, *provided the energies can vary continuously*. The situation becomes quite different if the energy can vary only by discrete jumps, each oscillator having a scale of equidistant values. In the search for a compromise between minimizing energy, *i.e.* staying at the lowest possible altitude of each scale, and maximizing entropy, *i.e.* occupying as many values as possible to increase messiness, the "choice" of the field will depend on its

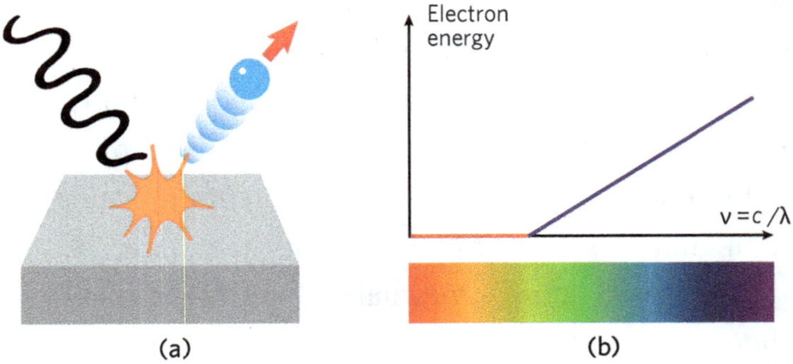

Figure V.3. The photoelectric effect. (a) Above a threshold frequency, light pulls electrons from a metal surface. (b) Einstein's theory predicts that the energy of ejected electrons depends only on the color of the light. From a null value at the threshold, this energy increases linearly with the frequency of the light wave. The slope of this variation is equal to Planck's constant h.

frequency and temperature. At low frequencies, there are many states with an energy on the order of $k_B T$, and the law of energy equipartition and the Rayleigh–Jeans formula derived from it are good approximations. But when $h\nu$ is on the order of $k_B T$, climbing the ladder costs too much energy for the system to increase its entropy by occupying many states while keeping its free energy low. In this case laziness outweighs messiness. High-frequency quantum oscillators remain frozen in their ground state, which consequently eliminates the ultraviolet catastrophe and results in Planck's law.

In the same article, Einstein showed that the notion of the photon explained another mystery of light: the strange properties of the photoelectric effect. When a metal is irradiated, it can emit electrons—elementary particles discovered in 1897. But this photoelectric effect has a strange characteristic: it occurs only when light has a frequency above a threshold that depends on the metal. Below this threshold, regardless of the light's intensity, no electrons are emitted. Above it, even very weak light extracts electrons.

Einstein explained this result by postulating that an electron from the metal must absorb a photon in order to escape, and that this photon must have a minimum extraction energy W, hence the frequency threshold. If the light does not have a large enough frequency, even if it's intense, none of its photons can extract an electron. By analogy, we can understand that a continuous heavy snowfall will have less of a disruptive effect on a fragile surface than a hailstorm in which the same frozen water arrives as discrete energetic entities.

Einstein also predicted a property that remained unchecked. If the frequency of light were increased above the threshold, the excess energy supplied would result in an increase in the kinetic energy of the extracted electrons, which varies linearly with the frequency of light. The slope of this variation should allow for a new measurement of Planck's constant. The verification of this result a few years later by the American physicist Robert Millikan led to Einstein's Nobel Prize in 1921, whose citation mentions his elucidation of the photoelectric effect, not relativity.

By accepting the notion that the quantization of energies was not just a mathematical trick but in fact a profound physical truth, Einstein took a giant step forward. Since Young's interference experiments and Fresnel's and Maxwell's works, it had been accepted that light was a wave, as Huygens had proposed as early as the seventeenth century, and Newton's corpuscular theory had been abandoned. In his 1905 paper on light, Einstein did not challenge the wave theory or Maxwell's equations, on which he had otherwise based relativity, but he made the revolutionary argument that light is a wave when it propagates and manifests itself through interference phenomena, and that it is formed of discrete entities when it interacts with matter.

Thus, reconciling Huygens and Newton in an unexpected way, Einstein introduced a principle of duality that has dominated physics ever since. The photon, a corpuscular manifestation of light, propagates at the limit speed c of information. Each photon of a plane wave of energy $h\nu$ is associated, according to special relativity, with linear momentum $h\nu/c$. If the light is circularly polarized, each photon also carries a unit of angular momentum equal to $h/2\pi = \hbar$ (see Chapter 1).

Quanta are generalized to matter

What made Einstein exceptional was his ability to draw conclusions from his discoveries without letting himself be blinded by preconceived ideas, and to generalize the results obtained in one field to other areas in physics. Having understood the need to quantize light oscillators, he extrapolated this result to the study of the mechanical oscillators that are atoms in a solid. The atoms vibrate around their equilibrium positions at frequencies that depend on the crystalline structure, the frequency of the vibrations being greater the more rigid the solid. Since the nineteenth century and the French physicists Pierre Louis Dulong and Alexis Thérèse Petit, it had been known that solids' atomic specific heat, *i.e.* the amount of energy per atom that must be contributed to increase their temperature by one degree, took, for a great many elements, a value equal to three times the Boltzmann constant.

This result was a consequence of the equipartition of energy, each atom corresponding to three elementary oscillators vibrating in the three directions of space. But there were exceptions to this rule—diamond, for instance, has a specific heat per carbon atom lower than this value. Einstein hypothesized that atomic oscillators were quantized, like electromagnetic field oscillators, and that they had to vibrate faster in diamond than in other elements to which the Dulong–Petit law applied. Consequently, as with thermal radiation, these high-frequency oscillators had to have their excitations frozen at room temperature. And experimentation confirmed that when diamond is sufficiently heated, its specific heat increases and satisfies the Dulong–Petit law. Conversely, Einstein predicted that at sufficiently low temperatures, below a value varying according to the solid, the specific heat per atom of metals should collapse and cancel itself out at the temperature of absolute zero on the kelvin scale.

Einstein presented this theoretical analysis at the first Solvay Conference in 1911, where he met Lorentz, Langevin, Marie Curie, and the other great physicists of the beginning of the century, and where the ideas of the nascent quantum physics were discussed for the first time. Ernest Solvay was a Belgian chemist and industrialist with a passion for physics who founded a series of conferences at which the most advanced ideas in physics have been analyzed regularly ever since. The official photograph of this first conference is famous because it shows the physics elite of the day, including those who were beginning to lift the veil over what was then called the *theory of quanta*, gathered around Solvay. Marcel Brillouin, professor at the Collège de France (his son, Léon, would be one of the founders of the quantum theory of solids twenty years later), was secretary of the conference along with the X-ray specialist Maurice de Broglie. The Collège de France still holds in its library the proceedings of the conference, mimeographed and annotated in pencil by Marcel Brillouin.

Einstein, then known mainly for his work on relativity, presented a paper entitled "On the Current State of the Problem of Specific Heat." Its subtitle was: "Relation between specific heat and the thermal radiation formula." The text detailed how Planck's formula could be applied not only to radiation but also to the material oscillators contained in matter.

Figure V.4. The 1911 Solvay Conference. *Seated from left to right:* W. Nernst, M. Brillouin, E. Solvay, H. Lorentz, E. Warburg, J. Perrin, W. Wien, M. Curie, H. Poincaré; *standing from left to right:* R. Goldschmidt, M. Planck, H. Rubens, A. Sommerfeld, F. Lindemann, M. de Broglie, M. Knudsen, F. Hasenöhrl, G. Hostelet, E. Herzen, J. H. Jeans, E. Rutherford, H. Kamerlingh Onnes, A. Einstein, P. Langevin.

From that moment on it was clear that Planck's constant was not merely a curiosity introduced into a mathematical model to make theory mesh with experimentation, but a fundamental constant of nature that intervened in all phenomena of the microscopic world. Its finite value indicated the discrete nature of microscopic physical quantities, and classical physics was to be treated only as an approximation when the thermal energies $k_B T$ involved were large compared to the quantum $h\nu$, a situation where h could then be considered to be zero. Thus, the revolutions in physics at the beginning of the twentieth century did not abolish Newton's and Maxwell's theories, but rather generalized them. Classical mechanics became a limit of quantum mechanics when h can be taken as 0, and a limit of the theory of relativity when the speed of light c becomes very large compared to the speeds involved in the observed phenomena. It can then be seen as infinite, and $1/c$ as 0.

Einstein's double stroke of genius lay in understanding what happens when the finite values of h and $1/c$ cannot be neglected. George Gamow, the young researcher who offered Bohr and Einstein his version of the photon box, was a great popularizer of physics, publishing works that owed much to his deep understanding of science and his sense of humor. These describe the fantastical adventures of his hero, Mr. Tompkins, exploring worlds where h and $1/c$ are big enough to not be neglected in everyday life. Gamow playfully introduces the concepts of relativity and quantum physics by showing the quirks of a world where one can reach a speed close to that of light, or travel like a tiny Lilliputian through quantum matter.

The discovery of the first quantum effects unlocked the idea that they occur in a privileged way at very low temperatures, which all the physics of the twentieth century would go on to confirm. It was in 1911, just after the Solvay Conference, that the superconductivity of certain metals—their ability to conduct electric current without resistance at very low temperatures—was discovered by the Dutchman Heike Kamerlingh Onnes, shown next to Einstein in the conference photo. Superconductivity is a quantum effect that disappears above a transition temperature. It was not fully explained until 1957. As we will see, however, in the 1920s Einstein would introduce a concept fundamental to the understanding of this phenomenon.

Ernest Rutherford, the man standing next to Kamerlingh Onnes in the conference photo, is also a player in our story. In 1911 as well, he performed an experiment on the transmission, through a thin sheet of gold foil, of α rays produced by the radioactive decay of unstable atomic nuclei. These rays are made up of helium nuclei carrying a positive electric charge double that of a proton. While most of the α particles passed through the foil without deviation, a small number were strongly deflected. Rutherford deduced that the atoms in the gold foil were made up of quasi-point positive charges, the atomic nuclei, which contained most of matter's mass. The rest was made up of much lighter electrons that did not disturb the trajectories of the α particles. These generally crossed the sheet in the empty space between the nuclei, and were strongly scattered by Coulomb repulsion only in the rare cases where their trajectories passed near the nuclei.

The idea that immediately followed was that the atom must be a miniature planetary system. The very small heavy nucleus concentrating most of the mass played the role of the Sun. The light electrons gravitating on orbits with very large radii compared to the dimensions of the nucleus were the planets. Here again, though, classical physics reached an impasse. According to Maxwell's theory, charged electrons revolving around the nucleus would radiate energy and fall rapidly onto the nucleus, making matter unstable.

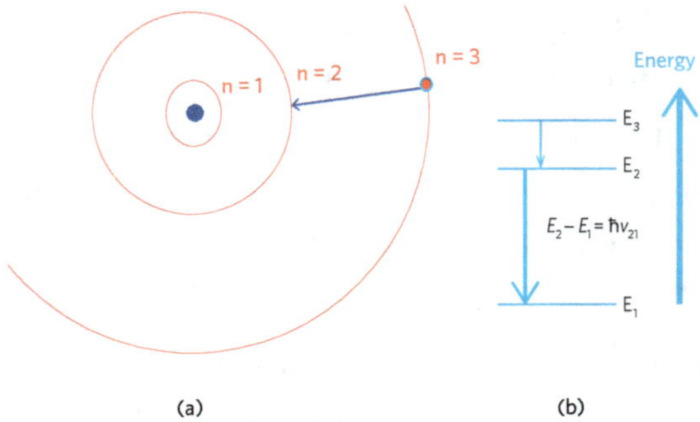

Figure V.5. Bohr's 1913 model of the hydrogen atom (circular orbits). (a) The electron is located on circles with the proton at their center, identified by their quantum number $n = 1,2,3...$ The radii of these orbits grow as n^2. The orbit of the ground state $n = 1$ has a radius of 0.53 angstrom (0.53×10^{-10} meters). This quantity is called the *Bohr radius*. The atom gains or loses energy when the electron jumps at a random instant from one orbit to another by absorbing or emitting a photon: a quantum jump. (b) Energy levels of Bohr's atom, showing the lowest fundamental energy level $n = 1$ and the first two excited levels $n = 2$ and $n = 3$, corresponding to the orbits of the figure on the left. The energies of the emitted or absorbed photons satisfy the energy conservation relationship of the atom + light system. The frequencies of the photons are thus proportional to the differences in energy of the transitions' initial and final states (the constant of proportionality is h). Quantization of the energy spectrum yields, one hundred years after Fraunhofer, an explanation of the discrete lines of the atomic spectra. This model, still based on the classical notion of trajectory, would soon be generalized to elliptical orbits by Arnold Sommerfeld, before being replaced in 1925 by the modern quantum theory describing the electron by its delocalized wave function. The complete theory shows that Bohr's energy levels split into several sub-levels of neighboring energies, corresponding to the various possible values of the electron's angular momentum (fine structure). The coupling of electron and proton magnetism adds an additional enrichment to the spectrum (hyperfine structure). Finally, the coupling of the electron with vacuum fluctuations adds a small displacement to the energy levels (the Lamb shift).

Niels Bohr, a young visiting postdoctoral student working with Rutherford, provided an answer to this problem in 1913. Bohr established the first quantum model of the atom by postulating that the energies of the electrons therein were quantized, like Einstein's oscillators. Electrons can occupy only certain orbits and never have intermediate energies. Once again we imagine steps on a ladder, this time with unequal distances between rungs. In this model the electrons jump from one orbit to another by emitting or absorbing a quantum of radiation, the energy $h\nu$ of the photon being equal to the difference in energy between the levels in question. On the lowest fundamental energy level, the electron cannot radiate, and the atom is stable. The transition from one level to the other is made by instantaneous quantum jumps.

Bohr's model made it possible to calculate the energy levels of the simplest atom, single-electron hydrogen, with a formula that brought back Planck's constant. The spectrum of frequencies emitted by the atom was in very precise agreement with this model, reinforcing the confidence quantum ideas were beginning to accrue in the scientific community.

Gregarious photons and atoms that mimic them

Einstein was pleased to see his ideas generalized to an understanding of the structure of matter. I will say more later about his relationship with Bohr and their passionate discussions about the interpretation of quantum physics. In 1913, however, as we have seen, Einstein had temporarily abandoned the study of photons to devote himself to the theory of general relativity. In 1916, once this theory was established, he returned to quanta. Now that the quantum nature of light and matter was better understood, was it possible to revisit the theory of black-body radiation by analyzing in detail the mechanisms by which the atoms in the partitions of the cavity interact with the photons contained therein?

Atoms can either absorb or emit photons, and thermal equilibrium results from the balance between these two mechanisms. Absorption occurs for each atom with a probability per unit time proportional to the light intensity in the modes resonating with atomic transitions, while the emission of an excited atom toward the ground level occurs, according to Bohr, following a spontaneous and random process in which an atom emits, at an unpredictable instant, a photon in one field mode or another according to a law of probability independent of the light that is present.

Einstein noted that these two mechanisms were insufficient to bring back Planck's law. To do this, it was necessary to add a third process whereby an excited atom receiving a photon emits a new one, in the same mode as the one it receives, with a probability proportional to the number of photons already present in this mode. This mechanism, called *stimulated emission*, is the exact counterpart of absorption. It amplifies light instead of attenuating it. This means that when photons are in the presence of excited atoms, new photons appear in the modes already occupied, all the faster as the number of photons already present increases.

Stimulated emission would be used in the lasers invented some fifty years later. These devices produce beams of amplified and coherent light, all their atoms emitting an extremely stable light in the same direction and with the same frequency and phase. These sources are very different from classical lamps, in which atoms independently emit photons in different modes, with random phases and frequencies.

We're familiar with the importance of lasers in modern technology. Gravitational antennas that detect minute variations in the distance of mirrors several kilometers apart exploit the extraordinary phase and frequency stability of extremely powerful lasers. Remarkably, then, the discovery of gravitational waves helped unite two essential discoveries by Einstein in 1915 and 1916: that of general relativity, which predicted the existence of these waves, and that of stimulated emission, which laid the groundwork for the lasers that would make it possible to detect them. But I'm getting ahead of myself by talking about lasers: in 1916, Einstein was far from envisaging their existence. Nor does he seem to have taken any particular note, at that time, that photons tended to accumulate in the same state. This property of gregariousness would lead him, a few years later, to make another important contribution to the emergent quantum theory.

But this would not happen until 1924, when he received a letter from an unknown young Indian physicist. Satyendranath Bose sent Einstein a short note in which he re-demonstrated Planck's formula of thermal radiation, starting simply from the principles of thermodynamics and considering the light field as a gas of photons exchanging energy with a heat reservoir. In this he followed the method Boltzmann had applied to an atomic gas. To calculate the entropy of the photon gas, he counted all the states available for it and assumed that at equilibrium all of them were equiprobable.

Bose made an implicit hypothesis in this calculation, that of the *indistinguishability* of photons. This hypothesis differed from the one advanced by Boltzmann, who had numbered the atoms and treated as two distinct configurations the case where atoms 1 and 2 had the positions and momenta x,p for atom 1 and x',p' for atom 2 and the case where these values were reversed: x',p' for atom 1 and x,p for atom 2. Bose, on the other hand, considered these two states to be the same for photons. In this case, the state of the photon gas is specified completely by the number of particles for each value of x,p, with no need to worry about numbering the particles. This apparently harmless change in counting method has an important consequence. It makes the presence of all photons in the same state relatively more likely.

To explain this property, we may note that if the particles are distinguishable, those configurations where they occupy different states can be attained in multiple ways by exchanging particles between states, whereas the configurations where they occupy the same state can be realized in only one way. Distinguishability thus makes configurations where the particles are in the same state rarer than those where their states differ. Indistinguishability, on the other hand, increases the relative probability of configurations where all particles occupy the same state—in particular the ground state, in which they all occupy the lowest energy state.

With this hypothesis Bose recovered Planck's formula, which he could not have obtained using Boltzmann's method. This simple fact validated his counting method, the basis of what is now called *Bose–Einstein quantum statistics*. Reading Bose's article, Einstein realized its importance. He may also have recognized that this fundamental indistinguishability of photons confirmed their gregarious nature, their tendency to accumulate in the same state—a property at which his discovery of stimulated emission had already hinted eight years earlier.

Einstein's extraordinary intuition was to take this paper, which merely confirmed a known formula for light, and extend its scope by applying Bose's quantum statistics to material particles, to a gas made of atoms considered fundamentally indistinguishable. In doing so he found the classical properties of Boltzmann's gas at room temperature, but at very low temperatures he predicted that all atoms would condense into the same quantum state, the system's ground state, not unlike the photons of a laser beam, which all occupy the same field mode. This hypothetical state of ultra-cold matter, now known as a *Bose–Einstein condensate* and imagined long before the invention of the laser, left Einstein's colleagues skeptical, since they saw it as no more than a curiosity resulting from an ultimately arbitrary method of calculation.

Once again, however, the rest of the story proved Einstein right. The Bose–Einstein condensate was first observed in 1995, in a gas of alkaline atoms cooled and trapped using lasers. When released from their traps, these condensates form laser-like beams of coherent matter in which photons are replaced by atoms. This new state of matter is now the subject of very active studies in many laboratories around the world.

The article introducing the Bose–Einstein condensate was much more general in its scope than the discovery of a particular state of matter at very low temperatures. By taking seriously the consequences of applying the Bose rule to the enumeration of states accessible to a gas of particles, Einstein guessed an essential aspect of quantum physics far beyond this particular case. The indistinguishability of particles of the same species is an important property of the quantum world, and plays a fundamental role in our understanding of the properties of matter and light. Einstein was the first to glimpse it at the end of 1924, before the corner of the veil still covering the mystery of the quantum world was lifted a few months later.

The veil is lifted on matter waves

At this point in our story, quantum physics had brought together a disparate set of observations and methods but was not yet a finished theory that could be described in a clear and simple mathematical way. This would change radically in the subsequent months. Einstein was not directly responsible for this last step, but he contributed to it nonetheless. Let's return once more to the photo from the 1911 Solvay Conference (page 325). On the far right, next to Einstein, we see Paul Langevin, and at the center is Maurice de Broglie, a specialist in X-ray physics. Maurice's younger brother, Louis, was not at the conference. A history student, he was not yet interested in physics, and is therefore—like the young Bohr—absent from this picture. It is said to have been Maurice who stirred Louis de Broglie's passion for physics by telling him about the discussions at the 1911 conference.

After studying classical physics at the Sorbonne and spending World War I in radio communications at the Eiffel Tower, Louis began a thesis in 1920 under Langevin's direction. Taking up Einstein's idea of wave–particle duality, he extended it to matter by postulating that material particles, like photons, should be associated with waves. To find their wavelength, de Broglie made an analogy between the corpuscles of matter of mass m and the light particles that are photons. The momentum $p = h\nu/c$ of a photon can also be written $p = h/\lambda$, where $\lambda = c/\nu$ is the wavelength of the radiation. De Broglie extended this relationship between wavelength and momentum to material corpuscles and associated any particle of momentum $p = mV$ with a wavelength $\lambda = h/mV$.

The details of this association remained unclear, but de Broglie predicted the possibility of observing the interference of these matter waves and was able to use a simple model to retrieve Bohr's formulas on the hydrogen spectrum. The electron matter wave revolving around the nucleus had to satisfy a resonance condition, closing in on itself after oscillating an integer number of times. After a simple calculation, this condition gave the energies of the authorized Bohr orbits. To arrive at this result, de Broglie largely took inspiration from Einstein's work: his papers on quanta, but also those on relativity.

Invoking relativity in the context of de Broglie's reflections helps us understand the importance of the concept of phase in quantum physics. Classical Fresnel optics, discussed in Chapter 3, introduced this concept by associating the phase of a light wave to the direction of a vector in an abstract plane (or, equivalently, to a complex number represented by a vector in this plane). To a particle moving freely with momentum p along direction x, de Broglie associated a matter wave of wavelength $\lambda = h/p$, whose phase should thus, at a given instant, be equal to $2\pi x/\lambda = px/\hbar$. The Fresnel vector representing this phase indeed turns by 2π when x increases by λ.

Instead of focusing on the spatial evolution of the phase of the matter wave, we may wonder how it evolves in time at a given point. Special relativity gives us the answer. We have seen that the quantity $Et - px$ is a relativistic invariant for a particle moving in direction x with energy E and momentum p. It is therefore natural to admit that if $px/\hbar$ describes the phase of the matter wave at a given instant, the quantity $(px - Et)/\hbar$ must be the spatiotemporal phase at point x,t of space-time. This phase has an absolute meaning. The event corresponding to the passage of the de Broglie wave at a maximum for a certain observer must correspond to the same event for another observer in another Galilean frame of reference. The phase of the matter wave associated with a free particle is thus a relativistic invariant. At a point x in space, this phase evolves as the angle $Et/\hbar$. It thus rotates with frequency E/h. Quantum physics therefore dually associates momentum p and position x on one hand, and energy E and time t on the other.

De Broglie's idea was revolutionary, and Langevin didn't know what to make of it. He sent the manuscript of the thesis to Einstein, who was impressed and replied that the young de Broglie had lifted part of the great veil hiding the laws of the microscopic world. Langevin was reassured and de Broglie was able to defend his thesis, with the atomic physicist Jean Perrin presiding over the jury. After the defense Maurice de Broglie allegedly asked Perrin what he thought of his brother, to which Perrin noncommittally replied: "I think your brother is very intelligent." The observation of the first electron interferences, by the American physicists Clinton Davisson and Lester Germer in 1926, proved that de Broglie was not just intelligent: he was also right.

A few months later, Erwin Schrödinger, a physics professor in Zurich, presented de Broglie's work to his colleagues in a seminar. One listener pointed out that if there were a wave, there must be an equation describing its propagation. What, then, was the equivalent for these matter waves of Maxwell's equations for photons?

Schrödinger tackled this problem, and in a few months of feverish work established the famous equation bearing his name. This associates a wave function to a particle that evolves over time under the effect of the force field to which it is subjected. This evolution is described by a differential equation linking the temporal and spatial derivatives of the wave function. Applying this equation to an electron evolving in the proton's electric field, Schrödinger found that his solution required the energies of the hydrogen atom to be quantized. His equation assigned them the same values as those of Bohr's model.

Independently of Schrödinger and almost simultaneously, a young German physics student named Werner Heisenberg's presented a mathematical theory of quantum physics based on a description of physical observables in the form of tables of numbers called *matrices* obeying a noncommutative algebra. Heisenberg too found the spectrum of the hydrogen atom whose quantization resulted from the non-commutation of operators describing the position and momentum of the electron. It was quickly recognized that Schrödinger's and Heisenberg's points of view were equivalent, describing the same physical reality in two formalisms that could be translated into one another. Quantum physics in its modern form had been born.

Wave function, quantum states, and the superposition principle

The deep nature of these waves, whose propagation was described by Schrödinger's equation, remained to be understood. What were they made of? In what medium did they propagate? Was it necessary to reinvent the aether Einstein had eliminated from physics twenty years earlier? The answer came quickly from Max Born, Heisenberg's teacher: they are abstract mathematical waves associated with probability distributions that don't need a medium to propagate. The square of the wave's amplitude at a point represents the probability of finding the particle at that point. This formulation revealed an essential aspect of the theory, its probabilistic character. In quantum physics we cannot speak of a particle's position, velocity, or energy before we have measured it. Before we do, all potential results are possible, their probabilities described by the Schrödinger wave or by the values of elements of Heisenberg's matrices. These probabilities are due not to our insufficient knowledge of the system, as in classical statistical physics, but to a fundamental uncertainty.

The principle of superposition is essential as well. A particle's matter wave, like a light wave, obeys the Huygens–Fresnel principle. To calculate its propagation from one point to another, we can break it down into secondary wavelets from virtual sources present on successive wavefronts, the global wave resulting from the interference of all these wavelets. As a consequence, if several matter waves describing the evolution of a particle satisfy Schrödinger's equation, then a combination of these waves obtained by summing their amplitudes with arbitrary coefficients is also a solution, describing another possible state of the particle.

To be a complete theory describing not only the behavior of an isolated particle but also the interactions between particles that lead to the formation of nuclei, atoms, molecules, or solids, the notion of a wave function representing a four-dimensional distribution of amplitudes in space-time must be generalized. Describing a set of *N* particles requires a function with $3N+1$ variables—each particle's spatial coordinates, plus time. This wave function can no longer be represented simply as a wave in ordinary space. It becomes a function evolving in an abstract multidimensional space, but the essential properties of a particle's wave function remain. The square of the amplitude associated with a configuration of particle positions always represents the probability of detecting all the particles in this configuration. And the principle of superposition remains. Multiparticle wave functions combine and can give rise to interference phenomena in the space of their configurations. This description privileges measurement of the particles' positions. We could also measure their velocities. For this, we would describe functions that take their values no longer from ordinary space, but from that of momentum. These functions can be deduced from the former by a Fourier transformation operation, whose properties are discussed in Chapter 3.

Physicists may also focus on parameters other than particles' position or velocity. For example, they can look at the polarization state of photons, the spatial orientation of the spin of electrons, or the angular momentum of atoms. In that case the spatial wave function is replaced by a more general mathematical object, the state vector of the system defined in abstract Hilbert space, as introduced in Chapter 1. Knowing the coordinates of this vector in this space makes it possible to calculate the probability of finding this or that result when measuring an observable of the quantum system being studied, whether it is the polarization of a photon, the angular momentum of an atom, or some other physical quantity. The spatial wave function is thus only a special case of a quantum state, adapted to the study of the variables of position or velocity in the particles under observation.

The principle of quantum superposition results simply from the rules of addition of state vectors in Hilbert space. If a quantum system can be represented by different vectors in different states, the sum of these vectors, with arbitrary coefficients, represents another possible state of the system. Note the analogy with the composition rules of Fresnel vectors describing the vibration state of light, which express the principle of superposition in optics.

As we saw in Chapter 1, transformations affecting quantum states play an essential role in the theory.

They can describe how these states are modified under the effect of rotations or translations in space of the observed quantum objects, or how they evolve in time. Each transformation is represented by an operator in the Hilbert space of states, a mathematical object described by a matrix of complex numbers. This matrix is used to calculate the coordinates of the state vector resulting from a given transformation once we know the coordinates of the initial state. In general, the product of two quantum operators is non-commutative, the result depending on the order in which they are applied. This property, discussed in Chapter 1, plays an essential role in the theory as well: it leads to the discrete nature of the possible values of certain observables, such as the internal energies of atoms or their angular momenta.

All of these notions, which remain necessarily vague for the non-specialist reader, came into focus between 1925 and 1930 thanks to passionate debates between the theory's founding fathers. We shall return to this later.

The particle family expands

In the 1930s and 1940s, quantum theory was supplemented to take relativistic principles into account. Schrödinger's equation is indeed a valid approximation for particles of low velocity compared to *c*. Quantum concepts, combined with those of relativity, showed that the electron should be associated with its antiparticle, the positron of the same mass but opposite charge. This particle was observed experimentally shortly after the theoretical prediction of its existence by Paul Dirac.

As we have seen, electrons and positrons can annihilate each other by producing gamma photons, transforming matter into pure electromagnetic energy according to Einstein's formula $E = mc^2$.

The theory of quantum electrodynamics, established in the 1940s by the Americans Richard Feynman and Julian Schwinger, and independently by Sin-Itiro Tomonaga in Japan, describes all phenomena that involve charged particles and light. Feynman's formulation of this theory is particularly telling. It analyzes the relativistic interactions between particles in the form of Minkowski space diagrams, giving a vivid representation of the physical processes involved.

Each particle is represented therein by segments of world lines interrupted by punctual events, or vertices, corresponding to processes of emitting or absorbing photons. Since quantum physics does not assign particles a defined trajectory, each line symbolizes an infinite number of possible paths for the particles to follow between two vertices. The theory fixes mathematical rules that assign a probability amplitude to the different paths and indicates how to add these amplitudes to calculate each diagram's contribution to the probability amplitude of the process under study. The summation of these complex amplitudes recalls Fresnel vectors. As in optics, only those paths whose quantum phases do not get blurred by destructive interference matter.

These diagrams describe, for example, how the electron of the hydrogen atom interacts with the proton via an exchange of *virtual photons*, so called because they do not appear outside the interacting systems and cannot be detected directly.

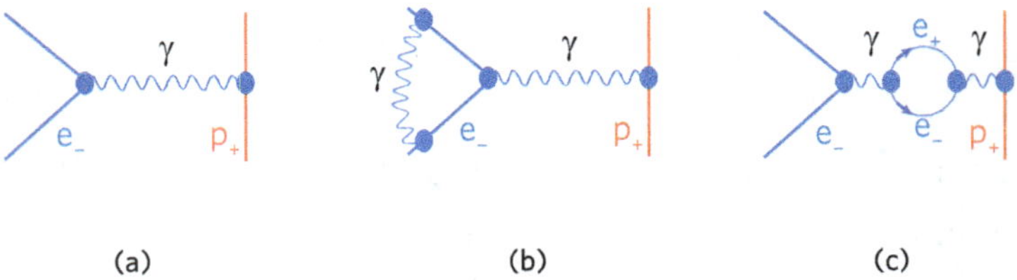

Figure V.6. Three Feynman diagrams illustrating the elementary processes of electron–proton scattering in the hydrogen atom. Each drawing represents the world lines of interacting particles. Time flows vertically and the horizontal coordinate represents space, reduced to one dimension to simplify the drawing. Bold lines describe the world line portions of material particles (electron e_-, positron e_+, and proton p_+), while those of the photons (γ) are represented by wavy lines. The intersections between the lines, called vertices, represent the processes of emission and absorption of light quanta. The photons that appear transiently in these processes are said to be *virtual*. (a) Diagram describing the elementary process of two-vertex interaction between an electron (e_-) and a proton (p_+) by exchange of a virtual photon. (b) Diagram with four vertices describing the main contribution to the Lamb shift (see Chapter 1): the electron emits and absorbs a virtual photon before and after its interaction with the proton. This elusive photon describes the effect of vacuum fluctuations on the electron scattered by the proton. (c) Another four-vertex diagram describing the contribution of vacuum polarization to the Lamb shift: the photon exchanged between the electron and the proton creates a transient electron (e_-)–positron (e_+) pair that annihilates itself to recreate a photon, which is finally absorbed. This process reduces the electron–proton interaction slightly and contributes 5 percent to the Lamb shift.

Each of these photon exchanges corresponds to an elementary scattering process. The permanent binding of the electron to the nucleus, in the atom's ground state as well as its excited states, results from the additive effect of all these processes that create a stationary state in which the electron binds to the proton. The summation of all diagrams' contribution makes it possible to precisely calculate the atomic levels' relativistic energies.

To the diagrams describing the effect of the direct photon exchanges between the electron and the proton, we must add those that correspond to the fast emission and absorption of virtual photons by the electron, which describe the effect of vacuum fluctuations. It is also necessary to account for the possibility of a fleeting appearance of electron–positron pairs between the electron and the proton, a manifestation of what is called *vacuum polarization*. These diagrams give the value of the Lamb shift of hydrogen energy levels with excellent accuracy. The coincidence, to nine or ten significant figures, of the calculated predictions and the atomic spectroscopy measurements makes quantum electrodynamics by far the most accurate theory in all of physics. Feynman was able to say that if this theory predicted the distance from London to New York, the result would be accurate to within a hair's breadth!

Even more fundamentally, quantum electrodynamics has served as a model for quantum field theory, which describes in a unified form the various interactions existing in nature. In this theory, the different particles are the quanta of their respective fields. The photon, whose field carries the electromagnetic interaction between electrons and protons, is the "first cousin" of the massive particles W and Z, vehicles of the weak nuclear interaction responsible for the radioactivity of unstable atoms. Electromagnetism and radioactivity were brought together in the 1960s under the theory of "electroweak" interaction. This theory, the culmination of the work of Steven Weinberg, Abdus Salam, and Sheldon Glashow, extended the process of unifying physics that Maxwell had begun a century earlier. It describes radioactivity and electromagnetic phenomena as the different facets of the same fundamental form of interaction between particles in the universe.

Going further toward unification, the photon also appears as the "second cousin" of gluons, particles carrying the strong nuclear interaction responsible for the cohesion of quarks, protons, and neutrons in atomic nuclei. Electroweak and strong interactions are now united in an extended family constituting the Standard Model of elementary particles, which accounts with a very high degree of precision for all forces of nature except gravity.

This model, developed by theoretical physicists Murray Gell-Mann, Gerald 't Hooft, Martinus Veltman, David Gross, Frank Wilczek, and David Politzer, was built gradually between the 1960s and the 1980s through a constant exchange between theoretical work and experiments carried out in particle accelerators. The high-energy collisions that occur in these machines reveal cousins of the electron—muons and neutrinos—not present in ordinary matter, but whose existence is essential to the model's coherence and to any account of the nuclear processes that occur within stars. Other particles, predicted by theory and discovered thanks to accelerators, are the cousins of the quarks present in atomic nuclei. They have poetic names, such as *charm* and *strange*, that testify to the imagination of the physicists who named them.

Only gravitational interaction still resists what would otherwise be the great unification of all forces of nature; for the moment, the graviton, the hypothetical quantum of the gravitational field, remains elusive. The difficulty of this final unification undoubtedly stems from the fact that gravitation is not a force like any other, but rather one that expresses the very structure of the fabric of the universe. Quantizing it, reducing it to discrete entities, raises questions about the discretization of time and space, at scales so small that to date no experiment to verify it can be conceived.

A fundamental identity

This rapid overview of the development of quantum theory over the last century would be incomplete without a return to the importance of the indistinguishability of particles. This aspect of quantum physics had been announced by Einstein's paper on Bose–Einstein condensation, and would later take on central importance in the theory. An essential principle of the theory consists in describing particles of a given kind as totally identical objects. In classical physics, in principle, we can always distinguish atoms of the same element, or electrons orbiting an atomic nucleus, if only by numbering them arbitrarily and following them theoretically on their trajectories over time.

In quantum physics this possibility, even theoretically, disappears. There are no more trajectories for particles, only wave functions describing their probabilities of being at different points. When two particles collide, the waves associated with them overlap and it is impossible, even in principle, to determine which is which once they separate after interacting.

From this impossibility physicists deduced, beginning with the advent of quantum theory in the second half of the 1920s, that nature had only two ways of distributing identical particles between different states. In one, which Bose had hypothesized for photons and Einstein had extended to the category of particles known as *bosons*, an arbitrary number of particles can be in the same state without being distinguishable from one another in any way. This rule explains, as we have seen, the gregariousness of bosons.

The other rule of nature, first glimpsed by the Swiss physicist Wolfgang Pauli and later made explicit by Enrico Fermi and Paul Dirac, concerns the category of particles called *fermions*. Also called the *Pauli exclusion principle*, this rule applies to electrons in particular. It says that at most one particle can be in a given quantum state. While the bosonic rule favors, at low temperatures, the gathering of a set of bosons in the lowest energy state, Fermi–Dirac statistics requires that fermions be distributed among all accessible energy levels without more than one particle ever occupying the same state.

The exclusion principle plays an essential role in explaining the structure of atoms and matter in general. Because two electrons cannot occupy the same quantum state, the atoms of the successive elements on the periodic table get larger and larger. Indeed, electrons cannot accumulate near the nucleus and must come to occupy increasingly excited orbits, moving farther away from the heart of the atom the more electrons there are to squeeze in. The same principle offers a profound explanation for the mechanical properties of solid matter. Although atoms are mostly "filled with nothing," since most of their mass is concentrated within very small nuclei, they cannot interpenetrate because their peripheral electrons "refuse," due to the exclusion principle, to occupy the same space. It is by virtue of the same principle that we don't pass through the floor on which we stand: the electrons in the soles of our shoes refuse to be in the same place as those in the ground.

The partition between bosons and fermions is an essential feature of the Standard Model. The fundamental particles that make up matter are all fermions. At the most microscopic level, these are the electrons that orbit around nuclei and quarks that combine in threes to form protons and neutrons, the constituents of nuclear matter. The particles that carry the forces—photons, W and Z particles of weak interaction, and gluons—are bosons, as is the hypothetical graviton.

Finally, there exists another fundamental boson, as the media informed the general public not long ago: the *Higgs boson*, named after one of the physicists who predicted its existence in the 1960s. This particle, discovered at CERN in 2011 after a long search, made headlines at the time. The field associated with it suffuses the entire space. By interacting with this field, most Standard Model particles, which initially had zero mass, became massive at a primitive stage in the evolution of the universe. The photon remained of zero mass, while its cousin particles, the W and Z bosons, became very heavy. This difference explains, as we will see, the great disparity in range between electromagnetic forces and weak nuclear interactions.

Bosons and fermions are distinguished by their intrinsic angular momenta. The spins of the former are integers; the latter's are half-integers. The photon in particular is a boson of spin 1 (its angular momentum along its direction of propagation can take the values $+\hbar$ or $-\hbar$), while the electron is a fermion of spin 1/2 (its intrinsic angular momentum is $+\hbar/2$ or $-\hbar/2$). The link between the statistics of elementary particles and the value of their spins was demonstrated by Pauli in 1940.

Composite systems formed of several particles bound together also obey the principle of quantum indistinguishability. They are bosons if the total number of elementary particles composing them is even (their spin, the sum of an even number of half-integer spins, is then an integer). They are fermions if the number of elementary particles is odd (their spin is then a half-integer). Protons and neutrons constituting atomic nuclei, each made up of three strongly bound quarks, are thus fermions.

The same rule applies to atoms, which are bound systems of quarks and electrons. The hydrogen atom (one proton and one electron, *i.e.* four elementary particles) is a boson, as is the helium-4 atom (two protons, two neutrons, and two electrons). Deuterium (hydrogen isotope formed of one proton, one neutron, and one electron) and helium-3 (two protons, one neutron, and two electrons) are fermions. The odd isotopes of alkaline atoms, which all have an odd number of quarks and electrons, and thus an even total number of fermions, are composite bosons. As we have seen, these atoms were the first to be prepared in Bose–Einstein condensates.

The existence of composite bosons formed by pairing even numbers of fermions in matter explains a number of quantum phenomena that were discovered even before the first experimental realization of the Bose–Einstein condensate. Condensed boson gases have very special quantum properties. Their particles behave in a collective manner and allow these systems to manifest superfluidity, flowing together without viscosity. This is the case of helium-4 liquid at very low temperatures, whose superfluidity was discovered in the 1930s. Superfluidity also manifests itself in the superconductivity of certain metals, which I have already mentioned. Their electrons are coupled in pairs of bosonic character that flow without resistance. Thus, by predicting bosonic condensation, Einstein provided an essential element in our understanding of superconductivity.

The Pandora's box of quantum physics

What was Einstein's attitude toward the development of quantum physics, which began with a question about light and went on to radically advance our knowledge of the structure of matter and the universe? He had introduced into physics a dualism incompatible with classical concepts. He recognized that this physics carried in it a fundamental indeterminism, describing the spontaneous emission of photons as a random process. He was also the first to understand the importance of quantum indistinguishability, an essential property of particles behaving like waves. We may add that Einstein believed in the reality of photons long before Planck, even before Bohr.

When we consider Einstein facing the quanta he had discovered, we can't help but think of the myth of Pandora's box—though we must immediately qualify and clarify this image. Mythology tells us that Pandora, by opening the jar she was supposed to leave closed, triggered a series of catastrophic events. In the case of the box that held the photon, the consequences were certainly immense: its opening brought us knowledge of the microscopic world and gave us the keys to all modern technology, which can hardly be called a catastrophe. Nuclear power plants, computers, cell phones, lasers, magnetic resonance imaging, the internet, atomic clocks, and the GPS are quantum technologies that have changed our daily lives over the last century. In light of the misuse of some of these inventions, we may be tempted to restore a sinister connotation to the mythological term. Indeed, we can think of the military applications of nuclear energy, or the negative aspects of the internet and the sometimes perverse use of social networks, as a consequence.

Einstein did not share this pessimistic view of knowledge and science. If there was a Pandora's box for him, it was not for this negative reason. Rather, this image with its almost tragic resonance applies to his discovery of quanta because the theory that finally emerged from it always made him uncomfortable, and until the end of his life he tried to go beyond it, to return to a vision of the world more in line with classical ideas. And, it must be said, despite its immense success, this theory continues to be the subject of many interpretations and to exert on the non-scientific public a fascination that extends far beyond physics.

To the classical mind, waves and particles are irreconcilable. Wave phenomena are delocalized and continuous. Waves spread throughout the space accessible to them. They add or subtract according to their respective phases, and obey the principles of superposition and interference discovered at the beginning of the nineteenth century by Young and Fresnel. Particles, on the other hand, are strongly localized objects, following trajectories defined under the effect of the forces applied to them. The principles of superposition and interference mean nothing to them. This opposition between waves and particles, illustrated by the ancient quarrel between Huygens's and Newton's supporters, dates back to the origins of modern science. Einstein had refused to walk the beaten path by siding with one camp or the other, claiming that at least for light the classical dichotomy between waves and particles, obvious as it seemed, could not be maintained.

He had already been an iconoclast when that same year he rejected another dogma of classical physics, that of the universality of time. As counterintuitive as the relativistic principles were, they were still less disturbing to the science of the era—and to Einstein—than the photon's escape from Pandora's box. The quantum revolution would indeed prove, for physics and for science in general, much more upsetting to our worldview than relativity had been. Step by step, quantum theory would reveal a strange world in which the continuous and the discontinuous, the certain and the uncertain, the fluctuating and the immutable, are in constant contact. This is a world that disconcertingly combines seemingly contradictory aspects, giving a preeminent place to chance, which classical determinism had sought to eliminate. It is a world of molecules, atoms, and elementary components of nature, among which photons, omnipresent particles traveling at the speed of light, play an essential role.

The founders of this theory—Einstein, Bohr, de Broglie, Schrödinger, Heisenberg, and others—discovered the quantum laws through thought experiments, as Einstein did for relativity. But for Einstein, at least, these quantum thought experiments took on a very different function from that of the trains, carousels, and clocks he had imagined in 1905. While the latter had served as a guide to support and confirm his theory, quantum thought experiments allowed him to play devil's advocate in his increasingly desperate attempts to prove that quantum theory couldn't be the last word in explaining the laws of nature. Quantum thought experiments were also subtler and more complex than their relativistic counterparts because they had to analyze counterintuitive and apparently contradictory phenomena, whereas the relativistic train or carousel experiments illustrated only one simple and rational idea: the equivalence of all reference frames for the description of the laws of physics.

From classical to quantum: a dialogue across centuries between Fermat, Maupertuis, and Feynman

If quantum physics seems strange to us, it's because its laws do not generally manifest themselves directly at the scale of the macroscopic phenomena to which our senses are sensitive. This is due to the extreme smallness of Planck's constant compared to the actions we perceive on a daily basis. Here I use the word *action* in its physical sense. An action is an energy E multiplied by a time t, or a momentum p multiplied by a length x. Actions are expressed in the usual unit system in joules × seconds (J.s).

The variation of the action of a particle with energy E and momentum p between two events separated by a small space-time interval dx, dt is equal to $pdx - Edt$. (For simplicity, our explanation here will involve only one spatial dimension.)

Relativity and quantum physics teach us that the matter wave associated with this particle has, at point x,t, a phase equal to its action $A = px - Et$ divided by Planck's constant $\hbar$. This phase is a dimensionless number, gigantic for a macroscopic body. Similarly, the angular momenta of the systems we encounter in everyday life, which also have the dimension of an action, are enormous compared to the spin of a quantum particle. The angular momentum of a bicycle wheel spinning at a frequency of a few hertz is typically equal to a few tens of J.s, while the unit of quantum angular momentum $\hbar$ is very close to 10^{-34} J.s, a fraction on the order of one in ten million billion billion billion!

In spite of the very high value of the frequency ν of light, on the order of 5×10^{14} hertz for a radiation of yellow color, the energy $h\nu$ of a photon is extremely small, on the order of 3×10^{-19} J. A 100-watt lamp radiates approximately 330 billion billion photons per second! These order-of-magnitude exercises help us appreciate the enormous distance separating the macroscopic world from that of atoms and photons. The energy exchanges in the macroscopic processes we observe in our daily lives involve gigantic quantum numbers. The granular aspect of quantum physics, then, is totally unobservable; all phenomena seem continuous to us.

Note that the orders of magnitude expressed in numbers involving high negative powers of ten have led physicists to choose more practical atomic units. The joule has been replaced by the electronvolt (eV): the work of an electron undergoing a potential drop of 1 volt, which is equal to 1.6×10^{-19} J. The energy of the yellow-light photon then takes the more practical value of 1.85 eV.

Let's now explore in a more fundamental manner the frontier between the classical and the quantum worlds by analyzing the dynamics of a particle based on Feynman's ideas. By replacing the increment dx with vdt, where v is the speed of the particle, we can write the variation of its action A on a small segment of a world line as $dA = (pv - E)dt$. The global variation ΔA of the particle's action when it follows a path between two events separated by any distance is the integral on this path of the increments dA. For a non-relativistic particle, E is generally the sum of its potential energy U (which depends for example on its height in a gravitational field) and its kinetic energy $mv^2/2 = pv/2$. The increment dA is then written $(pv - U - pv/2)dt = (pv/2 - U)dt$. The variation of the particle's action is thus simply the integral of the *difference* of its kinetic and potential energies along its world line. This integral obviously depends on the path followed.

Let us recall that the particle evolves according to the principle of superposition, "simultaneously" on all the paths connecting the two space-time points considered. The probability amplitude of finding it in $x + \Delta x$ at time $t + \Delta t$ if it was in x at time t is the sum of complex numbers whose phases are the quantities $\Delta A/\hbar$ associated with all the paths connecting these two points. This rule, which applies to the calculation of Feynman diagrams, is analogous to the Huygens–Fresnel principle, which prescribes how to sum the Fresnel vectors associated with the propagation of light between two points. Only the contributions of stationary phase paths, *i.e.* those for which ΔA varies only on the order of $\hbar$ on neighboring paths, count in this sum. For microscopic quantum systems, a large number of paths satisfy this condition. The particle follows all of them "simultaneously." It behaves like a wave and the quantum effects are considerable.

When the particle is macroscopic, on the other hand, its action ΔA measured in units of $\hbar$ is a huge number, on the order of a very high power of ten. Only then does it follow the extreme (maximum or minimum) action path, since the contributions to the probability amplitude of all those that deviate from it are blurred completely by destructive interference. Paradoxically, then, starting from the principle of quantum superposition, we find the principle of *least action* of classical mechanics, reminiscent of Fermat's principle in optics.

Quantum mechanics is thus to classical physics as the geometrical optics of light rays is to wave optics. Light's diffraction and interference effects become important only when the size of the obstacles presented to the light rays approaches the order of the wavelength. Similarly, quantum interference is observed only if the actions of the systems under study are on the order of Planck's action h.

I have followed here an argument developed by Feynman, who invented his diagram method in the 1940s based on a reflection on the importance of Planck's constant in defining the boundary between the classical and quantum worlds.

That said, of course, the classical *principle of least action* did not wait for Feynman, and its historical discovery far preceded quantum physics. It was first expressed in the middle of the eighteenth century by Maupertuis, the man who measured the shape of the Earth (as we saw in Chapter 2). On action, Maupertuis tells us:

> When any change occurs in Nature, the amount of Action used for that change is always the smallest possible.

The argument we see here is almost philosophical, close to Fermat's whereby *Nature acts in the shortest and simplest ways*. Maupertuis's remarkable intuition was confirmed a few decades later by the mathematician Joseph-Louis Lagrange, who showed that the principle of least action is equivalent to Newton's law. He expressed in integral form, by summing the contributions to the action of infinitesimal trajectories, what Newton had described with a differential equation, equating the derivative of a particle's momentum to the force it undergoes.

Deriving the classical principle of least action from that of quantum superpositions, as Feynman did, has the advantage of answering a question that had been troubling classical physicists and that we have already seen with respect to light and Fermat's principle: how does the particle "know" the path it follows is extreme? How does it "determine" that the other paths correspond to larger or smaller actions? Just as for light, the answer is that the particle is associated with a wave that allows it to "feel" the neighboring paths and thus "decide" which corresponds to the extreme action. I have put the verbs in this anthropomorphic analogy in quotation marks; of course the particle neither feels nor chooses anything at all. It simply obeys the laws of quantum physics, which naturally bring out the classical deterministic behavior from the wave theory when the action grows large compared to Planck's constant.

A journey in orders of magnitude

The smallness of Planck's constant entails that of the wavelengths of matter discovered by de Broglie. Estimating these wavelengths allows us to understand, more specifically than Feynman's theoretical argument allowed, why quantum phenomena generally escape our direct observation. An oxygen molecule, with mass $M = 5 \times 10^{-26}$ kilograms, moving in the ambient air at the speed $v = 500$ m/s, is associated with a wavelength h/Mv on the order of 2×10^{-11} meters, or 0.2 angstroms. This length is very small compared to the distances between molecules in the gas, typically on the order of 30 angstroms. The matter wave associated with a molecule oscillates over very small spatial distances and is scattered by its interaction with other particles randomly distributed at several tens of wavelengths from each other. The matter wavelets generated by all of these interactions have phases whose differences are completely random, so that all interference effects between these waves are blurred. The physics of the gas can then be analyzed analogously to classical small marbles colliding according to the laws of billiard balls. This was how Boltzmann and Maxwell, before the advent of quantum theory, established the kinetic theory of gases.

The wavelengths of matter increase as the mass of the particles and their velocity decrease, making quantum phenomena relatively greater. A helium gas, eight times lighter than oxygen, consists at room temperature of atoms with a wavelength close to 0.6 angstroms. This is still small compared to interatomic distances, but less far off. And if we compress and cool this gas, we reach, more readily than we would with oxygen, situations where quantum effects, linked to the fact that interatomic distances decrease to the order of wavelengths, begin to manifest themselves in a spectacular manner. Liquid helium-4 becomes superfluid at 2.17 kelvin and climbs the walls of its container, flowing without viscosity like a fountain of spontaneously escaping liquid. This is a rare macroscopic quantum effect, a strange "bosonic" property not observed until the 1930s.

Let's dig deeper into the quantum world by evaluating the wavelength of an electron—whose mass m is 7,300 times smaller than that of helium—orbiting in a hydrogen atom. In the ground state, the electron has a kinetic energy $E_c = mv^2/2$ = 13.6 eV. Its velocity $v = \sqrt{2E_c/m}$

(equal to 2.2×10^6 m/s) corresponds to a de Broglie wavelength h/mv of 3 angstroms, equal to the circumference of the orbit Bohr assigned to the electron in his 1913 model. It is thus clear that quantum effects are essential. The constructive interference condition of the electron wave—circulating around the nucleus and returning to the same phase after one turn—justifies Bohr's quantization rule.

Electronic interference also plays an important role in solid matter. The atoms in a metal—copper, for example—are distributed in a regular crystal lattice with a periodicity of about 2 angstroms. They each pool one or two electrons moving freely in the lattice. According to Pauli's principle, which prevents two fermions from occupying the same state, all of these electrons are distributed on a scale of quantum states, occupied up to a maximum energy level E_F, called the *Fermi energy*. Neglecting in a preliminary approximation the interactions of these electrons with one another and with the atoms, we find through an elementary calculation of level filling that in copper $E_F = 7$ eV. This energy corresponds to an electron velocity $v_F = 1.5 \times 10^6$ m/s and to a de Broglie wavelength h/mv_F of about 4 angstroms, greater than the distance separating the neighboring atoms in the lattice. Quantum interference between the electron waves scattered coherently by the metal atoms plays a key role in explaining its macroscopic properties, in particular its electrical conductivity.

Let's dive even further into the heart of the matter. In the nucleus of the atom, protons and neutrons of neighboring masses $M = 1.6 \times 10^{-27}$ kilograms are confined to the extreme over dimensions on the order of 10^{-15} meters and their velocities are close to that of light c. The dimensions of the nuclei are comparable to the wavelengths h/Mc of their constituent particles. Nuclear forces confine the particles in the nuclei as a barrier with a thickness on the order of h/Mc would. Material waves can leak through this barrier by a phenomenon analogous to what is called *frustrated total reflection* in optics. If two blocks of glass are brought together with their flat surfaces separated by a thin air gap on the order of the wavelength of light, a ray undergoing total reflection in one of the blocks can in fact pass into the other thanks to the evanescent wave established in the air between the two diopters (this wave was described in the nineteenth century by Fresnel; see Chapter 3). The photons pass from one block to the other, whereas they couldn't do so if the diopters were separated by more than a wavelength or two. A similar effect can occur with the matter waves of nuclear particles trapped in nuclei. This transmission of particles through the nuclear potential barrier that confines them is called a *tunnel effect*. This is how helium atomic nuclei (consisting of two protons and two neutrons) escape, through a quantum wave effect, from unstable atomic nuclei. This is the radioactivity α that was discovered at the beginning of the twentieth century. As we have seen, the α particles thus produced were used as projectiles by Rutherford to bombard a sheet of gold foil, leading him to discover in 1911 the atomic nucleus and the planetary structure of the atom.

In all quantum effects, the notions of state superposition and interference are essential. Thus, these concepts, introduced in optics at the beginning of the nineteenth century, were extended to the description of all physics with the advent of quanta. The fields of vectors that Fresnel had first imagined to describe light, and that Faraday and Maxwell had broadened to the description of electrical and magnetic phenomena, were extended in the twentieth century to the description of the fields associated with all fundamental particles. In this sense, it can be said that the scientific history of light carries in it the whole history of physics. At first described as a delocalized field, electromagnetic radiation acquired a corpuscular character thanks to Einstein, and this dualism quickly generalized to other fields, describing the behavior of all particles that make up the universe.

It's not surprising that this paradoxical aspect of physics first appeared in the exploration of optical phenomena such as thermal radiation or the photoelectric effect. Indeed, the photon is the particle among all others whose wavelengths cover the widest ranges of spatial dimensions. Because its mass is null, the photon's energy can take an extremely wide spectrum of values, from those on the order of micro- or nanoelectronvolts (radio waves) to gigaelectronvolts (gamma rays). The associated wavelengths, inversely proportional to these energies, range from kilometers to attometers (10^{-18} meters). At one end of this spectrum, the interference effects of large-wavelength waves are evident, while at the other limit the granular behavior of high-energy photons bombarding matter is clearly identifiable. Between these two extremes, in the optical domain, light subtly manifests the combination of wave and corpuscular effects that Einstein was the first to glimpse.

The quantum scene: individual objects or statistical ensembles?

Quantum effects at the scale of a gas, a solid, an atom, or a nucleus were revealed by indirect observation. The wave behavior of matter was reconstructed by subtle inferences based on the statistical behavior of macroscopic samples of particles. The quantum was not revealed directly; it was, so to speak, veiled. It was in order to establish its rules that physicists imagined what would happen if they could manipulate and observe isolated electrons, atoms, molecules, or photons protected from the disturbances of their environment. Even as they imagined these experiments, they thought they would forever remain impossible to perform. Schrödinger expressed this conviction in a 1952 text:

> We never experiment with just one electron or atom or (small) molecule. In thought-experiments we sometimes assume that we do; this inevitably entails ridiculous consequences.

This sentence may seem strange, as in 1952 there was no doubt as to the existence of particles. Their trajectories in bubble chambers were already being studied by accelerators that explored the structure of matter with increasing precision. Aware of this though he was, Schrödinger saw a major difference between these experiments and the ones he and his colleagues had imagined. In their thought experiments they envisioned what would happen if we could manipulate and observe quantum particles without destroying them, whereas in the accelerators' detectors the experimenters observed the traces left by quantum objects destroyed in high-energy collisions. To underline this essential difference, Schrödinger pointed out that

> we are not *experimenting* with single particles, any more than we can raise Ichthyosauria in the zoo. We are scrutinising records of events long after they have happened.

In other words, particle physicists behave like paleontologists trying to reconstruct past events from fossil evidence. They do *post-mortem* analysis, whereas thought experiments focus on imagining what would happen if we could observe isolated quantum systems *in vivo* without destroying them. Schrödinger, no doubt like his colleagues at the time, thought these dreams would never be fulfilled. When he speaks of "ridiculous consequences," he even seems to imply that the effects described by these experiments, the superposition of states and quantum jumps, will forever be observable only by their statistical consequences, their effects in systems made up of a large number of particles, and that it is futile to try to consider them at the level of individual particles.

The recent history of physics has proven Schrödinger wrong. We know now how to manipulate isolated atoms or photons without destroying them, thus directly revealing the properties of the quantum world, that which had previously been hidden in the complexity of the classical world around us. Thought experiments have become real, and light has played a great role in this revolutionary advance in physics.

It was, in fact, the laser that made it possible to explore matter at the most elementary level, to trap atoms or photons and study their behavior on a stage where they manifest their quantum properties directly. The laser is a technology born of quantum physics, exploiting the knowledge this physics has given us of the interaction between atoms and photons at the most fundamental level. Einstein had the initial idea for the design of this instrument when he described the phenomenon of stimulated emission. Implementation of this idea led to experiments that tested and confirmed the counter-intuitive principles of quantum physics Einstein and his colleagues had imagined without ever dreaming they could become real. This paradox illustrates once again the complementarity of fundamental and applied research. The former lays down the principles on which the latter can develop, and the resulting inventions allow us to explore nature with greater precision, confirming or correcting existing models in a fertile back-and-forth between experimentation and theory.

I took part in this adventure and contributed, through my research, to the domestication of atoms and photons that will perhaps, one day, be the instruments of new quantum technologies. I have been fortunate to be part of a vast global community of researchers committed to exploring the world of atoms and photons, first to understand it better, then to exploit it by developing new devices that increase our opportunities for action and information. This is the adventure I will discuss in the following chapters. But first, to shed some light on the context, we must analyze the passionate discussions that took place at the 1927 and 1930 Solvay Conferences, where quantum theory was developed.

Young's double slit revisited

No doubt the first thought experiment that came to mind for Einstein and Bohr was that of Young's double slit, the observation of interference fringes that seemed, at the beginning of the nineteenth century, to definitively vindicate Huygens by establishing the undulatory character of light. How to interpret this experiment if light was in fact made up of discrete particles called photons?

And how to explain this if we replace light with a flow of material particles such as electrons? According to de Broglie, these particles were associated with a matter wave and should also give rise to interference, as the famous experiment by Davisson and Germer showed in 1927, leading to the 1929 Nobel Prize for the young French physicist.

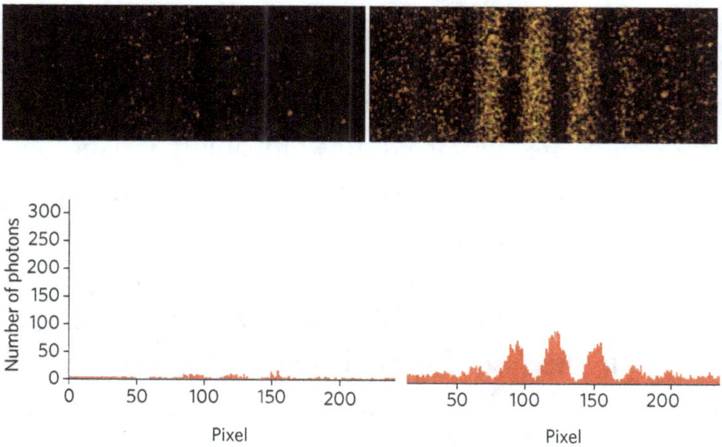

Figure V.7. Two-wave interference experiment with photons passing through the apparatus one by one (*top screens*): the first photons seem to reach the screen at random points (*left*), before a clear image of the fringes is formed (*right*). We see in this image the pointillism corresponding to the particles' individual arrivals on the screen. The histograms below the pictures show the number of particles detected at these two instants versus the abscissas of their positions measured in pixels (1 pixel = 25 microns). The two interfering waves come not from two slits but from the light passing through a Fresnel biprism, formed by two prisms joined at the base. This device, which Fresnel invented to study light interference, produces fringes similar to the ones he obtained with his two inclined mirrors. The principle is identical to that of Young's double slit (experiment performed by Jean-François Roch, École normale supérieure, Cachan).

Quantum interpretation is, at base, very simple. Matter waves have complex amplitudes. Those associated with the passage of particles through the two slits overlap and interfere. They become stronger when the path difference is an integer number of de Broglie wavelengths and cancel each other out with destructive interference when the difference is an odd number of half wavelengths. Where probability amplitudes are added, the probability of finding the particles is high; where they are subtracted, this probability is canceled. If the flow of particles is large, the distribution on a screen of the particles' impact points reproduces almost instantaneously the familiar interference fringe pattern that Young and Fresnel observed more than a century earlier. In the case of light, there seemed to be nothing new about this. But for electrons and other particles, the result was novel and opened fascinating perspectives for physics, extending the concept of interference to the properties of matter.

But there's more to this experiment. The very notion of physical reality is called into question if we seek to analyze what happens at the level of the individual particles passing through the device. Suppose the source of particles, photons or electrons, is so weak that it emits only one corpuscle at a time. The interference pattern will then appear only after a long exposure time, with the first particles arriving on the screen in a seemingly random way until an image of the fringes—first pointillistic, then increasingly precise—forms. This interference experiment with fringes, appearing gradually, could only be imagined in the time of Einstein and Bohr, but is now carried out with all kinds of particles, photons, electrons, atoms, and molecules, in various types of interferometers bifurcating and then recombining the paths of the particles coming from the same source. It always returns the expected result.

For interference to occur, of course, both paths must be open to the passage of each particle. If one of the two slits in Young's experiment is randomly closed, the image formed on the detection screen will be that of two diffraction spots simply added in the classical manner, without interference.

Measurement, complementarity, and uncertainty relations

For a mind trained in Newtonian physics, accustomed to reasoning in terms of orbits and trajectories, a question immediately arises. What happens at the moment the particle crosses the screen pierced by the two slits? How can it be that it never arrives on a dark fringe if it passes through one of the two slits, whereas it could if the other slit were closed? How does it "know" that the slit it's not passing through is open or closed in order to "decide" whether or not to go to a dark fringe? The answer given by quantum physics is… that the question is meaningless. It is not possible to ascribe a physical reality to the particle's passage through one hole or the other if no measurement is made to answer the question. There is a probability amplitude for the photon or electron to pass through each hole, and these probability amplitudes will interfere to give a result when a measurement of the particle's position is made *after* it passes through the screen.

If we try to determine where the particle is at the moment it passes through the slits, we will get a result (it passes on one side or on the other) but the wave function and the particle's subsequent evolution will be disturbed in such a way that the fringes disappear. In other words, knowing where the particle is at the level of the slits and determining where it is likely to fall on the detection screen afterward are two incompatible tasks. The same experiment cannot ask nature to answer both simultaneously. We must ask one question or the other, and modify the design of the experiment accordingly. Bohr expressed the impossibility of simultaneously measuring incompatible physical quantities in what he called the *principle of complementarity.*

This principle has fundamental consequences for our intuition of physical reality, first of all for the notion of trajectory. If we don't look at the electron passing through the device, there's no sense in asking what trajectory it follows. In a pictorial way, we can say that in the interferometer it follows both trajectories at the same time, passing through one slit *and* the other, in a superposition of the two states—suspended, so to speak, between two classical realities. In fact, a more exact expression of this theory would say it is the probability amplitudes associated with the two trajectories that are superposed, in the mathematical sense of the term. It is this mathematical superposition that forms the basis of the rules of calculation for the Feynman diagrams mentioned above.

For Bohr and Heisenberg this point of view was obvious, and they accepted it as an incontestable property of nature on a microscopic scale. Einstein, Schrödinger, and de Broglie saw it differently. "I know where the Moon is in the sky even if no one is looking at it," said Einstein, who found it difficult to give up ascribing a physical reality to a quantity simply because it was not explicitly measured. In an attempt to convince his elder, the young Heisenberg once told Einstein that he had only established quantum concepts by following the method Einstein had borrowed from the ideas of Ernst Mach. This Austrian physicist and philosopher argued that physics should be interested only in what was measurable, and that in order to define any physical quantity it was necessary to consider precisely how it was evaluated practically. After all, wasn't that what Einstein did when he questioned the notion of absolute time by closely examining the synchronization of clocks? Heisenberg had taken the same approach in establishing a theory that explicitly linked measurable quantities to each other. The question of the value taken by certain quantities, particles' positions or velocities, was futile when the experimental device could not determine them. Einstein opined that Mach's thought had helped him with relativity, but that "one should not repeat the same joke twice."

On the topic of Heisenberg, it's time for us to look at his famous uncertainty principle, which plays a significant role in quantum physics and is often used in common language to evoke quantum mysteries. It is important to discuss and demystify the principle at this stage, so closely is it linked to the notion of complementarity introduced by Bohr; it will also help us analyze in greater detail the thought experiments discussed at the Solvay Conferences of 1927 and 1930.

These famous uncertainty relations express that certain so-called *complementary* or *conjugated* quantities can be determined simultaneously in quantum physics with only limited precision. They link together the uncertainties Δx and Δp that must be respected when measuring the position and momentum of a particle, or the uncertainties ΔE and Δt with which one can determine the energy E exchanged between two quantum systems and the instant t of the exchange. The relation linking position and velocity, $\Delta x \Delta p \geq \hbar$, expresses the fact that, at the quantum level, increasing the precision of the measurement of a particle's position comes at the cost of increasing uncertainty about its momentum and therefore its velocity. The analogous relationship between time and energy, $\Delta E \Delta t \geq \hbar$, means that in order to accurately determine an energy one must renounce knowing with certainty when it was exchanged.

The smallness of Planck's constant appearing in these inequalities indicates, of course, that they become real limitations only at the level of the microscopic processes of atomic or nuclear physics. At the macroscopic scale, the products of the classical experimental uncertainties concerning x and p or E and t are much larger than $\hbar$, and the accuracy of the measurements is not limited by the Heisenberg relations.

These relations can be deduced qualitatively from the properties of the de Broglie matter waves. A given particle with momentum p has a well-defined wavelength. It thus extends in space over a theoretically infinite distance, corresponding to a uniform distribution of the probability of finding the particle anywhere. Extreme precision on a particle's momentum thus comes at the cost of a lack of information on its position. Conversely, when a particle is detected at a precise point, its wave function is perfectly localized at that point at that instant, and is made up of the superposition of an infinite number of plane waves associated with an infinitely wide distribution of momentum values. These are fundamental properties of the Fourier transform, which we discussed in the context of optics in Chapter 3.

Knowing a particle's position with infinite precision is thus paid for by total ignorance of its momentum. Between these two extremes, we can prepare a particle's wave function as a *packet of waves* whose wavelengths are dispersed around a central value, giving a width Δp to the distribution of the momentum and an inversely proportional width $\Delta x = \hbar/\Delta\mathrm{p}$ to the distribution of the possible values of its position. The product $\Delta x \Delta p$ of the two uncertainties is then equal to $\hbar$; this is the best possible compromise if we seek to optimize the accuracy of the two measurements.

An analogous reasoning, also based on Fourier analysis, helps us understand time–energy uncertainty relationships. A particle of well-defined energy E is associated with a de Broglie wave of a perfectly determined frequency E/h, which requires it to develop over an infinite time interval. It is then impossible to know when the wave appeared. Conversely, a particle created at a very precise moment yields total indeterminacy regarding its energy. Here again, the best possible compromise is to consider a wave packet that accepts a certain blur ΔE on its energy and a conjugated blur Δt on the creation time of the packet (or of its passage at a point), the product of these two uncertainties being equal to $\hbar$.

These uncertainty relations play an essential role in quantum physics. They give a qualitative explanation for fundamental phenomena in the world of atoms and photons. As we have seen, one of the mysteries Newtonian physics and Maxwell's equations were unable to explain was the stability of atomic matter. The electron of the hydrogen atom that revolves around the nucleus, attracted by its electric field like a planet around the Sun's gravitational field, should radiate electromagnetic energy and finally, after a very short time, fall into the nucleus. However, it remains at a distance of about 0.5 angstroms from the core. Bohr admitted this property as a postulate by assigning the electron privileged orbits, the lowest being precisely the one where the electron gravitates at a Bohr radius from the nucleus.

Uncertainty relations give this postulate a more general explanation. If the electron were to fall into the nucleus, its position would become infinitely precise; therefore its momentum, and consequently its kinetic energy, would increase indefinitely. By staying at a finite distance from the nucleus, it optimizes the product of uncertainties concerning its position and momentum so as to minimize its energy. This is the sum of its electrical potential energy, which becomes ever stronger and more negative as it gets closer to the nucleus, and its positive kinetic energy, which becomes ever greater. The size of the electronic wave function, on the order of 10^{-10} meters, corresponds to the minimum of the sum of these two energies evolving in opposite directions along with the electron's distance from the nucleus. If we gradually add other electrons to the atom in order to form the successive elements of Mendeleev's periodic table, we must—taking into account Fermi–Dirac statistics—place them in increasingly large orbits. We can thus say that the structure of matter can be explained by the combined requirements of Heisenberg's uncertainty relations and Pauli's principle of exclusion.

The time–energy uncertainty relationship also plays an essential role in quantum physics. When an electron in an atom is brought to an excited state, it falls back to its ground state after a certain time by emitting a photon whose frequency is related to the energy difference of the two states by Planck's formula $E_1 - E_2 = h\nu$. This emission occurs at a random time, as an instantaneous quantum jump. The average time τ that the atom spends in the excited state is related to the uncertainties ΔE and $\Delta\nu$ of the energy and frequency of the photon emitted by the relation $\Delta E = h\Delta\nu = \hbar/\tau$. This determines the limit of accuracy in measuring the frequency of light emitted by an atom at rest (an additional uncertainty is added if it moves, as we will see in the next chapter).

If we're looking for as monochromatic a light source as possible, for instance to build an atomic clock, we will have to choose to lock its frequency to an atomic transition between a ground state and an excited state of the longest possible lifetime, *i.e.* of the largest possible value for τ. In optical atomic clocks, the excited atomic states have a lifetime of a few hundred seconds, making it possible to define the frequency on the order of 10^{15} hertz to less than a thousandth of a hertz and to give the clock a relative uncertainty of about one part in 10^{19} (less than one second since the beginning of the universe!). Therefore, uncertainty relationships are not synonymous with fuzziness or imprecision. Quantum physics allows us to reach as high an accuracy as we want when measuring a parameter (*e.g.*, the energy or frequency of a photon) *on the condition* that we accept a large indeterminacy on the conjugate variable (in this case, the photon's emission time).

The time–energy uncertainty relationship can also be invoked to understand the quantum tunneling effect, analyzed by Gamow in the 1920s. We have seen with the case of α rays that a particle trapped by a nuclear force can escape its trap and move away from it indefinitely. Classically this is impossible, because crossing the barrier constituting the trap would violate the principle of energy conservation. Simply put, the trapped particle lacks the kinetic energy needed to jump over the barrier that confines it. In quantum physics, however, a transient process can cause an amount of energy ΔE to appear that does not conserve energy, as long as this process does not last longer than a time Δt on the order of $\hbar/\Delta E$. We can say that the particle, in order to escape by tunnel effect, borrows from the nucleus the energy that allows it to jump the barrier, and that this borrowing is possible because the jump lasts only a very short time.

A similar argument explains the great difference in scope between the "cousin" interactions that are electromagnetic forces and the nuclear interactions responsible for radioactivity. The former are long-range because they are conveyed by zero-mass photons, whose energy can be arbitrarily small. According to the time–energy uncertainty relation, the time between the creation and annihilation of a virtual photon exchanged by two charged particles can be arbitrarily long, all the more so as its frequency shrinks. This photon can thus propagate over large distances at the speed of light, which explains the long range of electromagnetic interactions. The weak nuclear force is carried by W and Z bosons, whose masses M_w and M_z are roughly a hundred times that of a proton. The appearances of these bosons in nuclear processes are thus extremely fleeting, not exceeding a time $\hbar/M_wc^2$ or $\hbar/M_Zc^2$, on the order of 10^{-25} seconds. In this time they can only propagate over distances on the order of 10^{-17} meters, limiting the influence of the weak nuclear force inside atomic nuclei.

Debates around imaginary experiments

To understand how Heisenberg's uncertainty relations illuminated Bohr's principle of complementarity, let us return again to Young's double slit experiment. Einstein had not given up on the existence of the trajectory of the particle crossing the screen pierced by two slits. Blocking one slit to ensure that the photon or electron passed through the other was obviously too crude a procedure, so he imagined a subtler method. The two slits would now be horizontal, one above the other. The bottom slit was pierced in the partition of the fixed screen, and the top slit was part of a moving contraption, a small blade suspended from a vertical spring left free to oscillate up and down. Einstein's idea was simple.

If the particle passed the device through the bottom slit, the top slit would not move. On the other hand, if the trajectory led it through the upper slit, it would be scattered by the slit's edges and see its momentum deflected slightly downward, causing by reaction an upward motion of the assembly. This assembly, retained by the spring, would start to oscillate and its motion would reveal the trajectory followed by the particle. The particle would then arrive on the detection screen and leave a trace there, as in the experiment in which the two slits are fixed. We would thus know the trajectory of each particle and observe fringes after a large number of impacts were detected. But could we be sure?

Figure V.8. Bohr and Einstein in 1925 in Leiden at the home of their friend, the Austro-Dutch physicist Paul Ehrenfest (© World History Archive/ABACA).

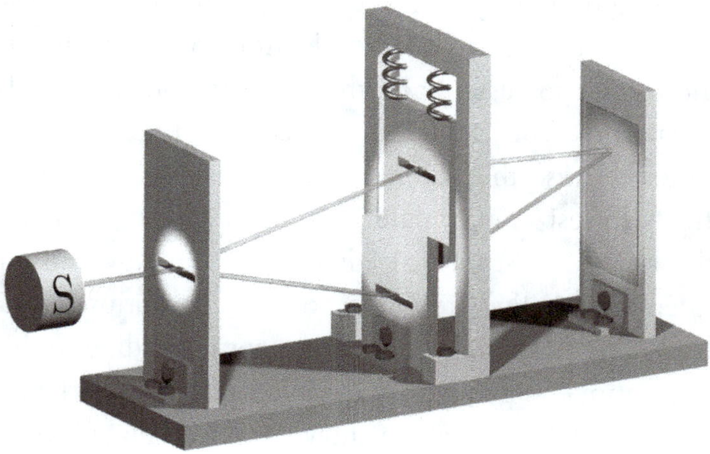

Figure V.9. Young's double slit thought experiment with a moving slit. The moving assembly, suspended from the spring, begins to oscillate if the particle passes through the upper slit (based on a drawing by Niels Bohr, reconstructed by J.-M. Raimond).

Bohr had no qualms with salvaging his principle of complementarity by calling Heisenberg to the rescue. In order to detect the very small change in momentum caused by the particle's passage, the initial momentum of the assembly had to be known with a very small uncertainty Δp, leading to an uncertainty regarding the vertical position of the slit at least equal to $\Delta x = \hbar/\Delta p$. This uncertainty would have the consequence of blurring the path difference between the two paths, clouding the interference pattern. The particle disturbed by the recoil it caused the moving slit would arrive on the detection screen anywhere, not necessarily on a bright fringe.

This thought experiment reveals a crucial property of a quantum oscillator, be it a small mass attached to a spring or a pendulum oscillating in the Earth's gravitational field. Just like the electron in the hydrogen atom, and for essentially the same reason, the oscillator cannot be located at a precise point at a given time, because then its momentum would be infinite. In its lowest energy state, the so-called *ground state*, the system must have residual kinetic energy and an uncertain position around its point of equilibrium. This blur of position and momentum, corresponding to the minimum energy state compatible with the uncertainty relations, defines what are known as the oscillator's *zero-point* fluctuations.

In the moving-slit thought experiment, these inevitable fluctuations make observation of the fringes incompatible with detection of the slit's motion, per Bohr's principle of complementarity. These limitations apply only to ultra-light, atomic-sized objects. For macroscopic bodies consisting of a gigantic number of atoms, the smallness of the Planck constant makes quantum fluctuations negligible. Interferometer slits do not move perceptibly and interference is observed without the particles' trajectory being known. The moving slit experiment remained an ideal thought experiment whose realization was unimaginable at the time. Things have since changed and versions of these experiments are now feasible, as we shall see.

Quantum fluctuations are also essential in describing the modes of the electromagnetic field that are also oscillators, as the study of thermal radiation has shown. In its ground state, in the absence of photons, a field mode of frequency ν presents fluctuations of the magnetic and electric fields whose energy is equal to that of a half-photon, $h\nu/2$. These so-called *vacuum field fluctuations* have detectable effects on the atoms immersed in this field. In particular, they are responsible for the Lamb shift of the atomic levels of hydrogen.

An oscillator in its ground state, whether it's a field or a mechanical system, has a totally random phase even though its energy is perfectly defined, since its number of quanta is zero. To acquire a phase, the oscillator must have an uncertainty regarding its energy, *i.e.* must be excited in a superposition of states corresponding to different numbers of quanta, distributed over an interval Δn around a mean value $\bar{n}$. In accordance with

Heisenberg's time–energy uncertainty relation, the larger Δn is, the more precise the phase of the oscillator. The fluctuation ΔE of the oscillator's energy is indeed equal to $h\nu\Delta n$, while that of the instant Δt when its phase passes through a given value introduces an imprecision $\Delta\varphi = 2\pi\nu\Delta t$ on this phase. We can then deduce from the relation $\Delta E\Delta t \geq \hbar$ that the product $\Delta n\Delta\phi$, equal to $\Delta E\Delta t/\hbar$, is greater than or equal to one. The product of $\Delta\varphi$ by the relative uncertainty $\Delta n/\bar{n}$ of the number of quanta is therefore *at least* $1/\bar{n}$. The state of the oscillator most closely resembling a classical system, having an optimally defined energy and phase *simultaneously*, corresponds to the compromise whereby the uncertainties concerning these quantities are both equal to $1/\sqrt{\bar{n}}$.

At the classical limit, the number of photons or vibration quanta is enormous and the quantity $1/\sqrt{\bar{n}}$ negligible. We return to the "certainty" of Newtonian physics, for which the amplitudes and phases of an oscillator can be defined with arbitrary precision. In the case of small numbers of photons or vibration quanta, quantum granularity is no longer negligible and the quantity $\Delta n/\bar{n}$ becomes appreciable, as do the quantum phase fluctuations $\Delta\phi$, on the order of $1/\sqrt{\bar{n}}$. We will see examples of this later.

Having lost the battle of uncertainty relations between position and momentum in 1927, Einstein returned, at the 1930 Solvay Conference, to the question centered on the relationship between energy and time. He presented to Bohr a thought experiment that seemed to violate Heisenberg's inequality $\Delta E \Delta t \geq \hbar$. Certainly this inequality was well respected in the case of a photon's spontaneous emission by an excited atom, but perhaps this was because the experimenter did not control the random emission of the light particle. Einstein conceived an ingenious device, the famous photon box described at the beginning of this chapter as Gamow built it, on view today at the Niels Bohr Institute. This box was, in principle, capable of releasing quanta of light at a well-defined instant.

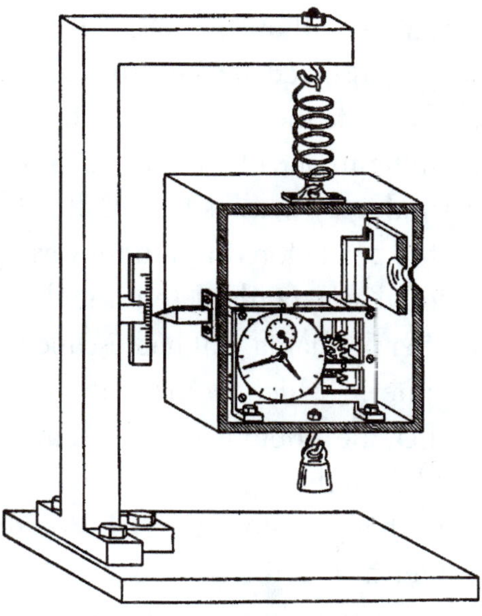

Figure V.10. The photon box designed by Niels Bohr. It was this box that Gamow "built" and offered to Einstein and Bohr for Christmas in 1930 (see Figure V.1) (Niels Bohr Archives, Copenhagen).

To measure the electromagnetic energy contained therein, the box was suspended from a spring in the Earth's gravitational field, so it sufficed to weigh the box like a common weight, observing the shift of a cursor attached to the box against a graduated ruler. For this, it was necessary to admit the equivalence of mass and energy and that of inertial and gravitational masses in general relativity.

The experiment thus combined the new quantum physics with the theory of relativity, a field of physics Einstein knew only too well. The box was supposed to have perfectly reflective inner walls that allowed the photons to bounce without loss, the energy of the system thus being conserved. To control the emission of a photon, a clock inside the box activated the opening and closing of a shutter in one of the walls for a small interval of time measurable by the position of the clock's hand. By weighing the box before the shutter was opened and after it was closed, one could determine the energy and thus the number of quanta that had escaped. There was an arbitrary time in which to make this measurement before and after the photons departed, and it seemed nothing would limit the accuracy with which one could determine the change of the energy E stored in the box during the interval t of the box opening, which could be as short as the experimenter chose.

Heisenberg's time–energy relation seemed to have a serious problem. Attendees of the conference recounted Bohr's restlessness when he could not immediately answer Einstein's question. It took the Dane a sleepless night to find the riposte that saved quantum physics: Einstein had forgotten his own theory of general relativity and neglected the dilation of time measured by the box clock!

Without going into detail, let us summarize Bohr's reasoning, which Gamow transcribed on the panel attached to his version of the photon box. The change of energy $E = mc^2$ contained in the box due to the photons' departure, equivalent to a change of mass, would lead to a force on the spring with an uncertainty ΔF proportional to that Δm of the mass we wanted to determine. This uncertainty on the force was related to the uncertainty Δp of the momentum acquired by the box, the relation between F and p being simply Newton's law whereby the variation of p per unit of time is equal to the force applied. Thanks to Heisenberg's relation between position and momentum, the uncertainty regarding p was linked to that Δz of the vertical position of the cursor measuring the box's weight.

A direct chain of relations thus linked the uncertainties Δm of the mass that we seek to determine, Δp of the momentum acquired by the box while the shutter is open, and Δz of the height of the cursor attached to the box. And it is at this point in the reasoning that general relativity intervenes. If the box's height in the Earth's gravitational field has changed, the time measured by the clock has changed as well. The pace of the clock measuring the time interval between the opening and closing of the shutter depends on z and the quantum uncertainty Δz affects the uncertainty Δt regarding how long it was open. A simple calculation then allowed Bohr to link Δm to Δt and find the uncertainty relation in the form $\Delta mc^2 \times \Delta t \geq \hbar$. General relativity had saved Heisenberg's relation! Witnesses to Bohr's triumph noted the irony of the situation: Einstein had been caught in an error in the theory of relativity, his beloved child, and it was this theory that allowed him to preserve the coherence of the very quantum physics that made him uncomfortable!

It may seem surprising *a priori* that quantum physics relied on the theory of relativity—the general one, at that—to be validated in this way, while there is still no model that combines these two theories in a general manner. But there is nothing shocking about it. In imagining his experiment, Einstein himself had defined the framework in which it was to be analyzed. The mass–energy equivalence of special relativity and the principle of equivalence between inertial and gravitational mass play an essential role here. Time dilation as a function of position in a gravitational field is an inevitable consequence, as is shown by the carousel thought experiment described in the previous chapter. Taking into account this effect on the time measured by the box's clock is therefore a perfectly justified approach, indispensable even if we want to ensure the coherence of physics. The fact that there is no theory yet available that describes the quantum aspects in the extreme gravitational conditions of black holes is not a problem for the experiment presented here, which depends only on the manifestations of general relativity in very weak field situations, where it can be described classically.

Quantum entanglement

The thought experiments of 1927 and 1930 contained the seeds of the notion of *quantum entanglement*, a concept that preoccupied Einstein for much of the 1930s and led to his final contribution to quantum physics. To interpret these experiments, we must describe as quantum objects not only the particles being studied, but also the experimental apparatus with which they interact. In the moving slit experiment, the slit and the particle form an inseparable whole whose physical parameters are strongly correlated. After the slits are passed through, the system evolves in a superposition of two distinct physical situations: one where the particle has passed through the lower slit without the upper slit moving, and one where it has been transmitted through the upper slit, which then begins to oscillate. Before the particle is finally detected downstream of the slits, at a distance from them that can be imagined to be arbitrarily large, these two parts of the system exist in a combined quantum state called an *entangled state*. For the classical mind, this entanglement has strange properties.

If the slit is observed, it will be found to be either stationary or in a state of oscillation, with equal probability between the two outcomes. In the first case we will know with certainty, even without having detected it, that the particle is described by the wave emanating from the lower slit; in the second case we will be sure it is associated with the wave emanating from the upper slit. This perfect correlation could even be ensured by modifying the experimental assembly. It would suffice, in the case of photons, to replace the detection screen by a lens forming an image of the two slits and to place detectors where these images form. We would then see that if the light slit remained stationary, the photon would always arrive on the detector coinciding with the image of the fixed slit, and the opposite if the light slit moved.

The intertwined state of two systems that have interacted and separated is very particular. Each part can no longer be described independently of the other. The quantum state of the whole carries information about the correlation, but has no information about each part considered separately. The particle will lead to one detector or the other with equal probability, and the slit will be observed to be moving or not moving with equal probability as well. Which of these results a measurement will observe cannot be predicted. The only information carried by the entangled quantum state is that these two results are always correlated. The particle *alone* is no longer in a superposition of states: it *shares* this superposition with the slit that carries information on its trajectory.

This sharing explains the disappearance of interference fringes. Even if no observer watches what happens to the moving slit, we can always assume that its state is recorded in a register that can be read at any time in the future. This simple possibility forces the particle to pass through one slit or the other depending on the result of this virtual measurement. We find ourselves in a situation not fundamentally different from that in which one slit or the other is randomly closed, in which case we understand that interference cannot occur. By the simple fact that the particle shares its quantum state with another system, the quantum superposition implying that it is on both trajectories simultaneously is replaced by a classical alternative. The particle is now associated with the wave from one slit *or* the other—no longer a superposition of one trajectory *and* the other—and the interference disappears.

Quantum entanglement thus plays an essential role in explaining the principle of complementarity. The undulatory aspect is observed only if no information on the particles' trajectory leaks into an environment apt to record this information. If such a leak has occurred, the particle under study is entangled with its environment and the interference effects are destroyed. The system is forced to follow a classical trajectory, determined by the information that has leaked into the environment. It is this phenomenon that explains why quantum particles leave traces of precise trajectories in bubble or spark chambers. Their continuous interaction with the liquid in the chamber or the potentials of the detector wires forces them to choose a trajectory and eliminates the undulatory aspect of their behavior.

Analyzing the 1927 and 1930 thought experiments led Einstein to wonder about entanglement, which resulted in a famous article written in 1935 in collaboration with his assistants Boris Podolski and Nathan Rosen, known simply as "the EPR paper." In it, Einstein and his colleagues imagine two non-interacting particles moving in opposite directions along the Ox axis. These particles are, at a given time, in a superposition of states with all possible opposite momenta. According to the laws of quantum physics and the properties of Fourier transforms, the system's spatial wave function exhibits a perfect correlation of the particles' positions with a fixed distance x_0 between them. The entangled pair describes a situation in which the information is contained in the correlations between the particles, while the position and momentum of each are undetermined. If one particle's momentum is found to be p_1, the other's is $p_2 = -p_1$. If a particle's abscissa is measured to be x_1, the other's is $x_2 = x_1 + x_0$. While the position and momentum of each particle cannot be determined simultaneously, due to the Heisenberg uncertainty principle, the quantities $p_1 + p_2$ and $x_2 - x_1$ are perfectly defined at the same time.

This situation can be compared to a classical one: consider two identical billiard balls approaching each other with the same speed before colliding head-on and moving away in opposite directions. As a consequence of the conservation laws of classical mechanics, the positions and momenta of the two balls in their rest frame remain opposite throughout their motion. The momentum correlation is the same as in the EPR situation. As for the correlation between the balls' positions, it is now defined not by the difference of their abscissae but by their sum ($x_1 + x_2 = 0$). This notwithstanding, the profound difference in the quantum situation is that, unlike the balls, which follow a well-defined trajectory, the EPR particles are, before being observed, in a superposition of fundamentally *nonexistent* states of positions and momenta. Indeed, the Copenhagen interpretation of quantum physics, supported by Bohr, asserted that positions and momenta had no physical reality before they were measured—that their values became actualized, gained meaning, only when a device equipped to measure either the velocity or the position of each gave a precise result. It was this argument that allowed Bohr, in his discussion with Einstein in 1927, to dismiss as meaningless the question of which slit the photon or electron in Young's experiment passes through.

In the EPR paper, Einstein tries again with a remark that seems sensible. If the particles are separated by a large distance and one is collected by Alice and the other by Bob, then Alice, if she wants to know the position of her particle, does not need to measure it. Without touching or interacting with it in any way, she simply asks Bob to make the same measurement on his particle and communicate the result to her by phone or radio. She will then know the position of her particle. The same procedure applies to the measurement of momentum or velocity. A measurement made by Bob, without Alice touching her particle, will give her the value of this parameter. Of course, Bob cannot measure position and momentum at the same time due to the uncertainty principle. Measuring one of the quantities will disturb the result obtained for the other.

Einstein no longer questioned this local property of quantum physics, but asserted that if the result of a measurement of any of these quantities on a particle could be known without interacting with it in any way, this value should be what he called an *element of reality*, truly existing prior to any measurement. Something must have happened at the moment of creation of the pair of particles to fix the values these parameters would take in a later measurement. The quantum uncertainty must, in one way or another, be reduced to a classical uncertainty, an ignorance of hidden parameters in the system that the measurement of one of the two particles had to determine without touching the other. This reasoning led Einstein to assert that the description quantum physics gave of the microscopic world was incomplete, that there must be hidden variables it was currently unable to describe.

The non-local behavior of physics involved in the Copenhagen interpretation strongly displeased him. It was already difficult for him to admit that "God plays dice" when it came to actualizing the result of a measurement at a point in space-time. This would be even harder to accept if the chance involved was also not local. That the random result of a measurement made by Bob should immediately affect the value of the correlated parameter observed by Alice seemed inconceivable to him.

And yet this was what Bohr asserted in an article responding to the arguments developed in the EPR paper. The physics of a quantum system made up of spatially separated parts formed an indissociable whole. The physical quantities correlated at a distance between them became real in the whole space only when a measurement was made on one of the parts of the system. This revelation of the result, the so-called *quantum collapse of the state*, was a phenomenon of non-local information acquisition. Whether the measurements were made simultaneously or in any temporal order, correlations had to be verified at all times without invoking the existence of hidden variables. Bohr intuitively understood that there could be no such variables, that their possible existence would call into question the entire edifice of quantum theory.

Not until three decades after the EPR paper was the question that divided Einstein and Bohr posed in experimentally verifiable terms. In 1964, after the deaths of both of the debate's protagonists, the Irish physicist John Bell reconsidered the question of hidden variables. Taking up the EPR thought experiment, he showed that if one measured not just two conjugated variables such as x and p, but a set of observable combinations of these quantities, one could show that a theory of hidden variables would be unable to describe the interference effects of probability amplitudes predicted by quantum physics.

To prove this, Bell established a mathematical inequality which a sum of probabilities of correlated measurement results had to satisfy if hidden variables existed, and which was violated by the predictions of quantum physics. A first attempt to test a Bell inequality was made in 1972 by the American physicist John Clauser, who measured the correlations between the polarizations of entangled photon pairs and found that the result was consistent with the predictions of quantum theory. However, the experiment was questionable because the photons' polarizations were fixed in advance, which did not exclude the possibility of a causal influence of one measurement on the other. A crucial improvement was made to this experiment in 1982 by my colleague Alain Aspect of the Université d'Orsay. Still operating on pairs of entangled photons, he randomly switched the orientation of the polarizers placed in front of the photodetectors in a time shorter than light takes to propagate from one detector to the other. His experiment confirmed that inequalities of the type envisaged by Bell were indeed violated, proving Bohr's point. This result has since been confirmed by multiple experiments of increasing precision, notably performed by the physicists Anton Zeilinger in Austria and Nicolas Gisin in Switzerland.

An argument sometimes used to deny the possibility of instantaneous collapse of the wave function at two points separated by a great distance is that this would violate the principles of relativity and causality. In fact, this is not the case; Einstein himself never used this argument to contest the Copenhagen viewpoint. The theory of relativity prohibits the transfer of information between two points at a speed greater than that of light. However, the totally random result of a non-local quantum measurement does not convey information. The information is contained entirely in the *correlations between the results* observed by Alice and Bob, and these correlations surface only when the two observers exchange their findings. This exchange must be done by classical means (*e.g.*, an electromagnetic message), and therefore respects the principle of relativity. The non-local correlations of quantum physics are strange—they constitute, as Einstein said, *spooky action at a distance*—but they do not violate any principles of physics.

Let's return to Young's double slit experiment by mentioning one last twist courtesy of John Wheeler, the American physicist who coined the expression "black hole." Reflecting on the strangeness of a world in which a physical system can be suspended between different realities prior to measurement, Wheeler imagined a version of the interference experiment where the photon does not have to choose between "going through one slit *and* through the other" or "going through one slit *or* through the other," even after the slits have been crossed! To look closer at this so-called *delayed choice* experiment, let's suppose that behind the fixed slits, both now open—there is no longer any moving slit—we place either the interference detection screen or a lens that forms the image of the two slits in a plane located further downstream, where we have positioned two detectors to count the photons forming these images.

The choice between the detector screen and the lens is made randomly and very quickly during the interval when the photons, having already passed through the two slits, propagate freely in the space inside the interferometer. In naive terms, the photons don't "know" when they cross the slits whether they should do so in a state of superposition (which will lead to the observation of fringes if a screen is later placed in their path) or according to a trajectory passing through one slit or the other (which will certainly lead them to one of the two detectors if the lens is substituted for the screen).

According to the Copenhagen interpretation, the wave function of photons doesn't have to choose. The dilemma of whether there are defined trajectories or waves does not exist until the photon is asked to answer a specific question by randomly choosing one device or the other, even after the photon has passed through the slits. Depending on the question asked, we will always observe either a dichotomous result (the photon shows up on one detector or the other) or an undulatory behavior (the photon arrives only on a bright fringe, never a dark fringe, of the screen placed in its path).

The experiment proposed by Wheeler in the 1980s was carried out in 2007 by Jean-François Roch in a different form, with a different type of interferometer, but following the same principle. It gave a result fully compatible with the Copenhagen interpretation. Giving a quantum particle a delayed choice never puts it in the wrong. It is always "ready," no matter the version of the device it will finally confront, to give a result in agreement with the strange predictions of quantum physics.

Schrödinger's cat and the classical-quantum boundary

We can't leave these thought experiments behind without mentioning the one that remains most famous of all: Schrödinger's cat. In a 1935 article in which he explicitly discusses the concept of entanglement for the first time (he was also the first to name it, *Quantenverschränkung* in the original German), the Austrian physicist imagines in passing the fate of a cat locked in a box with an excited atom. The atom can revert to its ground state by emitting a radiation, photon, or radioactive particle that triggers a device capable of killing the cat (opening a vial containing poison or shooting a gun at the poor creature, it doesn't matter). The fatal event can occur at any moment during the lifetime of the excited state, which we assume to be very long. During this interval, and before the box is opened to see what has happened inside, quantum physics describes the state of the atom as a superposition of the excited state and the ground state in the presence of the emitted particle. The cat interacting with the atom must therefore also be in a superposition of the state where it is alive and the state where it is dead due to the process triggered by the killer particle.

The question is whether the Copenhagen interpretation is valid in this extreme case. Must we admit that the cat's state has no physical reality before it is observed, that the poor animal is suspended between two very different classical realities that are liable, moreover, to give rise to interference phenomena? Schrödinger thought not, and described this experiment merely as an example of the ridiculous consequences that occurred when the predictions of quantum theory were taken at face value.

The situation of the woebegone cat is not fundamentally different from that of the moving slit in Young's experiment. Like the cat, it is suspended between two realities (it oscillates, and at the same time it doesn't) under the effect of its interaction with the photon passing through the two slits. To Schrödinger's credit, he associated quantum strangeness and the notion of entanglement not with the emotionally uninteresting fate of a banal oscillator attached to a small moving slit—but with that of a living animal, a cat, whose fate will surely be of greater concern to the reader.

Figure V.11. An illustration of the Schrödinger's cat thought experiment (Wikimedia Commons).

In addition to its usefulness in defining entanglement, the experiment raises the question of the boundary between the classical and quantum worlds. All of the thought experiments described in this chapter hypothesize that the devices used are sensitive to quantum phenomena, to the finite value of Planck's constant. To recoil at the mere touch of a photon or an electron, the moving slit of Young's device must be very light, and the same applies to the photon box if it is to recoil measurably when a single photon escapes it.

Beyond a certain size, quantum effects get blurry and the entanglement must somehow disappear, or at least be veiled. The same is true for the dramatic case of the cat. This erasure of quantum phenomena when the size of the systems under consideration increases is called *decoherence*. To understand it, we must analyze how quantum systems become entangled with their environment, which recent experiments have made it possible to study in detail, as we will see later.

The strange laws of quantum physics, as discovered through analysis of the thought experiments of the 1920s and 1930s, were finally accepted by the vast majority of physicists, who applied them to the study of the microscopic properties of matter, with the successful results we already know. Obeying the watchword "shut up and calculate"—as opposed to those who, like Einstein, were preoccupied with the quirks of the quantum world—the physicists of the 1930s all the way through to those of the 1980s didn't ask many questions about the interpretation of the theory. They were able to directly observe only systems containing a very large number of particles, whether atoms, molecules, or electrons. At these scales, quantum phenomena, such as quantum jumps or state superposition effects, appeared only statistically. Interpreting the wave function as a mathematical quantity providing information about the probabilities of measurement results on large samples of particles, as classical statistical physics has done since the nineteenth century, was generally sufficient.

The question of whether the notion of wave function remained relevant at the level of individual particles, as described in thought experiments, was not essential to understand the physics of atoms, molecules, or solids, or the processes of chemical reactions involving macroscopic quantities of reactants. And when it became possible to record the presence of individual particles in an experiment, it was in the form of sprays of trajectories in bubble or spark chambers, to which an obvious classical description could be attributed.

The subtleties analyzed by the thought experiments resulting from the interpretation we were trying to give the wave function of an isolated quantum system remained beyond experimental reach. This is no doubt what Schrödinger meant when he described the results predicted by quantum physics examining individual particles as “ridiculous consequences.” For many physicists, such experiments would remain impossible forever, and there was no need to further question what they really meant. The situation changed in the 1980s, when technological advances made it possible to manipulate isolated quantum objects in the laboratory, and the "naked" quantum revealed itself in experiments that will be the subject of the following chapters. The laser played a key role in these experiments. Once again, light illuminated physics and allowed it to move in new directions, directly revealing the strange world that thought experiments had made it possible to imagine.

Before turning to these experiments, let's pause for a moment, at the dawn of the modern age, to consider the history of light described in the last four chapters. The center of this history, we must note, has shifted over time. While the great discoveries of the nineteenth and early twentieth centuries took place mostly in Europe, the United States has become increasingly important over the years as the epicenter of knowledge building. The rise of America's industrial and commercial power after its Civil War was accompanied by a boom in science, illustrated by the creation of the National Academy of Sciences in 1865, two centuries after the Academies of London and Paris were founded.

The invention of the light bulb by Thomas Edison and that of the telephone by Alexander Graham Bell—as well as, at a more fundamental level, Michelson's interferometric experiments and Gibbs's theoretical work in thermodynamics—showed by the end of the nineteenth century that America was a force to be reckoned with in the advancement of science and technology. Closer to our concerns, Millikan's experiments on the photoelectric effect and Davisson and Germer's experiments on electron diffraction played an instrumental role in the emergence of quantum physics concepts. We must also add those of the American Arthur Compton on the scattering of X-rays and gamma rays by electrons in 1922, which finally convinced physicists that photons were real.

Nonetheless, in the 1920s, the preeminence of Europe, particularly Germany, remained quite strong. It was still a time when students and young researchers traveled eastward across the Atlantic to be educated in Europe. The father of magnetic resonance, the American Isidor Rabi, came to Hanover in the early 1920s to learn molecular beam technology from Otto Stern. A few years later a student of Rabi's, Robert Oppenheimer, who would become the father of the atomic bomb, made the same trip to Germany to work with Max Born and Werner Heisenberg.

The rise of Nazism and the persecution of Jewish scientists in Germany caused science's center of gravity to shift westward across the Atlantic in the early 1930s. Einstein's departure from Berlin and his move to Princeton in 1933 was a symbolic step; he was followed by many scientists, including Stern, Fermi, and Gamow, all of whom we have met in this chapter. The end of World War II began a relative reversal. America remains preeminent, but the European cradle of science has once again become a radiant center of research, on light as on many other fields. France, following the strong tradition it has developed in optics since Fresnel, is a major player in Europe.

Another field of competition is now emerging with the scientific development of countries in the Far East, with China in the lead. Modern science has been a borderless global activity since its birth in the sixteenth and seventeenth centuries. Its worldwide development is proof of this. From the outset, the means granted scientific research have reflected the economic power and ambitions of the states that support it. It is to be hoped that, in this increasingly fierce competition, the Old Continent on which science was born will have the will and the wherewithal to keep playing an important role in the great adventure that is the acquisition of knowledge.

CHAPTER 6

Lasers, photons, and giant atoms

In 2012, my colleague and friend David Wineland and I shared the Nobel Prize in Physics for "groundbreaking experimental methods that enable measuring and manipulation of individual quantum systems." The experiments thus recognized were laboratory realizations of some of the thought experiments imagined in the 1920s and 1930s by the founders of quantum theory, which I described in the previous chapter. The early studies by our two research groups, one in Paris and the other in Boulder, Colorado, date back to the 1980s. Half a century had elapsed since Einstein's and Bohr's discussions at the Solvay Conferences. Advances in technology, in particular the development of lasers, had made possible what the great physicists of the early twentieth century had only dreamed of: juggling with individual quantum particles in order to directly display the strange logic of the world on a microscopic scale.

For my part, the adventure had begun some twenty years earlier, during the formative years described in Chapter 1. As with all fundamental research, it is first and foremost curiosity and the desire to understand nature that have motivated my research group and myself. And the road we have followed has been one of twists and turns, lined with unexpected results, surprises, and sometimes failures. Even if the general direction of our research was clear from the outset—to use light to deepen our knowledge of the atomic world—the goal we finally reached, manipulating and measuring individual photons without destroying them, emerged only gradually after remaining uncertain for a long time. This is the beginning of the long quest that I describe in this chapter.

An atom dressed by photons

The adventure begins with my thesis work, supervised by Claude Cohen-Tannoudji. I outlined the experiments that initiated my research career in Chapter 1. I used the light of a classical lamp to orient the angular momenta (more simply called *spins*) of mercury atom nuclei and to study the way they danced when subjected to a radiofrequency field. Each spin has two states, + and -, of energy E_+ and E_-, corresponding to an orientation of the magnetic moment parallel or antiparallel to the direction of the static magnetic field B_0 applied to the atoms. The difference in energy $E_+ - E_-$ is proportional to the amplitude of the field B_0, and the frequency of the corresponding transition $v_0 = (E_+ - E_-)/h$, known as the *Larmor frequency*, is equal to 760 hertz for a field of 1 gauss or 10^{-4} tesla (the Earth's magnetic field being on the order of 0.6 gauss).

Another magnetic field B_1, rotating in the plane perpendicular to B_0 at the Larmor frequency v_0, induced the flipping of the spins, initially oriented in the + state toward the - state, then their return to the + state following a sinusoidal evolution known as *Rabi oscillation*. This periodic phenomenon is named for the American physicist who first observed it in the 1930s, when he discovered magnetic resonance. The Rabi frequency ν_R is proportional to the amplitude of the radiofrequency field B_1 and should not be confused with the Larmor frequency ν_0, which is proportional to the amplitude of the static field B_0.

The spin of a particle is a quantum observable that can take only discrete values for each atom, and the same is true of the radiofrequency field, which in these experiments consists of photons of very long wavelengths (300 kilometers for a frequency of 1 kilohertz). In spite of the fundamental discreteness of matter and field, the physics of the spins' dance could be described classically. Indeed, we observed billions of spins evolving together in our resonance cell, and the radiofrequency fields inducing this evolution transported billions of billions of photons per second. Any quantum granularity was totally unobservable. The magnetic resonance experiments carried out since Rabi's pioneering work had until then been described using classical images.

Claude, however, had an intuition that it was worth taking the granular aspect of our experiments seriously, and he asked me to describe them by making explicit the quantum states involved for both atoms and fields. Using Bohr's elementary image describing the energy exchange between an atom and the electromagnetic field, we focused first on what happens to a spin in the higher-energy + state in the presence of the photon vacuum. We noted the global state of the atom and of the field with the symbol |+,0>, the first term in the | > bracket referring to the state of the atom and the second to that of the field (0 photon). This notation is the one Paul Dirac introduced in the 1920s to define generally a quantum state. When the atom flips from + to - by emitting a radiofrequency photon, the system evolves from the state |+,0> to the state |-,1>, the notation being explicit.

The elementary Rabi oscillation corresponds in this case to a reversible flip-flop between the two states |+,0> and |-,1>. These are not well-defined energy states of the global atom-field system. If they were, they would remain in these states indefinitely, per the uncertainty relation between time and energy. This problem's stationary states, the so-called *energy eigenstates*, are the combinations $|H_0+> = |+,0> + |-,1>$ and $|H_0-> = |+,0> - |-,1>$, superpositions of the two states |+,0> and |-,1> with opposite relative probability amplitudes. The two states $|H_0+>$ and $|H_0->$ have energies separated by an interval ΔE_0 corresponding to what is called the *vacuum Rabi frequency*, $\nu_{R,0} = \Delta E_0/h$.

When we adopt this point of view, the Rabi oscillation appears as a quantum interference phenomenon. The initial state of the global system, |+,0>, is in fact the superposition of the two energy eigenstates $|H_0+>$ and $|H_0->$. Thus, at the initial instant, the atom–field global system is suspended between two quantum states of different energies, just as, in Young's double slit experiment, the particle is, at the moment it crosses the screen pierced by the two slits, in a superposition of two states of different classical positions. When we measure the spin of the atom at a later instant t, we get the result + or - with a probability resulting from the interference of the two amplitudes corresponding to the components $|H_0+>$ and $|H_0->$ of the initial state. These amplitudes, whose phases differ at time t by the quantity $2\pi\nu_{R,0}t$, lead to a probability of detecting the atom in the + or - state oscillating in time, just as in Young's fringes they gave rise to spatial oscillations of the probability of finding the particle at different points on the detection screen. Here spatial interference has become temporal interference, but the physics is similar and its mathematical expression is based on identical equations, with the position and momentum variables in Young's experiment replaced by time and energy variables in Rabi's.

We still had a problem to resolve. The frequency of the quantum oscillation between the 0 and 1 photon states depends on an essential parameter, the amplitude of the fluctuations of the vacuum field, which determines the Rabi frequency $\nu_{R,0}$. For our extremely low-frequency radiofrequency fields, the amplitude of these fluctuations was so small that the energy interval ΔE_0 and the corresponding Rabi frequency $\Delta E_0 /h$ were minuscule, leading to extremely long spin-flipping times that were totally unobservable at the scale of our experiments. What was making our Rabi frequency measurable? In fact, we were observing the evolution of atomic magnetic moments not in the vacuum field but in a radiofrequency field containing gigantic numbers of photons.

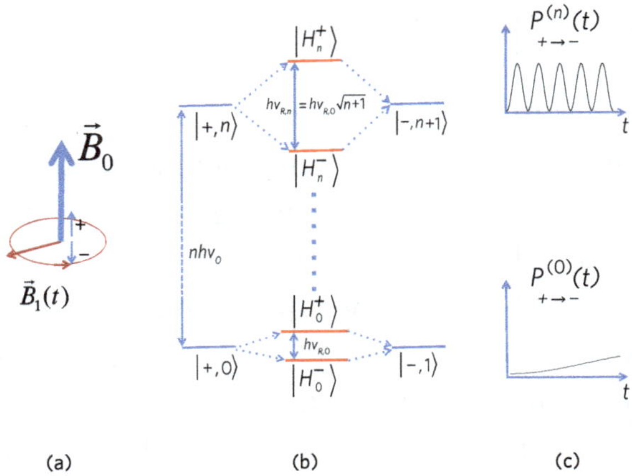

Figure VI.1. The Rabi oscillation in the dressed-atom theory. (a) Field configuration: the + and - spin states point in the directions parallel and antiparallel to that of the static magnetic field B_0. The circularly polarized radiofrequency magnetic field B_1 rotates in a plane perpendicular to B_0 at the Larmor frequency ν_0 and induces a spin oscillation between the + and - states at Rabi frequency ν_R. (b) Dressed atom energy levels: the coupling of the |+,0> and |-,1> states creates a doublet of "dressed" states $|H_0+>$ and $|H_0->$ in the ground state of the atom + photon system. The separation between these two states corresponds to the vacuum Rabi frequency $\nu_{R,0}$. The figure also shows a portion of the diagram of dressed-atom levels around the energy $nh\nu_0$, corresponding to a very large number n of photons. The energy splitting between the dressed states $|H_n+>$ and $|H_n->$ corresponds to the Rabi frequency $\nu_{R,n} = \nu_{R,0}\sqrt{n+1}$ (representation not to scale, as the amplification factor $\sqrt{n+1}$ of the Rabi frequency is gigantic). (c) The probability $P(t)$ of the spin flipping from the + to the - state as a function of time oscillates between 0 and 1 under the effect of the quantum interference associated with the evolution of the system in a superposition of the two dressed states in each of the doublets. In the case of the vacuum (*lower curve*), this oscillation is so slow as to be practically undetectable. Its period is observable, on the order of a hundred hertz, for radiofrequency fields containing very large numbers n of photons (*upper curve*; the time axes of the two evolutions represented here for $n = 0$ and for a large n are not on the same scale).

Given, to put it simply, a very large but well-defined number *n*, all of the previous analysis remains valid provided we now describe the evolution of the global system between the states $|+,n\rangle$ and $|-,n+1\rangle$, corresponding to the spin in the + or - state in the presence of *n* or $n+1$ photons. The energy eigenstates of the global system then become $|H_n+\rangle$ and $|H_n-\rangle$, superpositions of the two states $|+,n\rangle$ and $|-,n+1\rangle$. The energy difference of these states, $\Delta E_n = \Delta E_0\sqrt{n+1}$, varies proportionally to the square root of the number of photons increased by one unit—leading, for a very high number of photons, to a Rabi frequency $\nu_{R,n} = \nu_{R,0}\sqrt{(n+1)}$ on the order of a hundred hertz in the experiments we performed.

The increase of the Rabi frequency with the number of photons present in the field reflects the phenomenon of stimulated emission that Einstein discovered in 1916 (see Chapter 4). The larger the incident field on an atom, the greater the probability that the atom will subsequently emit an additional photon in this field. The formula $\Delta E_n = \Delta E_0\sqrt{n+1}$, in fact, expresses the phenomenon of field amplification, which in a different context is responsible for the laser effect.

By describing the atom and the field with which it interacted as a global quantum system, we introduced to magnetic resonance what is called the *dressed-atom* theory. The atom was said to be dressed by the cloud of radiofrequency photons surrounding it. Claude thought of this name via an analogy with quantum electrodynamics, which describes the properties of electrons and atoms as resulting from a *dressing* of these particles by the fluctuations of the vacuum field in which they're immersed (see Chapter 1). In the case of our spins, the dressing was not spontaneous—it was induced by the fields to which we subjected our atoms—but the idea was analogous. Claude had already evoked this analogy in his thesis, which he had defended a few years earlier, when he discovered that a light field non-resonant with an atomic transition and containing a large number of photons modified the position of atomic energy levels in the same way vacuum field fluctuations did in the *Lamb shift*. These *light shifts* were already a manifestation of the dressing of atoms by photons.

In developing this model, I was merely adapting it to radiofrequencies and generalizing it to various situations depending on the diverse orientations that could be given to the static field B_0 and to the polarization of the radiofrequency magnetic field B_1 dressing the atomic spin. The method was very general. For each configuration of the fields, we had to determine the proper energy levels of the dressed atom and to describe the experiments as interferences between probability amplitudes associated with the paths the system could follow by evolving in superpositions of states between these levels (the dressed atom was called in French *l'atome habillé*, hence the capital H that I use above in naming $|H_n+>$ and $|H_n->$ the energy levels of the global atom–field system). The synthetic image of the diagram of dressed atom energy levels immediately pointed to new effects that the classical description of magnetic resonance did not reveal as clearly. Guided by this model, I discovered several types of new resonances and field dressing effects on atoms that had never been observed before.

I defended my thesis in June 1971. Anatole Abragam, the French pioneer of nuclear magnetic resonance, was on the jury. Claude admired this elegant and caustic scientist, a professor at the Collège de France, who had taught him quantum physics and the principles of nuclear magnetism fifteen years earlier in the courses he gave at the CEA (French Atomic Energy Research Center). After the defense lecture, in which I presented my explorations of the atom dressed by radiofrequency photons, Abragam asked me with his characteristic ironic smile and elegant language why on earth I had bothered to develop the whole arsenal of quantum theory in order to describe effects explained perfectly by the classical treatment of the electromagnetic field, one he himself had exposed so well in his definitive book, *The Principles of Nuclear Magnetism.*

The question was largely rhetorical and asked for the benefit of the audience at the defense, since Claude and I had already pleaded the cause of the dressed atom before him, explaining how our synthetic approach, which treated atoms and the field symmetrically, had allowed us to shed new light on magnetic resonance, making fruitful connections with the theory of quantum electrodynamics, and how this viewpoint had guided our exploration of unexpected phenomena. Pleased that Abragam had given me the opportunity to repeat these arguments in front of the entire audience, I did so. Abragam then extended his question by asking if I could imagine new effects for which a quantum description of the field would not simply be a useful artifice, but prove necessary. In other words, were there intrinsically quantum phenomena, in which the granularity of the radiation quanta would manifest itself in a way that a classical description of the field could not explain?

For this, I answered, it would be necessary for the number of photons dressing the atoms to be reduced to a few units, such that the difference between n and $n+1$ photons became significant. The ultimate ideal situation corresponded to the case of Rabi oscillation between 0 and 1 radiation quantum, which at the time was totally theoretical. At least twenty orders of magnitude were missing if one wanted to make the coupling between the atom and the field strong enough to allow this oscillation to occur in an interval not exceeding the duration of a feasible experiment, even just the few years that could be devoted to a thesis.

I am reconstructing here an exchange that's almost half a century old and perhaps, in retrospect, rephrasing it under the influence of what I have learned and done since. Nevertheless, Abragam's challenge to explore the phenomena that make it necessary to quantize the electromagnetic field stimulated and guided my research from that day forth. I might summarize what followed in a lapidary way by saying that it has been, for me, a long hunt for orders of magnitude, for a way to shorten the period of Rabi oscillation in the vacuum from a few billion years to a few microseconds. The instrument that would make this exploration possible was the laser, which I had not used in my thesis and with which I thus had to familiarize myself. So, after my military service, I decided to go to Stanford University in California for a postdoctoral fellowship under Arthur Schawlow, one of the inventors of this promising new light source.

Introduction to lasers in California

Indeed, the laser would offer several decisive advantages, compared to traditional light sources such as the sun or incandescent or discharge lamps, to physicists wishing to explore the quantum world: the intensity and directivity of its radiation, its monochromaticity and temporal coherence, and finally its high intensity. Numerous fundamental physics experiments in the last half-century have exploited these properties, combined or in isolation, to gain ten orders of magnitude or more in the precision and sensitivity of their measurements, in their temporal resolution, or in the temperature range of the phenomena explored. Such quantitative leaps have led to the discovery of qualitatively new phenomena and opened up perspectives we could not imagine at the time.

The monochromaticity and coherence of laser radiation, its propensity to oscillate without amplitude or phase perturbation for an enormous number of optical periods, comes from the fact that the atoms or molecules of the light-amplifying medium produce photons in unison, in the same field mode. This phenomenon of stimulated emission, discovered by Einstein, is to be contrasted with the spontaneous emission of a conventional source, in which the atoms independently emit photons with randomly distributed frequencies, emission directions, and polarizations.

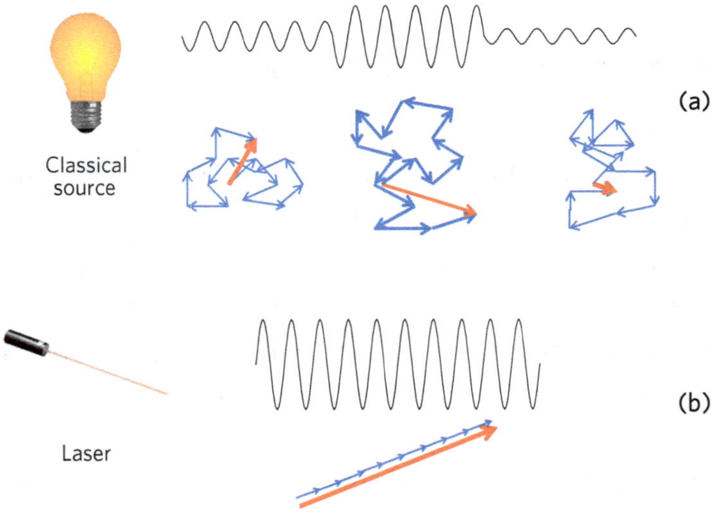

Figure VI.2. (a) Radiation from a classical atomic source: the vector representing the total field in the Fresnel plane is the sum of steps in aleatory directions, constituting a random walk. The atoms change phase rapidly and independently of each other (three successive configurations of the random walk are shown), producing rapid jumps in the amplitude and phase of the light. (b) Laser radiation: atoms emit in unison with the same phase. The field results from a succession of steps all oriented in the same direction, producing a field of large amplitude and stable phase.

To return to the representation introduced by Fresnel (see Chapter 3), the vectors associated with the amplitudes of the fields emitted by the different atoms of a laser are all oriented in the same direction in the Fresnel plane and thus add up by constructive interference, whereas in a classical source they are oriented in different directions, the global field resulting from the addition of elementary vectors constituting a random walk in this plane. This explains why laser light has a higher intensity than conventional light. As for the temporal coherence of the lasers, it is due to the fact that the atoms of the laser medium maintain the same phase of radiation for a long time, giving the Fresnel vector of the resulting field a stable direction, whereas in a conventional source the wave trains emitted by the different atoms change phase after a very short time, on the order of the lifetime of the excited atomic levels. This leads to a rapid change in the distribution of the vectors associated with the different atoms in the medium, and to random amplitude and phase jumps in the resulting field.

For the phenomenon of stimulated emission to become dominant, the number of atoms or molecules in an excited level must reach a critical threshold such that the amplification of light by the excited atoms outweighs the absorption by the atoms remaining in the ground state. The excitation is generally produced by an electric discharge or an intense classical light source illuminating the amplifying medium. To lower the laser oscillation threshold, this medium is usually placed between two reflecting mirrors facing each other, allowing the light to pass through the medium a large number of times and thus lengthening the path over which it is amplified. The mirrors constitute an optical cavity of length *L* in which the field is built up by constructive interference when *L* is equal to an integer number of half wavelengths.

Laser technology has evolved considerably over time, leading to the development of various sources seeking to optimize one radiation property or another for specific applications. Lasers with gaseous, liquid, or solid amplifying media have been developed and diversified. Today their size and power span an extremely wide range, from gigantic lasers several tens of meters long, which very briefly produce intense light outputs equivalent to those of several nuclear power plants, to the tiny laser diodes in our CD or DVD players, whose power is a few milliwatts.

Lasers were still in their infancy when I arrived at Stanford in 1972. Two qualities of these light sources were particularly sought after in atomic physics: their monochromaticity and their tunability, which made it possible to scan a radiation of a well-defined frequency over a wide spectral range in order to make the frequency of the light coincide with those of the transitions between levels of the atoms or molecules being studied. The first lasers offering this possibility had just been developed in Schawlow's laboratory by Theodor Hänsch, a young German researcher whose first name had been Americanized to Ted. These lasers used organic dye molecules, diluted in a small glass cell filled with alcohol, as an amplifying medium. The cell was illuminated by flashes of ultraviolet light produced by a pumping laser in which nitrogen gas was excited by brief electrical discharges.

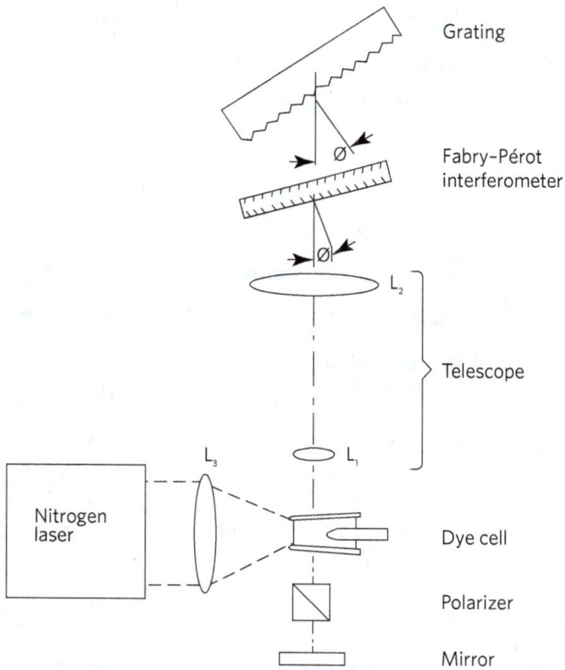

Figure VI.3. Diagram of Ted Hänsch's first dye laser. A nitrogen laser pumps the cell containing the dye. The laser cavity is closed at the bottom by a mirror and at the top by a grating that selects the wavelength. A telescope between the cell and the grating widens the beam incident on the grating, increasing its frequency resolution. A Fabry–Pérot interference filter narrows the spectral width of the laser. An optional polarizer defines the polarization of the light beam (after T. Hänsch, *Applied Optics* 11 (1972), 895).

One of the mirrors of the dye laser was replaced by a dispersive grating consisting of fine striations drawn on a glass substrate. The wavelength of the dye laser was roughly adjusted by rotating the grating around an axis perpendicular to that of the laser cavity with a micrometer screw, which changed the color of the light returned by the grating to the amplifying cell (the property of dispersion of light by a grating according to its color, discovered by Young, is discussed in Chapter 3).

To narrow the spectrum of the light, a thin blade with parallel and reflective faces was inserted into the laser cavity. This plate, called a Fabry–Pérot interferometer, filtered by interference effect the wavelengths whose half-integer multiple coincided with its thickness. As we have seen in Chapters 2 and 3, this chromatic property of thin plates was discovered by Newton, then explained by Young and Fresnel. When the grating was rotated to make a rough adjustment of the light frequency, this rotation had to be synchronized with that of the Fabry–Pérot filter to change the length of the light path through the blade and drive the filtered frequency continuously.

This produced a continuous sweep over a wavelength range of several tens of angstroms. Changing the dye made it possible to cover the entire visible or near-infrared spectrum. The laser's frequency could also be doubled by sending the light beam through a crystal that responded nonlinearly to light irradiation. This phenomenon was discovered shortly after the invention of the laser in the 1960s. Intense laser light at frequency ν induced oscillating dipoles with a Fourier component at the frequency 2ν in the atoms of this crystal, radiating a harmonic field in the direction of the incident laser beam.

The generation of this optical harmonic extended the laser's radiation spectrum into the ultraviolet, thus broadening the range of possible spectroscopy experiments. To record a spectrum, one could measure the intensity of the light transmitted by the cell containing the gas under study and detect the decrease in signal due to the absorption of light by atoms when the laser frequency coincided with an atomic transition. It was also possible to collect fluorescence light from the atoms in a direction normal to the laser beam, which increased significantly each time the light resonated with the atoms.

Ted masterfully juggled the various optical elements, mirrors, gratings, interference filters, and frequency-doubling crystals, adjusting them with precision and modifying their arrangement to reduce the spectral width of his lasers' light as much as possible. I admired his experimental skill, his imagination, and his intuition, all of which would lead him to make great discoveries over the next thirty years. These first tunable lasers were as primitive in comparison to today's instruments as the flying machines of the early twentieth century were to modern airplanes. All of their elements were controlled manually. The device was noisy, accompanied by a throbbing tick-tock from the flash of the pumping nitrogen laser. Breakdowns were frequent, and it was out of the question to let the instrument record data automatically in the absence of the students piloting the experiment, who had to constantly monitor that everything was working well—this being a time when computers were not yet capable of taking man's place in operating machines.

I was also struck by the close relationships established at Stanford between university researchers and engineers working at private companies to develop the first commercial lasers. This collaboration between academia and private industry brought together all the elements that would, a decade later, lead to the birth of Silicon Valley. It quickly resulted in the development of commercial lasers, which were more robust and more practical than the first do-it-yourself laboratory instruments. These devices were quickly perfected. The short-pulse dye pumping by a nitrogen laser was replaced with a continuous excitation provided by an argon atom laser. These continuously tunable sources produced a much more monochromatic radiation than the pulsed lasers of the first experiments, making the measurements all the more precise.

In this race for precision, Stanford researchers had a decisive advantage over their counterparts in Paris, for they were the first to benefit from the technical advances of the nascent laser industry. The first prototypes were sold to them at a discount, or simply given on loan. The spectacular experiments they carried out with these devices, then published in the most prestigious physics journals, served as excellent publicity. For both researchers and salespeople this collaboration was, as they say, a win-win situation.

I remember the enthusiasm and dynamism of those days in Schawlow's laboratory. He was a warm, good-natured, humorous man leading a team of young researchers who worked tirelessly under the pressure of intense international competition. But the atmosphere remained relaxed, and practical jokes often lightened the mood. To wit, Schawlow realized one day that the dye dissolved in the trembling gelatin that he considered a breakfast treat could be the amplifying medium for what he named the world's first edible laser. An experiment was soon conducted, and an article on the subject published in a serious scientific journal. This was unimaginable in the austere laboratory of Kastler and Brossel, where there was no lack of humor but where such an experiment would never have been considered. To be fair, we didn't have the same taste in food as Americans, and our imagination would never have led us to invent such an edible laser.

Beating the Doppler effect

Jokes aside, the Stanford researchers at that time performed many firsts in atomic and molecular spectroscopy. I will mention only those that achieved high resolution by eliminating the great enemy of atomic spectroscopy: the Doppler effect. For an atomic or molecular resonance spectral line to be narrow, it's not enough for the source that excites the atoms to be monochromatic. The atoms or molecules must also not move too much. If they move toward the light wave, the field seen from their reference frame is shifted toward the high frequencies. If they flee from the laser beam, on the other hand, they see a field whose frequency is decreased. This well-known effect was discovered in the nineteenth century by the German physicist Christian Doppler and studied in depth by Hippolyte Fizeau, who made the first earthbound measurement of the speed of light (see Chapter 3). The Doppler effect applies to waves of both light and sound: we've all heard a siren's pitch shift from high to low frequency when an ambulance or a fire truck passes us on the road.

The Doppler effect also has a simple corpuscular interpretation in terms of photons. In order to be able to absorb a quantum of light, the atom must undergo a process that conserves both the energy and the overall momentum of matter and field. If an atom of mass M moving in the direction of light absorbs a photon of momentum $h\nu/c$, its own momentum p must increase by this amount and its velocity $V = p/M$ increases by a small amount, $\Delta V = h\nu/Mc$. Its kinetic energy $E_c = (1/2\ (MV^2)$ then increases, in first approximation, by the quantity $\Delta E_c = MV\Delta V = h\nu(V/c)$. The atom, brought from the state of internal electronic energy E_1 to an excited state of energy E_2, thus sees its total energy increased by the quantity $E_2 - E_1 + \Delta E_c$. This energy must be equal to that *of* the absorbed photon $h\nu$. Thus we have the equality $E_2 - E_1 = h\nu - \Delta E_c = h\nu(1 - V/c)$, which can be rewritten, if V/c is very small compared to 1, as $\nu = [(E_2 - E_1)/h](1 + V/c)$. Thus, a photon propagating in an atom's direction of motion can be absorbed only if its frequency ν is higher than the Planck frequency $(E_2 - E_1)/h$ by a fraction equal to V/c, the ratio of the speed of the atom to the speed of light.

This is equivalent to saying that the atom "sees" in its reference frame the oscillating field at a frequency smaller than the one it would see if it were motionless, so it is necessary to increase this frequency by the quantity $\nu(V/c)$ to make the field resonant with the atomic transition. The absorption frequency shift changes sign and becomes negative if the photon moves in the direction opposite to that of the atomic motion. The atom is then slowed down each time it absorbs a photon, an effect that will prove essential for the implementation of laser cooling of atoms. I have simplified here to those cases where the laser beam is parallel or antiparallel to the speed of the atom. It can be shown simply, in fact, that the Doppler effect depends only on the projection of the atomic velocity along the laser beam. This effect is therefore null for atoms moving in a plane perpendicular to this beam.

For atoms moving with thermal velocities at room temperature, we have already seen in Chapter 5 that V/c is on the order of 10^{-6}. The Doppler shift for a light field of 10^{15} hertz, then, is on the order of a few gigahertz (10^9 hertz). This shift is positive or negative depending on whether the atoms move against or along the laser beam. When we use a monochromatic laser to illuminate an atomic gas made up of atoms whose velocities are random, the atoms' absorption lines then extend over a frequency range of a few gigahertz, centered at the frequency that would be absorbed by motionless atoms. The Doppler effect thus results in a broadening of the absorption line, which hides within a broad structure all the fine or hyperfine components of the atomic spectra due to the interactions of electronic and nuclear magnetic moments, which are typically on the order of a few tens of megahertz at most. One way to eliminate the Doppler effect and observe these structures was to illuminate the atoms of an atomic beam, in which all atoms propagate along a well-defined direction, by a laser oriented perpendicularly to this beam. The method is efficient but requires a rather complex apparatus, with a furnace to emit atoms and an enclosure emptied of any residual gas.

Ted preferred to start with another method, operating on the atoms of a gas contained in a small glass cell but arranging for the resonance signal to be sensitive only to those atoms whose velocity in the gas was perpendicular to the direction of the laser beam. For this he used a phenomenon called *saturated absorption,* which he had discovered a few years earlier while working on his thesis in Germany.

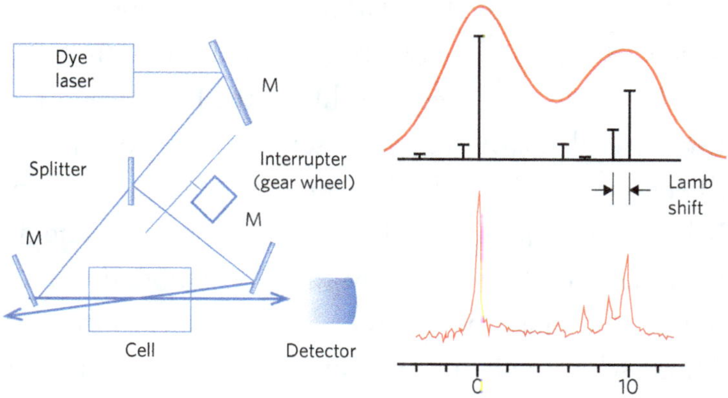

Figure VI.4. Saturated absorption spectroscopy of the hydrogen transition between the levels with principal quantum numbers $n = 2$ and $n = 3$. (a) Experimental setup: the beam of a red dye laser (tunable around 656 nanometers) is divided in two by a beam splitter so as to illuminate a cell containing a hydrogen gas H_2 with two waves propagating in opposite directions. An electric discharge dissociates the hydrogen molecule H_2 in its atomic form H and excites the atoms in the $n = 2$ level. The variation as a function of the laser wavelength of the intensity of the light transmitted by the beam propagating from left to right in the cell is measured by the photodetector. A gear wheel periodically interrupts the beam counter-propagating from right to left in order to modulate the signal and isolate by subtraction the component due to absorption saturation. (b) Spectrum observed by Ted Hänsch without a counter-propagating beam (*top*) and with this beam (*bottom*) (per *Nature* 235 (1972), 63). The vertical bars indicate the frequencies of the transitions calculated by the theory of quantum electrodynamics. While the fine and hyperfine structures are completely embedded in a Doppler width on the order of 3 to 4 gigahertz in the absence of a counter-propagating beam, saturated absorption spectroscopy clearly reveals these structures. For the first time the Lamb shift (separation of the $^2S_{1/2}$ and $^2P_{1/2}$ levels of hydrogen; see Chapter 1), on the order of 1 gigahertz, is directly observable on an optical spectrum.

The resonance cell containing the atoms under study is now illuminated by two beams propagating in opposite directions, obtained by splitting the laser beam into two waves that a mirror device sends along two opposite directions into the cell containing the gas to be studied. When the frequency v of the laser does not coincide with that v_0 of the atomic transition for stationary atoms, the light interacts with two classes of atoms in the gas. Those whose velocity projection on the laser axis is equal to $c\ (v\text{-}v_0)/v$ are excited by the beam traveling in one direction, and those whose velocity projection takes the opposite value $-c\ (v\text{-}v_0)/v$ are excited by the beam traveling in the opposite direction. If v is not equal to v_0, these two classes of atoms with opposite velocities interact with light independently of each other, contributing on the light transmitted by one of the beams to an absorption resonance line with the Doppler width.

When, on the other hand, the laser light has exactly the frequency v_0, the two velocity classes coincide in $V = 0$ and correspond to atoms propagating normally to the light beam. These atoms are then solicited by a luminous intensity twice as great as that seen by atoms interacting with only one beam. If the light intensity is high, the $V = 0$ atoms cannot absorb twice as much as atoms whose velocity along the laser beam is non-zero, resulting in a decrease in the absorption at line center. When the light intensity is so large that atoms are no longer able to respond proportionally to laser excitation, the matter–light interaction obeys the rules of what is called *nonlinear optics*, a field of light science born with the invention of the laser.

The nonlinear saturation effect of the response of the atoms moving perpendicular to the laser beam produces a slight dip in the center of the resonance profile transmitted by one of the beams. This is called the *Lamb dip*, since it was Willis Lamb, the discoverer of the Lamb shift, who first calculated it. The spectral width of this dip is ideally equal to the natural width of the excited atomic level, which as we have seen corresponds to the inverse of the lifetime of this level. This width, the one imposed by Heisenberg's uncertainty relation between energy and time, is typically smaller than the Doppler width by two or three orders of magnitude. To make the saturated absorption signal appear clearly, one can subtract from the light intensity measured in the presence of both beams the one that the cell transmits on one beam when the other is blocked. Then only the Lamb dip appears in the difference, revealing the details of the spectrum that the Doppler effect hides in ordinary spectroscopy.

Ted Hänsch was not the only inventor of this elegant method for getting rid of the Doppler effect. Christian Bordé, a young French researcher, had discovered it independently while working on his thesis at the end of the 1960s. I myself had been interested in saturated absorption and written a theoretical article describing it in collaboration with Francis Hartmann, a CNRS researcher who briefly welcomed me in his laboratory in Orsay in 1971, just after my thesis defense. I had done an internship there as a "contingent scientist" at a time when this possibility allowed young researchers to substitute work in a research lab for military service. So I was familiar with saturated absorption, at least theoretically, and fascinated by what Ted did when he had the opportunity to apply it to atoms by irradiating them with his dye lasers.

In particular, he carried out a magnificent experiment on saturated absorption of hydrogen—the simplest of all atoms, the one whose energy levels Bohr, Schrödinger, and Dirac had calculated in the early days of quantum physics. Hänsch showed clearly in an optical spectrum for the first time the fine and hyperfine structures predicted by Dirac and the famous Lamb shift, which had so far been detected only in radiofrequency spectroscopy experiments between excited atomic levels (see Chapter 1).

In this first hydrogen experiment, Ted had his pulse dye laser pumped by the flashes of a nitrogen laser. The laser's spectral width, on the order of a hundred megahertz, was limited by the inverse of the pulse duration, on the order of a few nanoseconds. Ted would later improve the method considerably by using commercial dye lasers that generated continuous light.

A few years later, he replaced saturated absorption in his experiments with another nonlinear optical method, *two-photon absorption without Doppler effect.* The experimental setup is the same as it is for saturated absorption, with an intense laser beam folding on itself, but its frequency is now tuned to coincide with *half* of the frequency corresponding to a transition between a ground and an excited state. The atom can then be excited by simultaneously absorbing *not one but two* photons in a process that conserves energy. The Doppler effect is eliminated when the atoms absorb one photon in each of the beams propagating in opposite directions. The Doppler effects of the two photons are then of opposite signs and cancel each other out. All atoms contribute to the resonance signal, not only those propagating in a plane normal to the light beams.

This method had been imagined in the early 1970s by Bernard Cagnac, a colleague of Claude's at the ENS, along with his student Gilbert Grynberg, one of my classmates during my thesis years. A Russian researcher, Veniamin Chebotaev, had also suggested it independently. Hänsch and Cagnac's students in Paris would go on to make remarkable use of it to refine the study of the spectrum of the hydrogen atom. Ted continued these experiments until very recently. To give an idea of the progress made since the time I'm describing here, the widths of the lines obtained by Ted and his team are now on the order of only a few hertz, for transition frequencies on the order of 10^{15} hertz. Not only is there no longer any Doppler effect, but the natural width of the resonance line is extremely small, because it corresponds to the excitation of a level whose lifetime is very long (on the order of a tenth of a second).

Quantum beats

In my year at Stanford I didn't just learn laser techniques and admire Ted's experimental prowess. Arthur Schawlow provided me a dye laser lab and asked me to mentor one of his students, Jeffrey Paisner. For the first time I was free to determine my own research topic, with the added responsibility of training a young researcher in my turn. I decided to observe the time dependence of the fluorescence light emitted by cesium atoms excited by the short pulses of my dye laser.

When a sample of atoms is brought at a given instant to a single excited state of lifetime τ, it is expected to return to the ground state by emitting light whose intensity decreases exponentially with a time constant equal to τ. This time is on the order of a hundred nanoseconds in the case of the excited cesium levels I proposed to study. These levels are in fact grouped in hyperfine structures, with separations between neighboring levels on the order of a few tens of megahertz. These structures are caused by the interaction between the electronic magnetic moments and the spin of the atom's nucleus. A laser pulse of duration T of a few nanoseconds has a spectral width at least equal to $1/T$ and simultaneously excites all the levels of the hyperfine structure in a window of energy $\hbar/T$. By observing the fluorescence light emitted following this excitation in a given direction and with a given polarization, we expect to observe a decrease in light intensity with modulations at frequencies corresponding to the energy differences between the simultaneously excited hyperfine levels. This is typically a quantum interference effect. After each laser pulse, the atoms follow several paths at once, passing through the different excited levels before falling back to the same ground state. The probability amplitudes associated with all paths interfere provided that no information is available about the path followed by the atoms. As in Young's double slit experiment, each atom is in a superposition of states at the moment of excitation, and the question of whether it is in one state or the other is meaningless.

Jeff and I set up the experiment quickly, and after a few feverish weeks we began recording the fluorescence signals on an oscilloscope that averaged thousands of traces induced by successive pulses of the exciting laser, repeating at a rate of ten excitations per second. At the beginning of each experiment we observed a simple decreasing exponential, a large and noisy trace. After a few minutes, we began to see a small modulation in the noise that gradually became clearer until, after half an hour, the modulated signal we had been waiting for appeared clearly. It is difficult to describe the joy of these first moments when a secret of nature is revealed. We had calculated the signal we were waiting for and the atoms in the small glass cell responded to the light pulses we sent them, exactly as predicted.

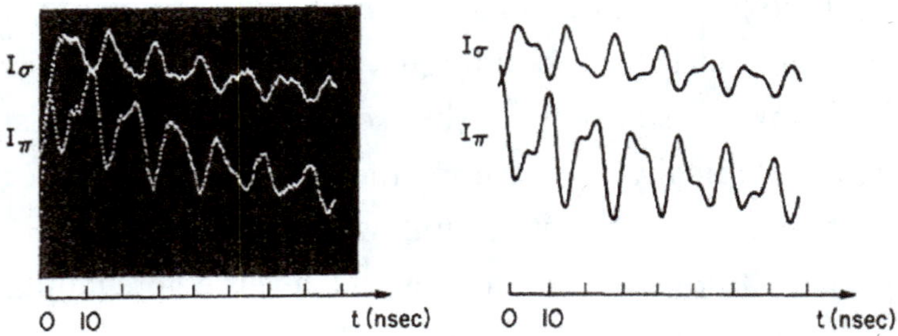

Figure VI.5. Quantum beats of the cesium atom. (*left*) Signals detected on the fluorescence light emitted with two orthogonal polarizations after a brief laser excitation preparing the atoms in a coherent superposition of three hyperfine sublevels. Each signal is the sum of three damped sinusoids. (*right*) The theoretically calculated signals. The excellent agreement between experiment and theory demonstrated that this method could be applied to the determination of unknown fine or hyperfine structures (per *Physical Review Letters* 30 (1973), 948).

The modulated signature that the atoms revealed to us was unique, just like an individual's fingerprint. It appeared as a sum of damped sinusoidal curves. Its Fourier transform, which a computer quickly revealed to us, allowed us to find the frequencies associated with the transitions between the different hyperfine levels excited by light. This was a demonstration experiment that taught us nothing we didn't already know about the cesium spectrum, as these structures had already been measured by other methods.

The *quantum beat* phenomenon, as it is called, was already known in other experimental contexts. We had simply shown that these *beats* could be easily induced by a pulsed dye laser, paving the way for a new method of optical spectroscopy that would prove useful later on in measuring other, previously unknown structures. I can't say exactly what led me to study this particular phenomenon. The reasons are probably multiple. While working on my thesis, I had the opportunity to chat with George Series, an Oxford professor interested in the light beats induced by classical lamps whose light was chopped to produce pulses of a few microseconds, a bit like in Fizeau's nineteenth-century experiments. The duration of the pulses allowed only very slow modulations of atomic fluorescence to be observed. Remembering these experiments, I thought that laser light, much more intense and concentrated in time, would be an ideal new source with which to take them up and give them a practical use.

I was also seduced by the idea that this spectroscopic method was complementary to the one Hänsch was implementing in the laboratory room next to mine. He was studying the frequency response of atoms excited by a laser whose wavelength he was scanning, while I was revealing their temporal response to a light percussion. The fact that the light from these early dye lasers was made up of short pulses was a flaw in Ted's experiments, since it limited the narrowness of the lines he was able to observe. For me, on the other hand, it was an asset, because I could cover a wider spectral range and excite more levels the shorter the duration of the light pulses.

Though this method was called *quantum beat*, it could still be interpreted by classical images because the signals observed were the average of the responses of billions of atoms excited in the cell. Each atom, however, obeyed the strange logic of quantum physics. The beat could indeed only be the sum of individual atomic phenomena. The light coming from two different atoms, flying at different speeds in the cell, could not interfere, if only because the Doppler effect, different for the two atoms, would scramble the corresponding beat. Finally, any attempt to determine which excited level an atom had passed through would make the modulation disappear. For example, one could try to place an interference filter in the laser beam to narrow the spectrum in order to excite only one hyperfine level at a time. This would have the effect of lengthening the duration of the light pulses, and the beat would disappear. The polarization of the light could also be adjusted to change the relative probabilities of excitation of the hyperfine levels involved. Here again the contrast of the modulations was affected so as to always respect the principle of complementarity.

The greater the information acquired about the path followed by each atom in the experiment, the lower the contrast of the observed oscillations. It was thus possible, by discussing the physics of this very simple system, to reproduce all the arguments of Young's double slit thought experiment. This study gave me confidence in my ability to explore nature in an original way. It was the first time I designed and conducted a personal research project. I had fulfilled my implicit contract with Arthur Schawlow and Jeff Paisner, and was beginning a long friendship with Ted.

Californian anecdotes

This first visit to California also reinforced my strong feeling of having joined a privileged community of researchers who, despite their different cultures and histories, were driven by the same curiosity. This community was rooted in the history of the physics of light and electromagnetism; this I could feel with great intensity at Stanford. During the convivial dinners hosted by Schawlow and his wife, Aurelia, I met Felix Bloch, the father (along with Edward Purcell) of nuclear magnetic resonance. He was also the man who, in the early 1930s, had explained the structure of metals through quantum physics. He was of Dirac's and Heisenberg's generation, and had skied with the latter in the Bavarian Alps at the end of the 1920s. He showed an extremely kind interest in me as a young researcher. It was most impressive for me to talk to one of the great actors and witnesses of the scientific revolution of the twentieth century, someone who had also experienced the tragedies of history, emigrating from Europe to America when Hitler came to power, just as Einstein had. Claudine, not a scientist, was less starstruck than I and thus more at ease talking with Bloch, which surely made our rapport easier.

Claudine also befriended Aurelia Schawlow, the sister of Charles Townes, who had invented the maser. This instrument, the first to exploit stimulated emission, was the precursor of the laser in the field of microwaves (*maser* being the English acronym for *microwave amplification by stimulated emission of radiation*). I remember an anecdote Schawlow told at one of his dinners: in the summer of 1951, Townes and Schawlow shared a room in a Washington, D.C. hotel. The two friends, soon to be brothers-in-law, had come to attend a conference of the American Physical Society. With two young children at home, Charles was used to getting up early. One morning at dawn, leaving Art sleeping in the room, he slipped away to think on a bench in Franklin Square, one of the quiet green areas of the U.S. capital near the White House. It was there that the idea of the maser suddenly revealed itself to him, exploiting the phenomenon of stimulated emission of ammonia molecules passing through a microwave cavity. This was the starting point of a great scientific and technological adventure that would lead, a few decades later, to a market for lasers worth between ten and twenty billion dollars annually. Townes was awarded the Nobel Prize in Physics in 1964, while Schawlow had to wait until 1981 to receive the same honor.

In 1973 Schawlow told this story with his infectious laughter, theorizing about the benefits of family life in fostering great discoveries. As he was a bachelor without children at the time, he had slept peacefully that early morning, and thus not shared this moment of exhilaration with Charles. Claudine and I had come to California with our two young children, Julien and Judith, and I could easily understand how waking up early to a peaceful morning could be conducive to new ideas.

A few years after that revelation, Townes and Schawlow co-wrote an article announcing how the maser effect could be generalized to light, calling the future instrument an "optical maser." They had tried to make the first device of this type, but had been scooped in 1960 by Theodore Maiman, an engineer at the private company Hughes Laboratories. As for the acronym *laser*, which simply swapped the M (*microwave*) for an L (*light*), it was coined by Gordon Gould, another colorful character in this story. As a thesis student at Columbia University in the 1950s, when Townes was teaching there, Gould came up with the notion of the laser but, instead of publishing it as Townes and Schawlow did, kept it secret by writing it in a draft patent. After the invention of the laser, he mounted a long legal battle to assert his prior right to the idea and its applications. Eventually, in the 1980s—long after the time I'm describing here—he won on some points in American courts and became rich.

Townes and Schawlow had had a difficult relationship with Gould, no doubt due to their different approaches to research. They had followed the academic route, disclosing their ideas openly in a scientific journal. Gould, sensing the laser's immense industrial potential, preferred the secret path of private research through a commercial company he created specifically to build and operate the new instrument. This was why he had abandoned his thesis at Columbia, drifted away from Townes, and finally lost in the race to build the first laser. This failure was more detrimental to him than it was to Townes and Schawlow, who had taken out patents on the laser as well but also published their work openly. Instead of submitting his ideas to the reviewers of a scientific journal, Gould had committed them to the obscure legalese of a patent application, for an audience of jurists and lawyers who could not judge their true significance.

The matter was further complicated by the fact that Gould had been a Communist in his early youth, during the era of McCarthyism, and was forbidden by the FBI to communicate with the scientists and engineers in the company he had founded, who were working unsuccessfully to construct the first laser. I found this story incredible and fascinating for its mix of political, financial, and scientific intrigue. Related by Schawlow with his humor (and his evident, understandable anti-Gould bias), it revealed to me aspects of science I had not suspected, far from what I had learned in Paris alongside idealistic scientists like Kastler, Brossel, and Cohen-Tannoudji. America was indeed an astonishing world, as was also demonstrated by the Senate hearings on the Watergate affair, which American television broadcast on a constant loop in the spring of 1973.

First major international conference

That summer, it was time to think about returning to Paris. But first I was to present my work on quantum beats in Vail, Colorado, at the first laser spectroscopy conference. In the majestic setting of the Rocky

Mountains, I met all the players in this new physics. My mentors Claude Cohen-Tannoudji and Jean Brossel were there, as was Christian Bordé, who had come from the nearby city of Boulder, where he was doing his postdoctoral fellowship in the laboratory of the great laser specialist John Hall. With his team from JILA, the Joint Institute for Laboratory Astrophysics, which I had visited four years earlier on my first trip to the United States (see Chapter 1), Hall had just made the ultimate measurement of the speed of light, giving it the value c = 299,792,458 m/s, with an uncertainty of three billionths. In principle, the method was very simple. JILA researchers had measured both the frequency ν and the wavelength λ of a laser locked onto a narrow absorption line in the near-infrared spectrum of the methane molecule CH_4, using the saturated absorption technique to eliminate the Doppler effect. They deduced the value of c as equal to the product $\nu\lambda$.

The measurement of λ was made through interferometry, by counting the number of wavelengths of the molecular transition of CH_4 in a meter, defined at the time as a multiple of the wavelength of a red spectral line of the krypton atom. This amounted to a simple interferometric comparison of two wavelengths. The uncertainty was about 10^{-9}, limited by the definition of the meter, which at the time depended on the accuracy of the krypton resonance line determination, itself limited by the Doppler effect in the cell containing this gas.

More delicate was the measurement of ν, the frequency of the molecular transition of CH_4, which is on the order of 10^{14} hertz. No instrument was capable of counting this frequency directly. Starting from a low frequency in the microwave range that could be measured directly using an atomic clock, the JILA experimenters gradually synthesized higher frequencies, multiples of the initial one, amplifying the successive signals by locking lasers to different harmonics until they reached a frequency large enough to be directly compared, by recording its beating, with that of a laser locked to the methane transition. This experimental tour de force, the results of which were presented in Vail, was carried out in a huge hangar housing all the elements of this frequency chain, which had to work simultaneously to bridge the gap between the microwaves and the optical frequency.

This was the final measurement of the speed of light. It was clear from that moment on that it was futile to try to increase its accuracy, since the length standard used for the measurement presented this uncertainty. Reversing the problem, metrologists decided a few years later to set the speed of light by convention to the value found at JILA and to define the meter as the distance traveled by light in 1/299,792,458 of a second. They thus linked the unit of length to the unit of time, which is determined by atomic clocks with much greater precision, the second being defined as an exact multiple of the period of a microwave transition of the cesium atom. We had come a long way since Römer first determined the speed of light in the century of Louis XIV (see Chapter 2). As we have seen, the error back then was 30 to 40 percent, the length standard used at the time being the crudely known radius of the Earth's orbit. Three centuries later, we had gained nine orders of magnitude in our knowledge of this fundamental parameter of physics. And the story didn't end there. In the fifty years since that conference in Vail, the measurement of optical frequencies has continued to make immense progress, resulting in the gain of nine more orders of magnitude, opening up fascinating new perspectives in optical instrumentation and metrology. The gigantic JILA frequency chain has been abandoned, advantageously replaced by a *frequency comb* laser that fits in a box less than a meter wide—an extraordinary invention developed by John Hall and Ted Hänsch, for which they were jointly awarded the Nobel Prize in 2005. We will come back to this later.

The *terra incognita* of giant atoms

Back in Paris in the fall of 1973, I set out on an original experimental project using the type of dye laser with which I had become familiar in California. The quantum beat experiments had shown me that this light source made it possible to excite atoms in levels far from the ground state. Using photons of increasingly short wavelengths, or performing a "stepwise" excitation in which the atom successively absorbed two photons through an intermediate excitation state, I expected to reach levels close to the atom's ionization limit.

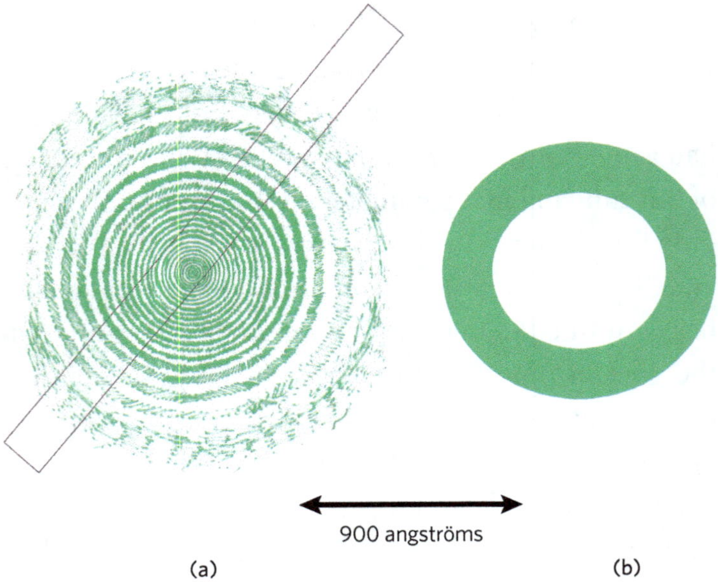

Figure VI.6. Distribution of the electron cloud in a Rydberg atom. (a) A drawing from an article I published in *La Recherche* in 1978. It represents the electron orbital of the 30S state of hydrogen of principal quantum number $n = 30$ and orbital angular momentum $l = 0$. The wave function is formed of thirty concentric undulations spaced increasingly as they move away from the proton, the invisible point in the center. The rectangle superimposed on the atom gives, for comparison, the dimensions of the tobacco mosaic virus. (b) Electron orbital of the same atom in the $n = 30$ level of maximum angular momentum ($l = n\text{-}1$) aligned along an axis normal to the figure. The electron of this circular Rydberg atom is located in a torus. It has a probability of maximum presence on a circle of radius equal to 900 Bohr radii centered on the proton (invisible at this scale). I have represented the torus surface on which the probability of electron presence is half its maximum value.

This limit corresponds to a binding energy of the peripheral electron tending toward 0. These very excited states were impossible to prepare with classical light sources, since the probability of a photon's absorption by the atom decreases very quickly as the ionization limit is approached. But lasers, with their high intensity concentrated in a small spectral range, seemed likely to make it possible to excite these atoms efficiently, opening a path to laboratory exploration in this *terra incognita* of atomic physics.

These very excited states are called *Rydberg states*, after the Swedish physicist who, at the end of the nineteenth century, established the formula giving the wavelengths of the transitions in hydrogen linking these levels to the atom's ground state. Rydberg's empirical formula, deduced from spectroscopic observations of hydrogen even when no theory could explain it, guided Bohr in his elaboration of the first quantum model of the atom. In a semantic shortcut, the atoms carried in these levels are referred to as *Rydberg atoms*. These are not new elements of the periodic table, but ordinary atoms that call for this qualifier during those moments when their peripheral electron orbits far from the atom's nucleus.

The existence of these states has been known in astrophysics since the 1960s. Intersidereal Rydberg atoms of hydrogen, helium, and carbon are formed in space when electrons are captured by ions of these elements, and are revealed by the microwaves they emit when the electron approaches the core of the atom, cascading from level to level while losing energy. After a long journey through space, these microwaves are detected by radio telescopes. Very fragile because their electron, weakly bound to the nucleus, can easily be disturbed by collisions with other atoms, these Rydberg atoms can survive a relatively long time in the near-perfect vacuum of space. Observing them in the laboratory would require, in addition to the use of lasers, special precautions.

Why did I take an interest in these atoms? Certainly the idea of doing atomic physics on an unusual scale played a role. When an atom is carried in a Rydberg state, the dimensions over which the electronic wave function extends become gigantic compared to what they are in the ground state. Since Rydberg and Bohr, the energy of these states has been labeled with a principal quantum number n (not to be confused here with the number of photons), the electron's binding energy with the nucleus decreasing as the function $1/n^2$. Conventionally, we speak of a Rydberg state when the integer n is on the order of 10 or more. The dimensions of the electron orbit increase as n^2 and there is, in principle, no limit to the value of n. A Rydberg atom with $n = 50$ has a "size" 2,500 times larger than an atom in its ground state, on the order of 0.25 micron. This is already a dimension approaching that of biological objects such as viruses or bacteria. With $n = 1{,}000$, we obtain atoms whose dimensions are on the order of a tenth of a millimeter, the thickness of a hair. If n reached 10,000, the atom would be the size of a pea or a cherry.

To keep me grounded in reality and prevent me from dreaming too much, I naturally realized that the binding energy of the peripheral electron in these artificially "inflated" atoms, varying by $1/n^2$, would become smaller the bigger the atoms were, which confirmed to me that special measures would have to be taken to protect them from disturbances. It was clear, however, that their very fragility would make them useful as ultrasensitive probes of their environment. I thus caught a glimpse of an adventure not unlike Gulliver's in the land of giants: the exploration of a world of unusual orders of magnitude, bound to hold many surprises.

Another factor attracted me to the Rydberg atoms: they were very simple quantum systems. Once the peripheral electron was promoted to a large orbit, the nucleus and the other electrons appeared as a near-point charge, almost identical to a simple proton. The fact that the atomic core is a complex system, consisting of a nucleus made of Z protons surrounded by $Z - 1$ electrons orbiting at about 1 angstrom, is no longer important to describe in first approximation the dynamics of the electron orbiting at thousands of angstroms, which simply sees the center of the atom as a point. The Rydberg atoms of all elements are thus, once n is large enough, similar to hydrogen in an excited state. All physics, in a good approximation, comes down to that of an electron moving in a central field decreasing as $1/r^2$. There are certainly small effects linked to the structure of the atomic core that differentiate the Rydberg states of different elements, but their effects are small, at least in first approximation.

I was also attracted by the analogy with astronomy, which was, as we saw in Chapter 1, my first scientific passion. The Rydberg electron orbiting the atomic core can be likened to a planet revolving around the Sun. Admittedly, it is a quantum system for which the notion of trajectory doesn't make much sense, but the classical image in terms of Bohr's orbits nonetheless remains useful for describing the system and evaluating the orders of magnitude of the different parameters that define it.

In astronomy, the planets obey Kepler's empirical laws, justified by Newton's calculations, based simply on the fact that the force of gravitational attraction varies as $1/r^2$, where r is the distance from the Sun to the planets. The same dependence with the distance to the atomic core characterizes the force the electron experiences in the atom. Kepler's laws must therefore also apply. The third of these laws says that the square of a planet's period around the Sun divided by the cube of the radius of its orbit is a constant, the same for all planets. Translated to atomic physics, this law tells us that the square of the period of revolution of the Rydberg electron on a Bohr orbit must vary as n^6, the cube of the radius proportional to n^2. We deduce that the period of the Rydberg electron varies as n^3. This period, evaluated by the Bohr model, is proportional to the wavelength of the photons the Rydberg atom emits while jumping from one Bohr orbit to the next, n changing from n to $n-1$. For $n = 50$, the wavelengths thus evaluated are millimetric, a few thousand times larger than the micrometric ones of the optical photons emitted on the transitions between deep levels of the atom. These fall in the microwave range, corresponding to frequencies on the order of a few tens of gigahertz. A simple calculation of order of magnitude, based on this astronomical analogy, made me understand that these atoms would be ultraprecise and ultrasensitive probes of microwave photons.

I was excited by the challenge of exploring atomic states that had never before been observed in the laboratory, and by the hope that these systems' unusual orders of magnitude would lead to the discovery of new effects. Perhaps there was also the sense that these systems were conceptually simple, and that the effects they manifested in their interaction with electromagnetic waves would clearly and directly illustrate the strange laws of quantum physics. But I'm most likely reconstructing what gradually emerged only after the fact. By the time I returned to Paris after my year in California, I had only a vague idea of what Rydberg atoms would bring.

To embark on this adventure, I had only to send Jean Brossel a two- or three-page handwritten letter from Stanford, explaining more succinctly than I have done here why the study of Rydberg atoms interested me. I promised no specific results, only to try and see where this research would lead me. For this I needed money for the nitrogen laser that pumped the dye lasers, as well as the optics and electronics required to build a few tunable lasers. I needed to add all the mechanics that would allow me to build an atomic beam, because the high-*n* Rydberg atoms could survive only when propagating along straight trajectories in a very good vacuum reproducing the conditions of space. I also had to be given the responsibility of supervising a student or two. Brossel trusted me and gave me what I asked for immediately, without consulting anyone, using credits the CNRS was allocating to his laboratory. Thus was I able to embark, with a free spirit, on an adventure that would last several decades.

I can't help but compare my situation back then with that of young scientists approaching research in France today. Without funding, even if they've been lucky enough to get a position at the CNRS or a university with virtually no recurrent budget, they must immediately submit a research project to be funded by a French agency (ANR, the National Research Agency) or a European agency (ERC, the European Research Council). This requires weeks of work: filling out dozens of tables, explaining in detail what they hope to find, detailing the stages of a research project that is by definition unpredictable, and imagining what their research could one day be used for.

Submitted to committees of experts and put in direct competition with hundreds of other projects, they have in principle only about one chance in ten of succeeding, and must therefore repeat this tedious exercise on average several times before they can truly get to work. This is enough to dampen the enthusiasm of the most passionate young scholars. Those who succeed are often forced to give up an ambitious, necessarily risky project for more conventional or directly "useful" research. Had I been obliged to overcome all these obstacles, which inevitably lead to overselling the project and promising results that can't be predicted in advance, I don't think I would have embarked on this adventure. Perhaps I would have chosen another job, or stayed a scientist in the United States.

Over the five years following my return to Paris, I learned to prepare and detect Rydberg atoms of alkalis (sodium, cesium, and rubidium), which possess an external electron orbiting around a core whose closed electronic layers have the compact, chemically inert structure of a noble gas. The pulsed optical excitation was performed in two successive steps, using two tunable lasers of different colors carrying the atoms in an intermediate level before reaching the final Rydberg level. I started by exciting quantum beats in levels whose principal quantum numbers ranged from 10 to 20, still moderate-sized atoms that survived well in a resonance cell. Detecting the fluorescence modulations of these atoms after their short pulse excitation allowed me to measure their fine structure intervals through so-called *spin–orbit* coupling. This is a relativistic effect, the electric field of the atomic core being transformed in the Rydberg electron reference frame into an orbital magnetic field coupled to the magnetic spin moment of this electron. Depending on whether this spin points in the direction of the orbital magnetic field or in the opposite direction, the Rydberg atom acquires a small additional positive or negative magnetic energy.

As a result, for each state of orbital angular momentum there are two levels of fine structure of slightly different energies. Their simultaneous excitation by the pulses of the dye lasers produced the fluorescence quantum beats that revealed the fine structure and made it possible to measure the magnetic field as seen by the Rydberg electron. The effect diminished as the size of the atom increased. This reflected the decrease of the electric field seen by the electron farther and farther from the nucleus, as well as the decrease of the speed of the electron moving slower and slower on a larger and larger orbit, making the relativistic effects smaller and smaller. The fluorescence of the atoms falling back to an unexcited atomic state also became weaker and more difficult to detect as we climbed up the Rydberg scale, the probability of emission of an optical photon by these states decreasing strongly with their degree of excitation. It became difficult to study atoms of principal quantum number greater than 20 with this method, so we had to turn to others.

We then prepared the Rydberg states by illuminating a beam of alkaline atoms with our dye lasers. The atoms were obtained by evaporating a small sample of sodium, rubidium, or cesium contained in a furnace with a small hole. This furnace was placed in an enclosure from which air was pumped. The atoms escaped the furnace in a straight line, forming a slightly divergent beam that propagated in a vacuum. The lasers excited the atoms in short pulses, which then flew to a detector consisting of two parallel metal plates between which an electrical potential difference was applied. The electric field between the plates ionized the atoms by tearing off the Rydberg electrons as soon as its amplitude exceeded that of the field attracting them to the atomic core. This field decreased with the degree of the atom's excitation, and this dependence allowed us to selectively detect Rydberg levels with different binding energies, with a sensitivity much higher than that allowed by the observation of their fluorescence.

The electrons escaping from the attraction of the atomic core were accelerated toward an electron multiplier consisting of successive cascading cathodes. Each Rydberg electron snatched several electrons from the first cathode, which in turn snatched several electrons from the second, and so on. This led to an important amplification of the current, finally recorded by a memory oscilloscope that stored the signals induced by the successive laser pulses, repeated at a rate of about ten per second. In our first experiments, we detected a large number of excited atoms at the same time. By decreasing the intensity of the laser light, we were gradually able to observe the signals produced by smaller and smaller numbers of Rydberg atoms. We finally managed to detect signals corresponding to the preparation of a single Rydberg atom, on average, per laser pulse.

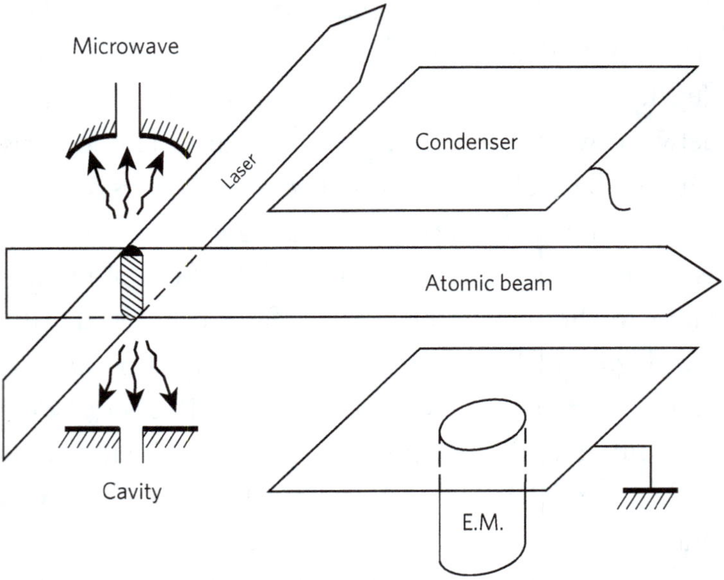

Figure VI.7. Sketch of the first experimental setup for the study of the microwave spectra of Rydberg atoms. Lasers excite an atomic beam in an open cavity formed by mirrors (*left*). After crossing the cavity, the atoms pass into a capacitor, where they are ionized by an electric field. The ejected electrons are counted by the electron multiplier (*E.M.*).

The birth of cavity quantum electrodynamics

We then began working on systematic microwave spectroscopy experiments. Between their excitation by the laser and their detection, our atoms were subjected to the field produced by a source radiating in the millimeter wave range, with frequencies on the order of a few tens of gigahertz. The microwave field was concentrated in a small cavity formed by copper mirrors in the shape of spherical caps facing each other. It propagated from the source to the cavity in a waveguide leading to a small hole drilled in one of the mirrors. The atoms moving with thermal velocities of a few hundred meters per second were subjected to the field stored in the cavity for the few microseconds they took to cross it.

The frequency of the microwave was swept along at the same time as the distance between the mirrors of the cavity, such that the field was created there by constructive interference (the distance L between the mirrors had to be a multiple of the half-wavelength of the field). We observed variations in the detection current of the Rydberg atoms when the frequency of the field coincided with that of a transition from the initial excited level to a final Rydberg level. Using a classical image, this experiment was equivalent to measuring the precession frequencies of Rydberg electrons on their Bohr orbits. As we have seen, these frequencies are close to those of an electron of the hydrogen atom on the corresponding orbits. The small differences from the frequencies of hydrogen come from the perturbing effect of the atomic core, which is a point charge only in first approximation. By systematically recording these spectra for alkaline atom Rydberg states of various principal quantum numbers n and orbital angular momenta l, we thus made a thorough study of the Rydberg electron's interactions with the atomic core.

Deviations from the hydrogen spectrum are expressed using a simple model. Instead of varying as $1/n^2$, as in hydrogen, the binding energies of the Rydberg electrons vary as $1/(n - \delta_l)^2$, where δ_l, called the *quantum defect*, is a number, small compared to n, that depends on the probability of the Rydberg electron penetrating the quasi-point core of the atom. This probability varies, in the Bohr image, with the ellipticity of the orbit, or the angular momentum l of the electron. Orbits of low l ($l = 0, 1,$ or 2) are very elliptical and the electron at its perihelion is very close to the core, hence a relatively large quantum defect δ_l (on the order of one). The orbits with large values of l ($l = 3, 4\ldots$) hardly penetrate the core, and their quantum defects are very small. They don't vanish completely, however, because even outside the atomic core the Rydberg electron is slightly influenced by it.

The analogy with astronomy can help us understand this effect. We know that the Moon, orbiting the Earth, induces the phenomenon of tides. The gravitational force it exerts on the Earth moves the oceanic masses, periodically deforming the shape of the Earth's geoid. Conversely, this deformation reacts on the Earth's gravitational field. The period of the Moon thus happens to be slightly different from what it would be if the Earth were a perfectly rigid and undeformable solid. Similarly, the Rydberg electron slightly modifies by its electric field the distribution of electronic charges in the core of alkaline atoms, and this modification affects its motion. It results in the model described above via the existence of a small residual quantum defect, decreasing when n and l increase.

One day in 1979, these experiments had a surprise in store for us. We were studying a transition between a Rydberg level of $n = 23$ and a less excited level of

$n = 22$ in the sodium atom, whose frequency was on the order of 340 gigahertz. We had not switched on the microwave field, and were very surprised to find that the atoms jumped from one level to the other *without* a microwave applied. All that was needed was for the distance between the mirrors to be properly tuned. The population transfer signal between the upper and lower levels passed through a resonant maximum when the distance L between the mirrors satisfied the resonance condition *for a field that shouldn't be there!* It didn't take us long to understand: we had unwittingly made a Rydberg atom maser. Just like the ammonia molecules in Charles Townes's maser, our Rydberg atoms radiated, by stimulated emission, a coherent field in our cavity. This field caused them to collectively switch to the lower transition state. It was a pulsed maser, the emission being repeated at the repetition rate of our excitation laser. We then evaluated the threshold of this maser by progressively decreasing the average number of Rydberg atoms prepared by each pulse. All that was needed was to attenuate the intensity of the excitation lasers.

To our own intense excitement, we found that the threshold of this maser was extremely low. The effect disappeared only when we prepared less than two or three hundred atoms per pulse. By comparison, it took billions of molecules in Townes's cavity for his maser to oscillate. This result was a revelation. The Rydberg atoms' extreme sensitivity to microwaves manifested in a way that allowed us to observe a system that was beginning to approach what I had dreamed of at the end of my thesis work eight years earlier. In our cavity, a few hundred atoms were interacting with a few hundred microwave photons.

This was not yet the ultimate goal of one atom and one photon, but it was beginning to get seriously close. Until then, we had paid little attention to the quality of our cavity. Now we understood that if we could increase its quality factor *Q*, we could reach a situation where the maser would need only one atom to work (the *Q* factor is a number proportional to the number of reflections a photon can undergo between mirrors before being absorbed). We replaced the copper cavity with two mirrors made of niobium, a metal that becomes superconductive below 9K. The theoretically infinite conductivity of these mirrors would considerably increase their reflectivity, and thus the *Q* of the cavity.

This new technology forced us to move to a cryogenic experimental setup, in which the core was cooled to the temperature of the liquid helium contained in a tank with which the mirrors were in thermal contact. Cryogenics had another advantage: lowering the temperature of the mirrors also reduced the temperature of the thermal field they spontaneously radiated. At $T = 4$K, Planck's law tells us, there is less than one thermal photon of 340-gigahertz frequency in a cavity. We could thus study the quantum effects of the interaction between the atoms and the field without being hindered by the thermal noise present at room temperature.

After adapting our apparatus to cryogenics, we were able to try the experiment in 1983. We found that a single atom was rapidly transferred from the upper to the lower transition level when the niobium cavity was properly tuned. We had reached the case where an atom interacts with a single photon, leading to a detectable signal. This was exciting, but our joy was still a few orders of magnitude from being complete. The photon emitted in the cavity disappeared into its walls before it could re-interact with the atom. The lifetime of the photon was about 300 nanoseconds, corresponding to a *Q* factor of about 7×10^5. We still lacked one or two orders of magnitude for the photon to stay in the cavity long enough for us to observe the Rabi oscillation in the vacuum field.

However, we knew what we had to do: extend the lifetime of the photons in our cavity, as well as that of the Rydberg atoms. Those we had prepared thus far, excited in a stepwise process induced by two or three laser pulses, corresponded to very elliptical orbits of low angular momentum, in which the Rydberg electron had a relatively short lifetime, on the order of a few hundred nanoseconds. To attain longer lifetimes, it was necessary to prepare Rydberg atoms of low ellipticity and high angular momentum, ideally the circular Rydberg level of maximum orbital angular momentum $l = n\text{-}1$. Increasing in parallel the lifetimes of photons and of atoms would require the construction of a new apparatus and the implementation of new techniques, which would gradually lead us to the experiments I will describe in the next chapter.

We weren't the only ones interested in the properties of Rydberg atoms in cavities in the 1980s. In the United States, Daniel Kleppner, an MIT physicist who had become a friend after we met at the 1973

Vail conference, performed a beautiful experiment in 1983 showing that the lifetime of a Rydberg atom was extended when it was confined in a cavity excluding the modes in which it could spontaneously radiate a photon. For this experiment he had used the circular Rydberg atoms I mentioned above, and invented the method to prepare them. This emission inhibition experiment was the opposite of the one we had just done with our single-atom Rydberg maser. In our case, the presence of the cavity increased the probability of spontaneous emission of the atom, whereas in the MIT experiment the cavity inhibited such an emission.

Modifying the lifetimes of excited atoms in the presence of cavities had been predicted in a brief 1946 note by Edward Purcell, the pioneer of nuclear magnetic resonance. He was interested in the modification of a spin's spontaneous emission under the effect of its coupling with the resonant coils of the circuits he used to induce magnetic resonance in solids or liquids. The context was different from that of our cavity atom experiments, but the effect was similar. The "dressing" of a quantum system by an environment of walls or wires changes the modes of the radiation vacuum in which the atom is immersed, thus modifying these systems' spontaneous radiation properties. Kleppner coined a name for this new physics: *cavity quantum electrodynamics*, or *cavity QED*.

Other researchers soon took up this physics. In Germany, Herbert Walther designed some beautiful experiments with Rydberg atom masers, in which atoms fed, one by one, the field of a superconducting cavity by successively emitting photons. Different groups in the United States and Europe generalized the concepts of cavity QED to the optical domain by developing microcavities confining visible light photons in very small volumes and coupling them to atoms passing through the cavity in their ground state.

Research and teaching on both sides of the Atlantic

I remember with pleasure and nostalgia this period of intense activity, surprises, and discoveries in Kastler's and Brossel's lab. With three PhD students and Philippe Goy, a colleague who had come from solid state physics to help me master the techniques of microwaves and cryogenics, we made a small group sharing not only a passion for the physics of atoms and photons being revealed before our eyes, but also much else, from music and painting to politics. Humor was never far off, allowing us to overcome the daily challenges we faced. My first students, Michel Gross and Claude Fabre, helped me set up the laboratory. The research I've just outlined was the subject of their theses, after which they continued their careers at the CNRS and the university. The next student was one of my great discoveries: a young Normalien, Jean-Michel Raimond, joined my group in 1978 and has not left since. He was my thesis student before becoming my colleague at the university. His deep intuition in physics, his vast knowledge of computer science, his great erudition in science and the humanities, and his devastating sense of humor contributed to the unique atmosphere of our group, a friendly rapport animated by the certainty that we were taking part in a great adventure, the exploration of the quantum world made possible by lasers.

We carried on this exploration intensely, through long nights in the laboratory running experiments requiring a vast accumulation of data. The rhythm of a researcher's life is not set by fixed bureaucratic schedules. Quite often atom and photon experiments require long hours of preparation and adjustment before nature is ready to reveal its secrets. Night has often fallen by the time data can be recorded. It is often in the special atmosphere of the laboratory, uncommonly calm and quiet, that the truth of an experiment is revealed. Then it is necessary to redouble one's attention, to combat fatigue and keep the senses alert. In such cases, working together in a friendly ambiance is an essential asset.

Writing articles or letters presenting the results of our work was also crucial. We had to find the right approach, reframing our work in the lineage of research to which it belonged while emphasizing its originality and its promising potential extensions. Having a manuscript accepted to the American Physical Society's *Physical Review Letters*, the top weekly journal for physics, was proof of quality work. Its title, introduction, and list of references—often the only parts that would be read—had to be particularly meticulous. In a few keywords, it had to attract the attention of colleagues all over the world working in the same field. The letter, which had to be four pages long, could not sacrifice precision for concision; in principle, it had to allow anyone to reproduce the experiment or redo the calculations. It was an art in which we indulged by writing parts together, encouraging and amiably critiquing each other until we were finally satisfied with the result.

But research work wasn't everything. Having started as a researcher at the CNRS, I was recruited in 1975, at the age of thirty, as a professor at Pierre and Marie Curie University. I was able to divide my teaching time between undergraduate students at the University and Normaliens and those who joined them at the ENS in the graduate program directed by Brossel, whose student I had been a decade earlier. I was happy to leave the CNRS for the university, because I've always enjoyed teaching. Trying to explain the subtle concepts of modern physics as well as possible taught me a great deal myself. The pedagogical effort frequently helped me understand things more deeply, and at times gave me ideas for experiments that I wouldn't have had if I were only a researcher. If the load isn't too heavy, teaching also helps the researcher change his perspective and escape from the difficulties inherent in research, the better to return to them with a rejuvenated mind.

At the beginning of the 1980s, I faced a difficult choice. Should I stay in France, in the laboratory that had trained me and given me all the freedom I could hope for to do the research I loved, or should I give in to the call of America, where I was being offered professorships at prestigious universities? My research on Rydberg atoms had been recognized, and I was beginning to receive invitations to give seminars and lectures at international conferences all over the world. I had the profile of one of those promising young researchers whom Harvard, MIT, or Stanford fight over, ever hoping to raise the scientific caliber of their research departments.

These offers were tempting. I had enjoyed the competitive but relaxed atmosphere of my postdoctoral year in California, and the idea of returning, this time to lead a research group, appealed to me. I was also seduced by the great mobility of American researchers, the intense competition between universities in the U.S. pushing them to move around among diverse and constantly renewed research experiences. This, I felt, must enrich them and broaden their vision of research, whereas permanent membership at the École Normale Supérieure could lead to a certain intellectual confinement, limiting the possibilities of my research. At the same time, I felt it would be difficult to find across the Atlantic the spirit of my group in Paris. Even back then, research in the U.S. was already a business that required not only the qualities of a researcher but also those of an entrepreneur, adept at raising funds and defending his interests against those of his colleagues and his institution's administration. This was very different from what I was accustomed to from working under the protective wing of an enlightened laboratory director like Jean Brossel.

So as not to have to choose definitively between France and America, I proposed to Yale University that I be granted a part-time position, allowing me to teach and research there for one semester each year. I would return to Paris for the other semester, where I could continue to teach and above all not cut myself off from the group that had brought me so much, with whom I wanted to pursue my dream of manipulating isolated quantum particles. I don't regret the semesters I spent at Yale between 1983 and 1992, which allowed me to live the life of an American professor in the special atmosphere of an Ivy League university. Relations between professors of different disciplines, in science and the humanities, were closer there than they are in France. The graduation ceremonies each year, which Americans curiously call Commencement Day, where professors and students show up dressed in academic gowns; the weekly department meetings, where the recruitment of future professors is discussed; the receptions at the university club, where the president and deans welcome their visiting colleagues in a simultaneously formal and relaxed manner—all of these events allowed me to experience traditions that had more or less disappeared from the French academic world.

More seriously, I enjoyed teaching, at Yale, students who were very different from the Normaliens. Coming from all over the world and selected according to a wide variety of criteria, they were a less homogeneous group than their ENS counterparts. As they were generally less knowledgeable in mathematics, they had less of a grasp of the basics of physics than my students in Paris did, but the American educational system pushed them to express themselves more, to ask questions without being embarrassed to reveal their ignorance. Paradoxically, their limitations made them more daring, and the classes were all the livelier for it. I also conducted interesting research at Yale after creating a small experimental team with two colleagues coming from Europe like me, Edward Hinds and Dieter Meschede. I won't describe this work here, as it is not directly related to the work that was later recognized by the Nobel Prize.

In spite of all the advantages of this double position on both sides of the Atlantic, it became obvious after a few years that I could no longer remain in "coherent superposition" between two situations so far removed from one another. It was becoming ever harder on my family as well. Claudine, a social scientist at the CNRS, could easily follow me to Yale and continue her work in that university's magnificent library, but Julien and Judith, teenagers at the time, needed stability and couldn't follow the rhythm of my travels. They spent the school year in Paris, which posed logistical problems for family life. In 1992, I finally chose to return full-time to Paris, where our work on atoms and photons was taking a promising turn. During one of my stays at Yale, Jean-Michel Raimond had recruited an exceptional young student, Michel Brune, who would remain with us and round out our team for the adventure that was to follow.

The laser cooling revolution

The enthusiasm and excitement we felt at the time was spurred on by an accelerated flow of laser-related discoveries and inventions. The 1970s had been a time of tremendous progress in laser spectroscopy. In the decade that followed, the laser also emerged as an ideal instrument for controlling the motion of atoms, stopping them in their tracks, trapping them, and studying them individually. Because stopping atoms means decreasing their kinetic energy, thus lowering their temperature, this field quickly became known as *laser cooling*. Our experiments were more concerned with controlling and manipulating the particles of light that are photons, but even though we weren't directly involved we followed the progress of our colleagues closely as they cooled and trapped atoms and ions. Manipulating atoms and manipulating photons have many points in common, and the advances in one field benefit the other, as is often the case in fundamental research. The physics of laser cooling was developing on both sides of the Atlantic, and my frequent stays in the U.S. allowed me to keep abreast of the experiments being carried out there and the intense international competition between American and European researchers in the field.

Cooling and trapping atoms to the point of freezing their motion to the extent allowed by quantum physics quickly became the ultimate goal of spectroscopy, the way to be totally free of that enemy of metrology, the Doppler effect. The nonlinear optics methods of the 1970s certainly made it possible to eliminate this effect to some extent, but they required subjecting atoms to intense light fields that caused perturbing effects on energy levels, displacing and widening the spectral lines by quantities that had to be corrected for. On the other hand, these experiments involved samples containing large numbers of particles that collided with one another, adding other perturbations responsible for widenings and shifts of atomic transitions. It soon became apparent that the way to avoid all these inconveniences was to stop the atoms and, if possible, keep them isolated from each other. Lasers would be able to do this. The initial goal, to further increase spectroscopic accuracy, was achieved beyond metrologists' wildest dreams, leading to other fundamental and totally unexpected discoveries.

Laser cooling relies on the radiation pressure exerted by light on matter. As we have seen, by absorbing a photon, an atom takes on the momentum of the particle of light. If the atom moves toward the laser source, its speed is reduced. After absorbing the photon, the atom returns to its ground state by spontaneous emission. Since this emission occurs in a random direction, the momentum received by the atom during this process is on average zero, because the probability of spontaneous emission in two opposite directions is equal. Subjected to the intense light of a resonant laser beam directed in the direction opposite their speed, the atoms in a beam undergo a large number of absorption and emission cycles per second, losing on average a small fraction of their speed with each cycle.

In order for the process to be efficient and lead to the atoms being slowed almost to a stop, they must be kept in resonance with the laser light and thus compensate for the variation of the Doppler effect. The latter causes the atoms to "see" light of a higher frequency than that of the laser in the laboratory's reference frame. Moreover, this apparent frequency decreases as the speed of the atoms does. To prevent the laser from coming out of resonance and the effectiveness of the radiative braking from diminishing, we can exploit a property of atomic magnetism. Atoms, which, as we saw in Chapter 1, carry small magnets, see their energies displaced by a quantity proportional to the amplitude of the magnetic field to which they are subjected. This is called the *Zeeman effect*, after the Dutch physicist, a colleague of Lorentz's, who at the end of the nineteenth century studied the influence of a magnetic field on atomic spectra.

In order to use the Zeeman effect in the laser cooling process, the beam of atoms is sent into a solenoid consisting of a winding of electrical wire on a cylinder. This solenoid creates an inhomogeneous magnetic field aligned along the beam. The number of turns of wire per centimeter decreases along the axis of the solenoid so that the atoms are subjected to a decreasing magnetic field as they lose speed. The Zeeman effect then compensates precisely for the variation of the Doppler effect, keeping the atoms in resonance with the light along their path.

This clever device, known as a *Zeeman slower*, was invented by William Phillips, a former student of Daniel Kleppner, working at the National Institute of Standards and Technology (NIST) in suburban Washington, D.C. In 1983, Phillips succeeded in stopping the atoms of a sodium beam excited by a yellow-light dye laser. He was the first to observe with the naked eye, and to photograph, a small fluorescent cloud of nearly stopped atoms at the exit of his solenoid.

Two years later, another American, Steven Chu at Bell Laboratories in New Jersey, generalized the one-dimensional cooling method demonstrated by Phillips to all three dimensions of space. An atomic beam of sodium, initially pre-cooled by a counter-propagating laser and a Zeeman slower, was irradiated with three pairs of laser beams directed along three orthogonal axes. Each pair was made of two laser beams propagating in opposite directions. At the intersection of these laser pairs, the atoms were thus subjected to light slowing them down, going against their speed in all three directions, but also to photons accelerating them by pushing them in their direction of propagation. In order to make the slowing effect greater than the accelerating one, Chu shifted the common frequency of the six laser beams slightly toward the red. Thanks to the Doppler effect, the atoms always saw the light propagating in the direction opposite to their speed closer to resonance than the one propagating in the same direction. They were thus slowed down more than they were accelerated, and finally, in a fraction of a second, were brought to a very low speed in all three directions.

The idea of exploiting the Doppler effect in this way to slow down atoms in a configuration of six laser beams had been proposed in 1975 by Hänsch and Schawlow, but the realization of the experiment had to wait about ten years for the technological advancement of lasers and the experimental skill of Chu. The atoms of the cooled gas moved slowly at the intersection of the six laser beams under the effect of light irradiation, undergoing a force opposite and proportional to their speed. This force is analogous to the force of viscosity that opposes the motion of a marble falling into a pot of honey, hence the name given to this device: *optical molasses*. The residual kinetic energy of a molasses atom should, per the accepted theory, be equal on average to the uncertainty in energy of the fluorescence photons. According to Heisenberg's relations, this uncertainty is equal to $\hbar/\tau$, where τ is the lifetime on the order of 10^{-7} to 10^{-8} seconds of the excited atomic state. The theory predicted a temperature T of the cooled gas on the order of $\hbar/k_B\tau$, *i.e.* a few hundred microkelvin, corresponding to an average atomic velocity of a few meters per second.

Phillips, taking up Chu's experiment, set about measuring this temperature. He suddenly switched off the laser beams of an optical molasses, let the atoms fall in the dark for a few milliseconds, then turned the light back on to measure, thanks to the detection of the fluorescence they emitted, the atoms' final positions. He deduced the distribution of their initial velocities in the molasses and determined the temperature of the gas. To his great surprise, it was on the order of a few microkelvin, a hundred times smaller than expected!

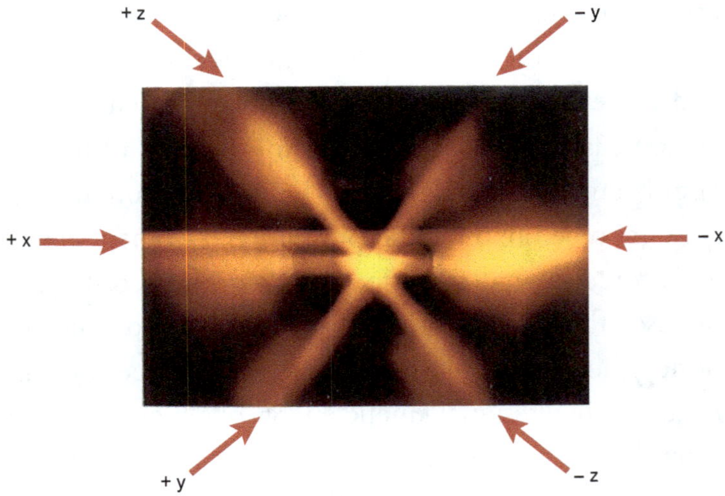

Figure VI.8. Optical molasses of sodium atoms photographed in William Phillips's laboratory at NIST. The six laser beams propagating in both directions along three orthogonal axes are visible thanks to the light scattered by the fast atoms they encounter on their path. At the intersection of the beams, a more intense scattering spot reveals the presence of roughly one billion cold atoms, moving randomly in Brownian motion with a velocity of a few tens of centimeters per second. The horizontal light beam appears double. Above the horizontal molasses lasers, we see the laser beam that pre-cooled the beam of sodium atoms.

A few months later, Claude Cohen-Tannoudji and his student Jean Dalibard gave an explanation of this happy surprise. The theory, they understood, had overlooked the fact that the sodium atom has two spin states in its ground level, making the exchange of momentum and energy between atoms and lasers more complex than a simple Doppler cooling. The superimposed laser beams of molasses form interference fringes. By "dressing" the atoms, they induce a spatial modulation of the energies of the different spin states. The atoms thus move in a landscape of hills and valleys of potential energy, depending on the state they are in.

Polarization of the interfering beams also varies spatially and favors the absorption of a laser photon when an atom in a spin state reaches the top of a potential hill. It then falls back by emitting a photon in the other state, at the bottom of a valley, and must go up the next hill before it absorbs another photon. Laser cooling, combining exchanges of energy and angular momentum between atoms and light, is thus much more efficient than simple braking by radiation pressure acting on atoms that would have only one ground state. Claude and Jean named this atomic braking mechanism *Sisyphus cooling*, recalling the mythological character condemned to endlessly push a boulder up a slope. For these discoveries from the 1980s, Chu, Phillips, and Claude shared the Nobel Prize in Physics in 1997. I was a privileged witness to this work, which took place in the laboratory room next to mine in Paris and, on the East Coast of the United States, in research institutes I had the opportunity to visit while teaching at Yale.

In research on laser cooling, shifting the energy levels of an atom subjected to non-resonant light would come to play an important role. This effect had been discovered, as we have seen, by Claude during his thesis work. The light shifts, tiny and simple curiosities when produced by classical low-intensity lamps, became very large with laser light. By superimposing laser beams of appropriately chosen directions and polarization, we can create by interference optical field configurations with periodic structures alternating maxima and minima of light intensity. The energies of atoms subjected to this light become spatially dependent, like a ball which, rolling on a surface with troughs and bumps, has its potential gravitational energy spatially modulated.

If the ball's kinetic energy is too small to allow it to move from one trough to the next, it will remain trapped in one of the pits. Similarly, laser-cooled atoms can be trapped in optical potential well arrays created by non-resonant laser beam configurations interfering with one another. Immersing a gas of cold atoms in these optical structures creates arrays of trapped atoms. The study of these trapped atoms began to develop in the early 1990s, considerably enriching experimental possibilities.

A temperature of 1 microkelvin corresponds to atoms moving at speeds of a few centimeters per second. To stop them from falling into the Earth's gravitational field, they must be subjected to a force opposing gravity. Dalibard invented an ingenious device combining the effects of radiation pressure with those of a spatially varying magnetic field. This trap, which equipped a large number of laboratories from the 1980s onwards, is called a *magneto-optical trap*, or MOT. Other types of purely magnetic atomic traps were also invented, allowing atoms to remain in the dark without suffering any light disturbance. In these traps, which resemble "magnetic bottles" with immaterial walls, the configuration of the fields exerts on the spin of the atoms a magnetic force that confines them to a small region of space.

Carl Wieman and Eric Cornell, two researchers at the JILA laboratory in Boulder where John Hall had measured the speed of light in 1972, set out in the early 1990s to further cool an already very cold boson gas suspended in a magnetic trap. Their goal was to reach the threshold temperature of Bose–Einstein condensation, obtained with atoms moving at speeds on the order of millimeters per second, corresponding to de Broglie wavelengths on the order of microns. According to Einstein's prediction from seventy years earlier (see Chapter 5), the bosons had to gather in the ground state of the trap, described by a single wave function for all atoms.

To reach this temperature, on the order of a hundred nanokelvin, laser cooling was insufficient. Wieman and Cornell subjected their atoms to another mechanism, called *evaporative cooling*. By gradually lowering the magnetic barrier that held the atoms in their trap, they allowed the hottest atoms, those with the greatest kinetic energy, to escape. The remaining atoms were then brought back into equilibrium by colliding with each other at a lower temperature. The process is analogous to the one we intuitively use to cool a hot liquid: by blowing on it, we accelerate the evaporation of the fastest molecules near the surface. Those remaining in the liquid acquire a lower average kinetic energy after thermal equilibrium, and are therefore cooled.

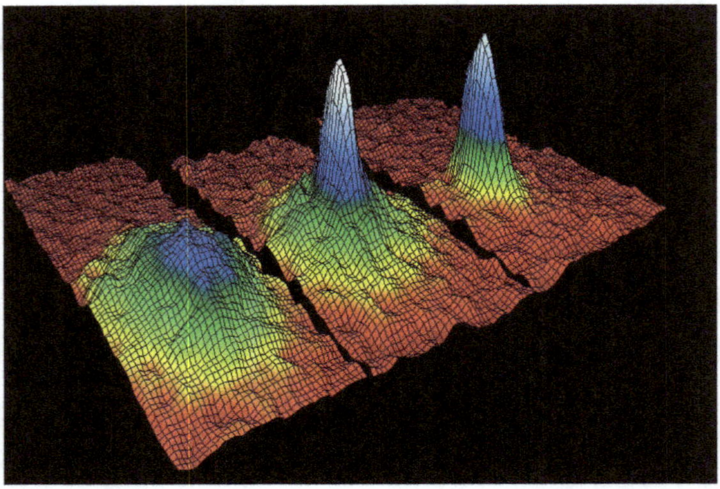

Figure VI.9. First image of a Bose–Einstein condensate. A gas of ultracold rubidium atoms, trapped in a magnetic bottle, is cooled by evaporation. To measure the distribution of the atoms at different stages of cooling, the trap is suddenly removed and, after a delay, the position of the atoms in the expanding cloud is detected by absorption of a laser probe. This gives the distribution of atomic velocities at the time the trap was switched off. This distribution is represented for three temperatures, decreasing from left to right, in a false-color 3D image, atomic density being represented in the vertical dimension. On the left, we observe the relatively wide thermal distribution of the atoms above the critical condensation temperature. When this temperature is reached (*middle image*), a narrow peak appears in the center of a wider pedestal of non-condensed atoms. At a lower temperature (*right image*), this pedestal fades away as most of the atoms fall into the condensed phase (Mike Matthews, JILA).

It took until 1995 for Wieman and Cornell to finally reach the threshold of Bose–Einstein condensation of a rubidium gas, thus creating a new state of matter with fascinating properties. A few months later, Wolfgang Ketterle at MIT succeeded in condensing a sodium atom gas. Wieman, Cornell, and Ketterle shared the 2001 Nobel Prize in Physics for this discovery.

Trapped ions and quantum jumps

To complete this panoramic vision of research on cold atoms, I must also mention the physics of trapped ions and electrons. An ion is an atom from which an electron has been torn off, either by photoionization or by collision with other particles. It thus carries a positive elementary charge and undergoes a force in an electric field. Various configurations of static or oscillating fields produced by electrodes and current loops make it possible, by acting on ions' charge and magnetic moment, to retain them around a position of equilibrium. The forces thus experienced by an elementary charge are much more intense than the radiation pressure acting on neutral atoms. Ion traps are therefore much deeper and more stable than the optical traps described above. They can retain particles whose temperatures, on the order of 1 kelvin or more, are thousands of times higher than those of atoms captured in light traps. The same trapping technique also applies to electrons and positrons, which are much less massive than ions but also carry an elementary charge on which field configurations exert restoring forces.

The motion of the trapped particles induces oscillating currents in the metal of the electrodes or wires that make up the structure of the trap; detection of these currents enabled the study of these particles' dynamics as early as the 1950s. This physics, pioneered by the German physicist Wolfgang Paul, reached an extraordinary degree of sophistication in the experiments carried out in the 1960s and 1970s on electrons or positrons. In a series of veritable technological tours de force, Hans Dehmelt, another German physicist who had emigrated to Seattle, managed to trap an isolated electron, then an isolated positron, and keep them in his apparatus for several months.

He used the electrical signals that these particles induced in the walls of the trap to measure their magnetic moments with extraordinary precision. This was the first study of quantum particles isolated from their environment, finally assembling the ideal conditions of the thought experiments of the 1920s.

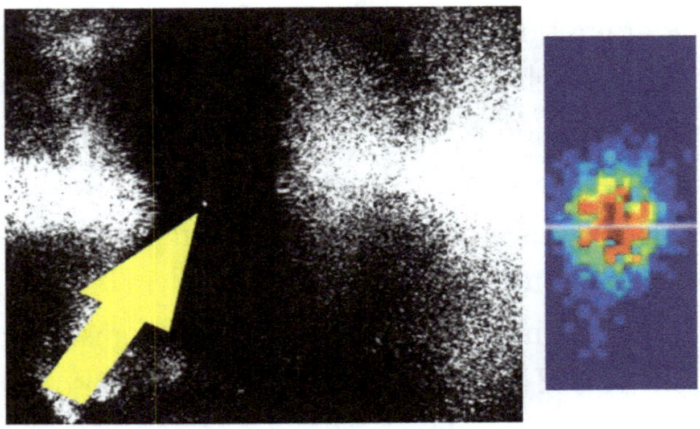

Figure VI.10. First optical observation of a single atom: a barium ion detected by its fluorescence light. The exposure time was ten minutes for this analog photograph on Kodak film, taken in 1978 by Dehmelt and Toschek. The stray light scattered by the electrodes of the trap did not quite eclipse the small bright spot indicated by the arrow (per *Physical Review* A 22 (1978), 1137). (*right*) The digital image of a single magnesium ion, now routinely obtained in a fraction of a second in modern trapped ion experiments.

These experiments allowed the theory of quantum electrodynamics to be verified with great precision, finding agreement up to the tenth significant digit between the theory and the experimental results.

When lasers became operational in atomic physics, Dehmelt quickly understood how they could be used to observe and manipulate trapped ions. In 1978, with a German colleague, Peter Toschek, he performed the spectacular first experiment to optically detect an isolated ion. Illuminating a small cloud of barium ions with a resonant laser on a transition coupling the ions' ground state with an excited state, they observed the fluorescence light of the atomic sample with a microscope. The ions, constantly re-excited by the intense laser light, scattered millions of photons per second toward the microscope lens. By attenuating the electrical discharge that created the ions before they were trapped, they managed to decrease the filling rate of the trap until only one ion remained, its fluorescence appearing as a small, faintly shining star in the center of the trap. The pictures they made of it were the first to show an isolated atom suspended in the vacuum, a quantum particle that could be observed with light without being destroyed. Thus, Schrödinger's prediction that this kind of experiment could lead only to ridiculous consequences was resoundingly contradicted.

Soon lasers were used not only to observe trapped ions but also to manipulate them. In 1975, along with David Wineland, who was doing a postdoctoral fellowship in his laboratory after finishing his physics thesis at

Harvard, Dehmelt imagined a method of laser cooling of ions similar to the one Hänsch and Schawlow had proposed the same year for neutral atoms. By illuminating the ions oscillating in the trap with a laser beam detuned slightly below one of their resonant frequencies, they could be made to preferentially absorb photons propagating in the direction opposite their motion, causing them to lose vibrational energy. The motion of an ion in its trap can be described in first approximation as that of a small oscillator vibrating around its equilibrium position. The energy levels of this oscillator are quantized, like those of an electromagnetic field mode. The energy scale of the vibrating quanta is made of equidistant bars, like that of photons in a cavity. The lowest of these levels is the ground state of vibration in the trap.

The particle trapped in the ground state is then as motionless as quantum physics allows. Its position and residual momentum present fluctuations that obey Heisenberg's uncertainty relation. By applying the laser cooling method adapted to trapped ions, Wineland, who went on to work in Boulder after his internship in Seattle, was finally able to cool a single ion in this state.

I was quite interested in Wineland's experiments, having met him at several conferences in the 1980s. His ion oscillating in a trap had a strong analogy to a Rydberg atom interacting with microwave photons in our cavity. In both cases, the state of the system was described by two quantum parameters, one associated with the internal electronic state of the atom or ion (which can be described in many experiments as a spin evolving between two long-lived + and - levels) and the other with the state of a quantum oscillator (described by the number of photons for the Rydberg atoms in the cavity and by the number of vibration quanta for the oscillating ion).

To clarify this analogy, let's consider an ion in its ground state of vibration prepared in the - state. We make it interact with two laser beams the *difference* in whose frequencies $\nu_1 - \nu_2$ is tuned to coincide with the difference in energy of the + and - states divided by *h*. An exchange of photons must then occur between the two laser modes. This process is called *Raman scattering*, after an Indian physicist who was one of the founders of molecular spectroscopy in the 1920s and 1930s. The ion switches from the - state to the + state, absorbing a photon in one laser beam and emitting a photon by stimulated emission in the other. This reversible phenomenon is analogous to a Rabi oscillation. Raman scattering causes the ion to oscillate between the - and + states without changing its state of vibration or the total number of photons exchanged between the laser beams. Now we adjust the difference in frequencies $\nu_1 - \nu_2$ so that it is equal to $(E_+ - E_-)/h$ plus the vibration frequency of the trapped ion. The conservation of energy then requires that the flipping of the ion from - to + be accompanied by the excitation of a vibration quantum. The ion must now oscillate between the states $|-,0\rangle$ and $|+,1\rangle$. Similarly, the Rydberg atom we were studying in our experiments in Paris should, if the cavity could hold the photons long enough, oscillate between two electronic states correlated to the presence in the field of 0 or 1 photon.

The physics of the Rydberg atom and that of the trapped ion obeyed the same equation! In both cases, a Rabi oscillation should be observed. In our experiment, it had to occur spontaneously under the effect of the coupling between the Rydberg atom and the vacuum field of the properly tuned cavity. In the case of the trapped ion, it was induced by laser light. David Wineland and I were both struck by this analogy long before we were able to clearly observe these phenomena in the laboratory. In the 1980s, we were moving by different means toward the same goal. On many occasions, the similarities of our two systems guided us to imagine new experiments inspired by our respective projects.

Finally, I must mention a trapped ion experiment from the 1980s that stoked my imagination and that I then sought to extend to our cavity QED experiments. We have seen that an isolated ion in its trap, irradiated by a resonant probe laser with a transition coupling the ground state g to an excited state *e*, scatters a large flux of photons per second. This fluorescence is observable with a simple microscope.

Let's now consider another excited level of the ion, which I will call *d*. A second laser resonant with the transition between levels *g* and *d* is also focused on the ion. The coupling with this second laser is much weaker than with the first one, resulting in a probability of excitation of the ion from *g* to *d* much smaller than the probability of absorption of a photon on the transition from *g* to *e*. What do we then observe on the ion's fluorescence at the frequency of the probe laser? The microscope receives a large flux of photons so long as the atom has not undergone a transition from *g* to *d*. As soon as the second laser has brought it to level *d*, the ion's intense fluorescence is suddenly extinguished, signaling that it has left the *g* level capable of scattering the light of the probe laser. The sudden disappearance of the fluorescence thus marks a quantum jump of the ion from the scattering state *g* to the "dark state" *d*, incapable of scattering the light of the probe. After some time, the atom spontaneously falls from *d* back to *g* and the fluorescence light reappears as suddenly as it disappeared.

This experiment was performed on a mercury ion in 1986 by Wineland and his group in Boulder. Dehmelt in Seattle and Toschek in Germany did similar experiments at the same time, directly observing the sudden jumps of a single ion between two levels. I remember a short film presented at an international conference that year: the audience, hypnotized, saw a single ion, isolated in a trap, glittering like a small star that flickered off and on again at random times. For the first time, the quantum jumps predicted by Bohr, and whose existence had never been accepted by Schrödinger, were directly visible.

This experiment made waves in the atomic physics community, which was divided between those who accepted the existence of these jumps and those who,

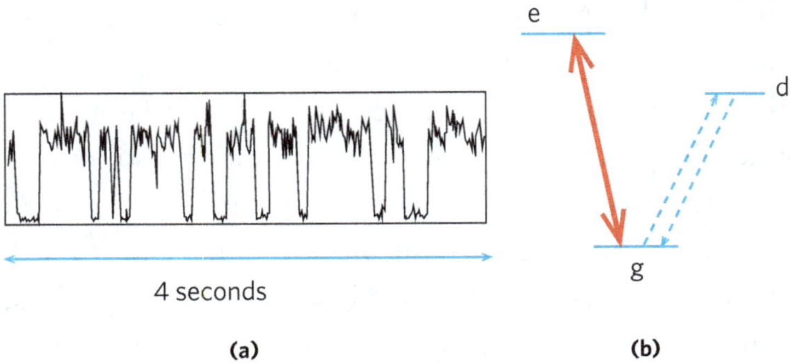

Figure VI.11. (a) Quantum jumps observed on the intermittencies of the fluorescence light emitted by a trapped mercury ion. (b) A diagram of the energy levels of the ion: the fluorescence is induced by an intense laser beam recycling the atom rapidly on the transition between the *g* and *e* levels. At the same time, a laser excites the transition between g and *d*, a very long-lived level. The fluorescence stops when the atom is excited from g to *d* and reappears abruptly when the ion falls spontaneously back to g (per *Physical Review Letters* 57 (1986), 1699).

like Schrödinger, didn't believe in them. The latter group's skepticism stemmed from the fact that until that point only large samples of particles had been observed, their statistical evolution masking the existence of individual atomic jumps. Theoretically, Schrödinger's equation does not predict any discontinuity in the evolution of a quantum system's wave function. As long as this theory is confronted with experiments on large ensembles of particles, the notion of jumps is not necessary. The observation of an isolated particle radically changes the situation.

The question of the reality of these jumps can be asked in simple terms. In the absence of the laser probe that measures the presence of the ion in the "bright" level *g*, does the atom jump discontinuously into the "dark state" *d* as a result of its interaction with the laser exciting the transition from *g* to *d*, or does it not? In other words, is there a quantum jump of the initially prepared ion to the state *g* if we don't "look" at this state? The Copenhagen school's answer is simple: there is a jump only if the system is observed. In the absence of coupling with a measuring instrument, the ion starting from the *g* state evolves continuously in a coherent superposition of the *g* and *d* states under the effect of the Rabi oscillation induced by the resonant laser with the transition from *g* to *d*, and it's meaningless to say it's in one state or the other. It is at the moment when the probe laser is turned on that the ion must randomly "choose" whether it is in *g*, in which case it starts to scatter light and no jump has occurred, or in *d*, in which case it does not fluoresce and remains invisible, indicating that a jump has occurred. If the laser probe continues to measure the state of the atom, there is a high probability that just after being found in *g* the ion will remain there for some time because the weak excitation by the laser carrying it in *d* only slowly builds up the probability amplitude of finding it in this "dark state." The ion's fluorescence is therefore observed over a certain time interval. However, the ion will eventually jump to *d*, at which point the fluorescence of the "bright state" *g* will suddenly extinguish.

This analysis contains the same arguments invoked to interpret Young's double slit experiment. It makes no more sense to ask whether the ion is in one state or the other in the absence of a probe beam than it does to ask whether a particle has passed through Young's double slit interferometer via one slit or the other, when no device is capable of answering this question. In both cases the quantum system is in a coherent superposition of states, and remains there as long as we don't try to remove this quantum ambiguity by measurement.

Ion traps used light beams to manipulate isolated particles of matter non-destructively. With my team, beginning at the end of the 1980s, I sought to reverse the roles of matter and light. We wanted to use Rydberg atom beams to manipulate photons trapped in a cavity. The extreme difficulty was that light quanta are much more fragile than ions or atoms. In captivity, they tend to disappear quickly on contact with matter. To observe them continuously without destroying them, to detect quantum jumps of an electromagnetic field, was to attempt the impossible. It took us fifteen years to achieve this and to carry out the experiments described in the following chapter, which led us to explore, with photons, the boundary between the quantum and classical worlds.

CHAPTER 7

Taming Schrödinger's cat

Watching is the most natural activity in the world. We perceive the reality of everything around us thanks to light, emitted by radiant sources and scattered toward our eyes by objects in our environment. This is how we receive most of the information our brain processes to tell us about the world in which we live. All of this information comes by way of photons, elusive and ubiquitous particles that travel through space at the speed of light. By interacting with matter—that of the objects around us, of our retinas, and of our detectors of visible and invisible radiation—these particles make us aware of the world. Photons inform us about what is directly perceptible, making it possible to form an understanding of our environment. They also give us, thanks to their detection by increasingly sophisticated instruments, knowledge of the universe, its origin, and its evolution.

The information carried by photons depends on their different attributes. Their color, an essential parameter, is analyzed by spectroscopy, which measures the wavelengths of light. Spectral lines are the signature of the different elements present in stars and in interstellar space. Their frequencies change according to the relative speed of the source and the observer. The Doppler effect tells us about the motion of atoms, whether in our laboratories on Earth or driven by the motion of the stars from which they send us their messages. Color also changes according to the rhythm of time's flow in the gravitational fields from which the photons we receive are emitted. The time dilation effect of general relativity informs us about the curvature of space-time. The polarization of photons, whose plane revolves around the direction of the magnetic fields along which they propagate—this is the Faraday effect discussed in Chapter 3—tells us about these fields' distribution around stars and planets. Finally, photons' wavelength changes due to the universe's expansion, which lengthens distances over cosmological eras and accounts for the cooling of fossil radiation, whose wavelengths, originally very short, now spread out in the radio-wave spectrum.

An essential part of the information that photons convey lies simply in the intensity of their flux. This is the number of photons arriving per second, from different directions in space, on the crystalline lenses of our eyes, the lenses of our cameras, or the parabolas of our telescopes, which create the images of the world we are observing. Another, subtler component of information is contained in the phase of the electromagnetic field, which can be detected only by observing an interference phenomenon. The principle is to compare the phase of the wave coming from an object with that of a reference wave that we've superimposed on it. This superposition presents spatial variations of amplitude. The information contained in the phase of the object wave is thus transformed into information about intensity, collected simply on a sensitive plate.

Simultaneously exploiting information on intensity and on phase is an art that has greatly enhanced the performance of our optical and imaging instruments in the domains of visible and invisible waves. Take the example of holograms. A photographic plate records the interference between the light of a laser beam and that which an object illuminated by the beam scatters on the plate. It thus collects information that makes it possible, when the plate is illuminated by a reading laser, or even by natural light, to reconstruct a three-dimensional image of the object—whereas ordinary photography, which exploits only the light's intensity, yields flat, two-dimensional images. Generalizing the principles of interferometry and holography to the observation of the sky, astronomers have considerably increased the resolution of images of the planets and stars by reducing in various clever ways the phase fluctuations of light caused by the turbulence of our atmosphere.

The principles of imagery I have just outlined can be described, with the exception of relativistic effects, as classical. They call exclusively upon laws of geometric and wave optics and phenomena known since the time of Fresnel, Doppler, Fizeau, and Maxwell. But we must not forget that the chameleon particle that is the photon is both a wave and a particle. Einstein's recognition of this dualism in 1905 was the starting point for quantum physics. There is a part of the information photons convey that cannot be reduced to corpuscular or wave aspects alone. The ambiguity whereby the photon does not "choose" whether it is a wave or a particle prior to measurement is a purely quantum characteristic that has been the subject of the debates, recalled in Chapter 5, illustrated by the thought experiments of the 1920s and 1930s. Laboratory study of this ambiguity on trapped photons was my goal and that of my team as soon as it appeared possible to do so by applying the methods of cavity QED.

There were two main difficulties to the task. One had to do with the photons' fragility. In free space they are eternal, as is proved by the fossil radiation photons that have traveled the universe for over thirteen billion years. When we try to capture them, though, they vanish easily on contact with matter. The most reflective mirrors and the most transparent media absorb them quickly. Even the most efficient optical fibers attenuate light over tens of kilometers, which photons travel in less than a millisecond.

The other difficulty is a consequence of this fragility. The standard detection of photons, whether it takes place in our eye, a photomultiplier, a photocell, an infrared camera, or a radio antenna, destroys them. The photoelectric effect, analyzed by Einstein in 1905, describes this phenomenon well. The photon disappears by yielding its energy to an electron ejected from the metal on which it has fallen. The resulting click signals the death of the photon. The effect is analogous to the detection of traces left by particles studied in accelerators, which are revealed by observing the debris resulting from their collisions.

In this sense, in a very different energy range, photon detection recalls the image of Schrödinger, who saw the trajectories observed in a bubble chamber as *post-mortem* signals left by vanished particles. The same applies to the clicks recorded by the standard photodetectors, including the eye. The detected photon is a vanished particle. So to observe photons *in vivo,* without killing them, we had to build a cavity to keep them in captivity for a long time and detect them using ultrasensitive atoms that are capable of extracting information without destroying them.

This new way of controlling, manipulating, and observing light would lead us to fundamental experiments. Seeing photons without demolishing them allowed us to dress light in strange new clothes. We managed to prepare it in superpositions of states where it oscillates with one phase and the opposite phase simultaneously. We could also generate states where it illuminates the mirrors of the cavity which traps it and at the same time leaves them in the darkness. We highlighted the extreme fragility of these exotic field states and showed how they become, in a flash, ordinary classical light. In this way we explored the boundary between the microscopic quantum world and the macroscopic world of our daily experience. By developing methods to preserve quantum strangeness in our cavity for as long as possible, we began to explore the possibilities of exploiting it in innovative information-processing devices. This is the adventure I discuss in this chapter.

The photon box

The photon box we finally managed to build doesn't look much like the one Gamow gave Bohr and Einstein as a reminder of their debates at the 1930 Solvay Conference. As we will see, however, it shares certain elements and can house a very special clock, which allowed us to determine the moments photons entered or left the box. Without changing the geometry of the cavity from our first experiments, made of mirrors in the shape of facing spherical caps, we gradually improved its quality, making the metal surfaces increasingly reflective by machining them with optical precision in polished copper and covering this metal with a thin layer of superconducting niobium. The surface irregularities of our current mirrors do not exceed a few nanometers, making microwave scattering outside the cavity extremely low. The almost infinite conductivity of the thin superconducting layer also minimizes absorption of photons into the metal.

The progress made between 1990 and 2006 can be measured by comparing two numbers. While photons could bounce off our first superconducting mirrors cut in solid niobium only five thousand times before disappearing, in our niobium-covered copper mirrors they reflect over a billion times. We take advantage of both the high mechanical quality of copper, which is much easier to machine than niobium, and the superconductivity of this metal covering the copper to a thickness of about ten micrometers.

The field mode in this cavity forms a standing wave with nine crests between the mirrors, which are 2.6 centimeters apart. The nodes between these crests, the points at which the cavity field amplitude cancels itself out, are 2.95 millimeters apart, half the 5.9-millimeter wavelength of microwave radiation that oscillates at 51 gigahertz. The photons bouncing between the mirrors of our most effective cavity survive on average 130 milliseconds and travel a folded path on the order of 40,000 kilometers, equal to the Earth's circumference. To speak of trajectory for these photons is to invoke a classical image that has no meaning in quantum physics. The photons are in fact completely delocalized in the standing wave, whose squared amplitude at each point is proportional to the probability of detecting them if we could insert a tiny photodetector there.

We can get an idea of the quality of this cavity by imagining an experience we've all had in an elevator with reflective walls. As we turn toward these mirrors, we see a few replicas of ourselves that are quickly lost in a distant blur due to imperfections in the reflective surfaces. If we were in an elevator with flat mirrors as perfect in the frequency range of visible light as those in our microwave cavity, we could see more than a billion copies of ourselves—a human chain as populous

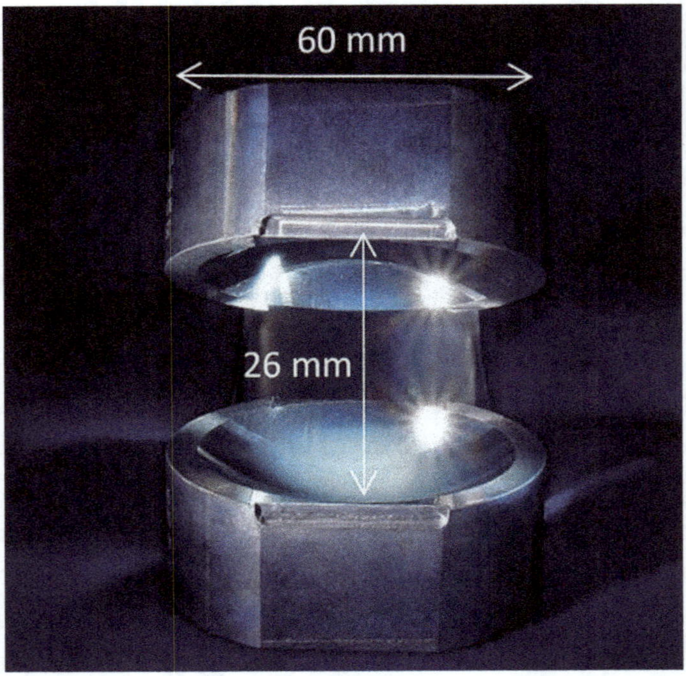

Figure VII.1. The ENS photon box. The microwave field at 51 gigahertz is stored between two copper mirrors in the form of spherical caps coated with a thin layer of superconducting niobium. The photons bounce over a billion times off the mirrors and remain trapped for an average of 130 milliseconds. The losses are caused by residual absorption in the metal and microwave diffraction on surface imperfections. The cavity is maintained at 0.8 kelvin. At this temperature the average thermal photon number is 0.05. For some experiments a small coherent field is injected into the cavity, containing on average a few photons. This field is radiated by a small horn close to the mirrors, coupled to a classical microwave source by a waveguide (not shown in the figure). It is coupled to the cavity mode by scattering at the mirrors' edges.

as China or India! True, our situation wouldn't be very comfortable, as these mirrors must be kept at a temperature of no more than 0.8 kelvin, making the surfaces' conductivity almost infinite and practically eliminating their thermal radiation.

We arrived at this ideal box only after a long period of trial and error, many attempts leading to dead ends, and experiments strewn with various mishaps. We were assisted in our quest for the ideal cavity by researchers and engineers from the Atomic Energy Commission (CEA) in Saclay, who had great expertise in the development of microwave cavities for particle accelerators. However, the constraints of their cavities were very different from ours, and we had to adapt their fabrication procedures to our experiment.

Though we knew that in theory nothing prevented us from making the cavity we wanted, it was not clear that we would be able to do so in practice. The slightest scratch on the copper substrate, the tiniest dust deposit on the mirrors from handling, or a tiny adhesion defect of the niobium film would be enough to shorten the photon lifetime by several orders of magnitude, rendering the experiments we were aiming for impossible. I remember our joy and relief on the spring day in 2006 when Michel Brune came into my office with a big smile to tell me that our students had finally managed to measure a photon lifetime of over one hundred milliseconds. That day we knew the experiments we had dreamed of were within reach.

The circular atom

To perform these experiments, a cavity alone, even one with near-perfect mirrors, would not do. To probe its field, we needed special atoms excited in very long-lifetime levels and extremely sensitive to microwave photons. The probes we used are *circular Rydberg atoms*, of principal quantum number n on the order of 50, in which the external electron rotates around the ionic core on a circle 2,500 times larger in diameter than that of an ordinary atom's orbit in its ground state.

These giant circular atoms are different from the ones we used in our early experiments. Those had a low angular momentum corresponding to very elliptical classical orbits of the external electron, similar to the elongated orbits of the comets in the solar system. These atoms were prepared by absorption of two or three laser photons, each of which could bring only one unit of angular momentum to the electron. In these states, the Rydberg electron approached the atomic core at the perihelion of the elliptical orbit, and its great acceleration in the vicinity of the core gave it a high probability of radiating a photon by falling to a less excited state. The lifetime of these elliptical states, thus limited to a few microseconds, was too short to make good probes of the microwave photons in our cavities. In a circular state, on the other hand, the external electron rotates on a circle and always remains far from the atomic core. It then undergoes a minimal acceleration, giving the atom a lifetime of about thirty milliseconds—on the order of that of the photons in our cavity.

The preparation of circular atoms, first performed by Daniel Kleppner at MIT in 1983, is a two-step process. First laser excitation brings the atom into a very excited state of low angular momentum. Then a circularly polarized radiofrequency field is applied to the atom, from which it absorbs about fifty very low-frequency photons, each bringing one unit $\hbar$ of angular momentum to the external electron, increasing its energy only very slightly. This second step recalls adjustments to the orbit of an artificial satellite whose ellipticity is changed by applying small thrusts from a booster rocket. Circular atoms are in fact the most classical atoms, for which the spatial distribution of the electron wave function is closest to a classical orbit. The region of space in which the electron has a non-negligible probability of presence is a torus, not unlike a truck tire whose thickness is about seven times smaller than the radius.

This torus defines a slightly blurry circular trajectory. The impossibility of completely specifying the orbit's radius reflects Heisenberg's uncertainty principle between electron position and momentum. The state of the electron is described by a de Broglie wave function rotating in the torus volume. The orbit's stability condition requires that there be an integer number of wavelengths on its circumference. This quantization condition is identical to the one Bohr discovered empirically in 1913, in his first quantum theory of the hydrogen atom. The principal quantum number n of the

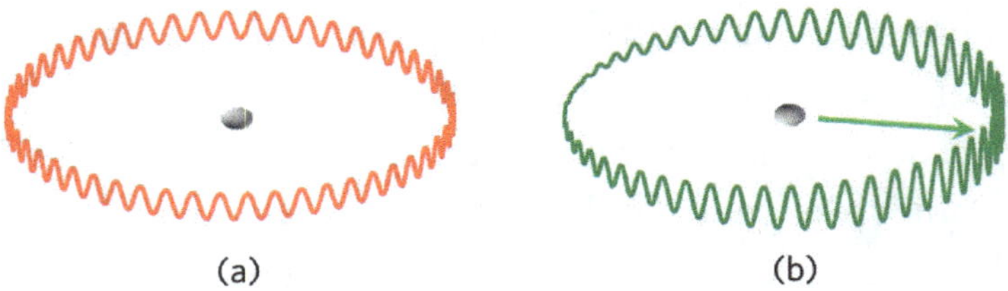

Figure VII.2. Representation of a circular atom by the de Broglie wave of its external electron. (a) The circular orbit $n = 50$ is associated with a wave revolving around the atomic core on a circle whose circumference is equal to 50 de Broglie wavelengths. The wave's amplitude is constant around the circle and the electric dipole of the atom is zero. (b) The superposition of the circular orbits $n = 50$ and $n = 51$, performed by a microwave $\pi/2$ pulse, prepares a circular wave with a crest at one point of the electron orbit and a node at the opposite point. The atom acquires an electric dipole (*arrow*) that rotates around the atomic core at the Bohr frequency of the transition between the $n = 50$ and $n+1$ levels (about 51 gigahertz).

circular atom simply measures the circumference of the orbit in units of de Broglie wavelengths.

In a circular Rydberg atom of quantum number n, the energy of the external electron is perfectly defined. Consequently, its position on its orbit at a given instant can't be, due to Heisenberg's uncertainty relation between time and energy. Quantum physics thus describes a particle that rotates at a very precise frequency, but with a completely indeterminate phase along its orbit. Since the electronic charge has equal probabilities of being at two opposite points on the orbit, the center of the electronic distribution coincides with the position of the atom's positively charged core. The electric dipole, which measures the separation of the centers of gravity of the atom's positive and negative charges, is thus null in a circular Rydberg state.

It is possible, however, to specify the electron's location on its orbit and to give its rotation a well-defined phase by creating a non-zero electric dipole in the atom. To do this, one can prepare a superposition of two neighboring circular states, of quantum numbers n and $n+1$, for example $n = 50$ and $n = 51$. A microwave pulse tuned to the frequency of the transition between these two states induces a Rabi oscillation between them, which carries the atom in a superposition of the two states. We can thus prepare a superposition whose probability amplitudes can be controlled by adjusting the duration and phase of the microwave excitation. For example, suppose we interrupt this excitation after a time equal to a quarter of the period of the Rabi oscillation (called $\pi/2$ *pulse* in magnetic resonance terminology because it lasts a quarter of a complete Rabi oscillation of 2π radians). The atom is then carried in an equal-weight superposition of the two circular states of main quantum numbers n and $n+1$.

The de Broglie electron wave revolving around the atomic core becomes the superposition of two waves presenting 50 and 51 oscillations along the orbits of the two states with almost identical radii. If the two waves are in phase at one point of the orbit, then they are in phase opposition at the diametrically opposite point, since there is a difference of exactly half a wavelength between the paths traveled by the two waves along a half-circumference.

The interference between the two matter waves is then constructive on one side of the orbit and destructive on the other. The superposition thus creates a crescent-shaped wave, with a crest and a node at two diametrically opposed points. This packet of two de Broglie waves, localized by interference, revolves around the ion core at the frequency of the transition between the two states of the superposition, *i.e.* 51 gigahertz in the case of the states $n = 50$ and $n = 51$ considered here. A rotating dipole is thus created on the atom, similar to a planet rotating around the Sun or to the hand of a clock rotating on a dial. The phase of this rotation can also be controlled by adjusting the phase of the microwave pulse that created the superposition of the two states. Finally, one can slightly vary the frequency of the electron rotation by applying a static electric field to the atom perpendicular to the plane of the orbit. This field slightly shifts the energies of the two Rydberg levels, and thus the frequency of the transition that separates them.

Circular Rydberg atoms are very simple quantum systems in theory, but very delicate to prepare and protect from perturbations of their environment in practice. In order to drive the atoms from their ground state to the circular state, we had to adapt the method invented by Kleppner to our experimental setup, combining laser and microwave pulses with the application of perfectly calibrated electric and magnetic fields and maintaining the atomic beam in a cooled chamber at a very low temperature, less than 1 kelvin. We learned to control the speed and position of these atoms as well. By exciting them in steps with spatially and temporally separated laser pulses, we managed to select in the atomic beam those atoms with well-defined velocity, within a few units between 200 and 300 m/s. This velocity filtering allowed us to know the position of the atoms at each instant, from the zone where they are carried in the circular state upstream to their detection after crossing the cavity downstream. In recent iterations of the experiment, we've also prepared very slow circular atoms, moving at speeds on the order of just a few meters per second. These are obtained by exciting previously laser-cooled atoms in a vertical atomic beam.

The development of these complex manipulations kept us busy for several years, in parallel to our work perfecting the mirrors of our cavities. During this period we verified a principle of complementarity that has nothing to do with quantum physics: the more we try to simplify an experiment on the conceptual level by controlling all its parameters as clearly and precisely as possible, the more complex the procedures become on the technical level. Conceptually, we wanted to establish a maximally simple situation: an atomic electron rotating on a circle that interacts with a few photons oscillating between two mirrors. The two interacting systems are both waves and particles, revealing, depending on the measurements made, one aspect or the other of their quantum duality.

Quantum ping-pong

The crucial test that showed us that the atom-cavity system was behaving ideally for our cavity QED experiments was observing the Rabi oscillation in the vacuum field. I had dreamed of this since the days of my thesis and my exchange with Anatole Abragam in 1971. It was easy to compute the evolution of circular Rydberg atoms coupled to the fluctuations of the vacuum field between the mirrors of our cavity. The dressed-atom theory (discussed in the previous chapter) indicated that an atom prepared in the circular state $n = 51$ and entering the initially empty cavity should, by oscillating between this state and the circular state $n = 50$, periodically emit and absorb a single photon. This phenomenon had to occur at the vacuum Rabi frequency $\nu_{R,0} = 50$ kilohertz, whose value depended only on the radius of our giant atoms and the volume occupied by the field in the cavity.

For this oscillation to be observed, the frequency of the atomic transition between the states $n = 51$ and $n = 50$ (henceforth e and g) had to be exactly equal to that of the cavity field, a resonance condition that could be achieved by adjusting the amplitude of the electric field between the mirrors. This oscillation corresponds to a periodic evolution of the atom-field system between the states $|e,0>$ and $|g,1>$, describing respectively the atom in state e in the presence of the vacuum and in state g in the presence of a single photon. It can be interpreted, as we have seen in the previous chapter, as a quantum interference effect between the probability amplitudes associated with two paths followed by the system in the energy diagram of the atom dressed by the cavity field. To observe this oscillation, the photon must survive between the mirrors for a long time compared to the period $1/\nu_{R.0}$, equal to 20 microseconds. We were thus able to observe it as early as 1996, with a still-imperfect cavity that kept the photons for only about a hundred microseconds.

Each circular atom crossed the cavity following a horizontal trajectory. We injected it in the *e* state and allowed it to interact with the field in the cavity for a fixed time *t* before interrupting the atom-field coupling by abruptly modifying the electric field between the mirrors. As the resonance condition was no longer satisfied, the evolution of the system was frozen at this instant and each atom at the exit of the cavity was detected by ionization in the state it had reached at time *t*.

By repeating the experiment a large number of times on atoms crossing the cavity successively, we obtained the probability of finding them at time *t* in the *e* or *g* state. The instant of observation was then changed and the accumulation of data started again at another instant *t'*. The probability of finding atoms in *e* or *g* as a function of time was finally reconstructed. This probability presented the typical oscillation signal for which we had eagerly waited since the Rydberg atom maser experiments we had performed about ten years before. The coherent exchange of a quantum between the atom and the cavity could be seen as a game of quantum ping-pong, which fascinated us upon first observation.

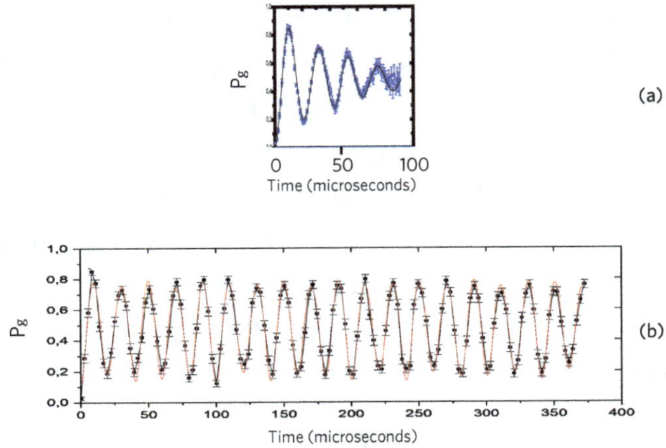

Figure VII.3. Rabi oscillations in the vacuum field. (a) The signal observed in 1996 with circular Rydberg atoms crossing the cavity in about 100 microseconds. The lifetime of the photon in the cavity is 220 microseconds. By injecting the atoms one by one in the *e* state and repeating the experiment a large number of times for increasing values of the atom-cavity interaction time, we observe that the probability of finding the atom in the *g* state as a function of time oscillates regularly, revealing a game of quantum ping-pong: the atom and the field reversibly exchange a quantum of energy. At the maxima and minima of the oscillation, the atom-field system is in the states |g,1> and |e,0> respectively. Between the two, the system is in an entangled state between the atom and the field (from Brune et al., *Physical Review Letters* 76 (1996), 1800). (b) The same signal observed with "cold" circular Rydberg atoms in a cavity keeping the photon for 10 milliseconds. For this experiment the atomic beam is vertical. The rubidium atoms are cooled by laser irradiation in a MOT trap located under the cavity. They are injected into the cavity by a vertical laser pulse at a velocity of 8 m/s. We then prepare them in the circular Rydberg state inside the cavity by subjecting them to optical excitation followed by interaction with a circularly polarized radiofrequency field produced by electrodes surrounding the mirrors of the cavity. The quantum ping-pong game is now much longer, with about twenty exchanges instead of three or four (experiment by Fréderic Assemat and Sébastien Gleyzes, 2019).

We then observed only three or four oscillations. Their number was limited by the lifetime of the photon and also by the time, on the order of a hundred microseconds, that each atom, flying at a thermal velocity on the order of 200 to 300 m/s, took to cross the cavity. We have since observed much more numerous vacuum Rabi oscillations by making cavities of high quality factor interact with slow laser-cooled atoms. These atoms, injected vertically from the bottom of the cavity to the top, cross it in half a millisecond, which now allows us to observe up to twenty complete oscillations.

The Rabi oscillation in the vacuum field creates a quantum entanglement between the atom and the field. We can set the interaction time between these two systems so they are at the atom's exit from the cavity in the state $|e,0> + |g,1>$, which presents the non-local quantum correlations Einstein analyzed in his famous EPR paper (see Chapter 5). Generating this entanglement requires only a quarter Rabi oscillation, so we were able to carry out experiments that demonstrated this phenomenon as early as 1996.

Quantum knitting

The possibility thus opened to us of studying the entanglement of atom and field was exciting because it brought us closer to a current of research developing very actively at the time: quantum information, which sought to exploit quantum logic to renew methods of computation and communication. The manipulation of individual quantum systems, then, was seen in a new light.

The ability to embed information in quantum systems that could be manipulated and detected by the methods being developed in laboratories by researchers working with ions, atoms, or photons stimulated the imagination of computer scientists and mathematicians, who invented algorithms to exploit quantum strangeness in order to compute or communicate faster or more efficiently than conventional devices.

Conversely, the appearance of these algorithms pushed atomic physicists to orient their research toward quantum information by presenting their experiments as demonstrations of elementary quantum computation or communication procedures. Artur Ekert, an English quantum information theorist, played an important role in the mid-1990s in bringing the community of atomic physics experimenters closer to that of computer scientists. He once explained to me at a conference that our experimental arrangement of Rydberg atoms in a cavity was an ideal device to make a quantum gate, the basic element of what could one day be a quantum computer. I came out of this exchange with the impression that I had been making quantum information without knowing it!

To some extent, the basic research we loved so much was taking a more applied turn. I must say my group and I resisted this trend somewhat, as we felt there were many fundamental points that needed to be clarified before we turned to applications. We also felt that by presenting our work as having applications too early we risked "overselling" it—a recurring problem in research.

Nonetheless, while continuing to improve our cavities, we followed the quantum information trend and oriented some of our work in this new direction by exploiting the characteristics of Rabi oscillations in the vacuum field. We were particularly interested in the properties of the entangled states of the form |e,0> + |g,1> prepared by this oscillation when limited to a quarter Rabi period.

By detecting the atom at the exit of the cavity, we caused the *collapse at a distance* of the field entangled to the atom. If we found the atom in the *e* state, the field was in vacuum. If we found the atom in *g*, the field became a one-photon state. If we mixed the atomic states *e* and *g* with a microwave pulse after the atom's exit from the cavity, the final measurement of the atom prepared the field in the state |0> + |1> or |0> - |1>.

We could also prepare a coherent superposition of photonic states by sending into the cavity an atom in the superposed state |e> + |g> before subjecting it to a half-period Rabi pulse in the vacuum field. The result of such an operation was easy to predict thanks to the superposition principle. The atom initially in the |e> state had to end up in the |g> state after depositing a photon in the cavity. If it was initially in |g>, it had to remain there, leaving the cavity empty. If it was prepared in the superposition of these two states, we had to find at the output the superposition of the two final situations, a state where the atom came out in |g> with a field prepared in the state $|0> + |1>$. Conversely, this state could be retransformed into an atomic superposition by sending a second atom through the cavity in the g state, where it also underwent a half Rabi oscillation. These experiments demonstrated the operation of a quantum memory that writes an atom's quantum state in the cavity field and then reads it back with another atom.

By refining these experiments, we were able to entangle two or even three atoms following each other in the device through their interaction with the field. These "quantum knitting" experiments exploited the combination of Rabi pulses lasting a quarter-period (Rabi $\pi/2$ pulse), a half-period (π pulse), and a full period (2π pulse). They could be seen as demonstrations of elementary steps in quantum information processing. The expertise acquired during the 1990s in the art of juggling atoms and photons was of great use to us in preparing the fundamental physics experiments we were planning with a cavity that trapped photons for more than a tenth of a second.

How to see photons without destroying them

Throughout these years, our dream was still to juggle with photons and atoms to control light without destroying it, and to prepare exotic field states in our cavity. In 2006, this dream became a reality. Before describing our work, let's return for a moment to the virtual world of thought experiments. This will allow us to introduce simply the principles that guided us in developing our methods of non-destructive light control. The guiding idea is that it is absolutely necessary to avoid basing the measurement of the field on a simple energy exchange between light and matter, in which we know the photons are destroyed. It is therefore necessary to extract information from the photons in a more delicate manner. The first idea that comes to mind is to detect not the energy of the photons, but their momentum.

Imagine that our facing mirrors are free to move in space, in an inertial reference frame, far from any gravitational mass, and that captive photons between these mirrors go back and forth without any loss in reflection. With each bounce off the mirrors, the photons' momentum changes sign, and the conservation of this quantity implies that the mirrors see their own momentum increase. Under the effect of this radiation pressure, the cavity thus undergoes a continuous expansion, the separation of the mirrors increasing over time. By measuring the force on the mirrors, we could thus in principle count the photons bouncing in the cavity without destroying them. This is a virtual experiment, of course, even more impossible to realize in practice than Einstein's photon box.

Note that energy conservation plays an important role here. The acceleration of mirrors is accompanied by an increase in their kinetic energy. This can only be taken from that of the photons, whose frequency must decrease after each bounce off the mirrors. This redshift of the light captive in the cavity is in fact a Doppler effect. An observer straddling one of the moving mirrors would see the light coming toward him shifted toward the long wavelengths, and this shift would be doubled after the bounce. The expansion of the cavity is thus accompanied by a spreading of the wavelengths of the photons it contains, much as the expansion of the universe caused the cosmological redshift of radiation from the Big Bang.

As maintaining a floating cavity in interstellar space is impractical, let's imagine a slightly more realistic experiment. Our mirrors are now fixed in our laboratory on Earth, still trapping a set of visible light photons bouncing losslessly between them. Suppose we insert a thin plate of perfectly transparent glass between the mirrors. The photons' propagation speed is divided by a factor equal to the glass's refractive index, which changes the wavelength of the radiation inside the plate. But, like a violin string fixed between two points, the field mode must always contain a whole number of crests between the mirrors, which have not moved. To compensate for the longer propagation time inside the plate, the field wavelength must decrease outside of the plate, which requires the field frequency to shift slightly. The insertion of the plate is therefore accompanied, once again, by a small energy exchange between photons and matter.

Analyzing the experiment more closely, we can see that the photons undergo a slight deviation if they pass the edge of the plate as it is entering the mode. This deflection changes their momentum and, in return, that of the plate in the opposite direction, according to the principle of action and reaction. The work of this force acting on the plate modifies its kinetic energy slightly.

If this change could be measured, a non-destructive photon counting would still occur. This thought experiment thus recalls Young's moving interferometer slit, as described in Chapter 5. It's just as infeasible in practice, but it brings us closer to reality.

Now let's replace the macroscopic transparent slide that doesn't absorb visible light with a single Rydberg atom *non-resonant* with a microwave field stored in our cavity. The force the atom undergoes from the field will, as happens to the virtual transparent slide, result in a small change in the atom's energy. By detecting this change, we will be able to extract information about the number of photons present in the cavity without destroying them. Of course, the atom must be extraordinarily sensitive to the electromagnetic field, a property possessed by our circular Rydberg atom.

Note, however, that even if it doesn't destroy them, the measurement disturbs the photons observed in this manner. As in the virtual experiment with the transparent slide, the photons' frequency is slightly altered while they interact with the probing atom. This effect will allow us to control the field phase in the cavity, to prepare it in a quantum state reminiscent of Schrödinger's famous cat, suspended between two classically different states. Counting photons without destroying them and studying the ambiguous states that the Austrian scientist considered as burlesque theoretical curiosities, then, are related experiments that reveal complementary and fundamental aspects of quantum physics.

Life and death of a photon

The simplest experiment we performed in the summer of 2006, as soon as we could trap the field in our super-cavity, consisted in continuously detecting a photon for the first time without destroying it. We used circular atoms that were non-resonant with the field, playing the role of the transparent plate mentioned above. Sending the atoms one by one into the cavity, whose mirrors were cooled to 0.8 kelvin, we measured the residual thermal radiation of the cavity mode. Planck's law predicts that at this temperature the field in the cavity will be in vacuum (with a probability of 95 percent) or contain a single photon (with a probability of 5 percent). The number of photons in this case is therefore a binary parameter, taking the values 0 or 1, and only exceptionally does the cavity contain two or more photons. So we had to execute an optimal procedure whereby the atom, also a binary system, was found at the exit of the cavity in the circular state e ($n = 51$) if there was a photon in the mode, and in the circular state g ($n = 50$) if the cavity was empty.

The cavity field had to control the state of the atom without being absorbed in the process. If an atom signaled the presence of a photon, other atoms following each other in the cavity had to confirm this result. In physics jargon, this is called a *quantum non-demolition* procedure. This type of field measurement differs fundamentally from ordinary photodetection, which, after a click indicating the presence of a photon, would leave the cavity empty for the next probe atom.

We ran the experiment for the first time on September 11, 2006, on my sixty-second birthday. In the middle of my birthday dinner, I received a phone call from Stephan Kuhr, the German postdoctoral fellow piloting the experiment that evening. After a day of delicate fine-tuning, he told me, he was ready to start the measurements. Before digging into my cake, I rushed to the laboratory in time to see the long-awaited signal on the screen of the computer controlling the experiment: a telegraphic sequence of vertical red bars indicating that the atoms were coming out of the cavity in level *e*, saying that the cavity contained one photon, followed by a sequence of blue bars corresponding to level *g*, telling us that the cavity was empty.

The experiment was repeated every second, each time showing a different trace, always with red intervals interspersed with blue. The red sequences, coming from hundreds of atoms passing through the cavity one after another, showed clearly that the method was nondestructive to light. Statistically, the average duration of the red intervals occupied 5 percent of the total observation time, and that of the blue intervals 95 percent. All the properties of Planck's law were unveiled right before our eyes.

After so many years of effort, the joy of seeing nature reveal one of its fundamental properties was unspeakable. I went home to deliver the good news to Claudine and my family, who had joked while they waited for me about physicists' mania for long work hours. The champagne that flowed that evening was as much for the success of the experiment as it was for my birthday. I later brought some to the ENS, where we drank it—moderately—while continuing to record, throughout the night, these first data from which we could not look away.

How do we acquire photon-number information about the state of the Rydberg atoms passing between mirrors? This cannot be explained by Rabi oscillation, since the frequency of the transition between the *e* and *g* levels is slightly out of tune with the frequency of the

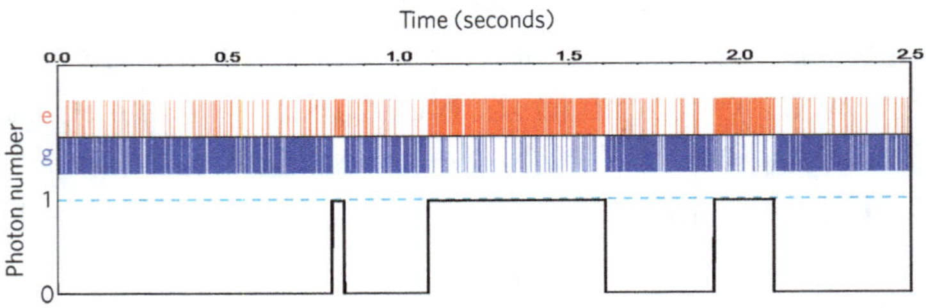

Figure VII.4. The life and death of a photon. The atoms, following each other in the cavity at a rate of two hundred per second, non-destructively measure the number (0 or 1) of thermal photons trapped between the mirrors (at the temperature of the experiment, the probability that the cavity contains two or more photons is negligible). The red and blue bars correspond respectively to atoms detected in the states e (n = 51) and g (n = 50). The signal is composed of sequences of atoms detected mostly in the same state, interrupted by quantum jumps corresponding to the birth or death of a photon. The detection of atoms in the level opposite that of the majority is caused by experimental imperfections. The photon detected in the middle of this trace survived for more than half a second (three times the average photon lifetime in our cavity) (per *Nature* 446 (2007), 297).

cavity mode. A frequency detuning δ of 100 kilohertz is sufficient to make any photon absorption by the atom negligible. Each atom, thus made "transparent," modifies the field frequency by an index effect, just as a glass microscopic plate inserted in the cavity would. This frequency increases or decreases slightly, by an amount $\Delta\nu$ on the order of a few kilohertz, depending on whether the Rydberg atom is in the *e* or the g level. The value of this frequency shift depends on the frequency detuning δ between the atom and the cavity mode, and can be finely adjusted. The total energy of the cavity when it contains *N* photons thus changes by $\pm Nh\Delta\nu$. By the conservation of energy, the atom between the mirrors sees the energy of its levels changed by the same amount, proportional to the number of photons. This effect is none other than the light shift discovered by Claude Cohen-Tannoudji, which, as we have seen, plays an important role in atomic laser cooling experiments.

The frequency shift of the field and that of the atomic energies are thus two complementary effects, expressing in a way the principle of action and reaction in the non-resonant interaction processes between matter and light. Measurement of the number of photons is thus reduced to measurement of the energy $Nh\Delta\nu$ transferred from the field to the atom during this non-resonant interaction. This energy is extremely small compared to the total energy $Nh\nu$ of the field. Indeed, the relative frequency shift of the cavity induced by an atom, $\Delta\nu/\nu$, is on the order of 10^{-7}, or one ten-millionth. The price paid in energy to determine the number of photons in the cavity is thus tiny compared to the cost of a measurement by photoelectric effect, which would absorb the photons in the cavity.

In fact, upon closer inspection, non-destructive measurement ultimately costs nothing in energy. The energy $Nh\Delta\nu$ is exchanged between the field and the atom only transiently. When the atom leaves the cavity after crossing it, matter and light stop interacting and each of the two systems regains its initial energy without changing the number of photons. It is therefore more accurate to say—if we wish to maintain our trading metaphor—that the atom takes a very small fraction of the field energy for a short time, then returns it before being measured. This brief borrowing allows it to record valuable information about the number of photons, which we must then decode.

A classical way to do this decoding would be to measure the direct effect of the force exerted on the atom by the field, which temporarily slows or accelerates it in its motion depending on whether its energy increases or decreases as it passes between the mirrors. Instead of making this direct measurement, which is practically impossible, we implemented a quantum interferometry method based on the measurement of the phase of a superposition of atomic states passing through the cavity. Before each atom enters the cavity, it is subjected in a small auxiliary cavity R_1 to a microwave $\pi/2$ pulse carrying it from the g state in a superposition of equal weights of the two circular Rydberg states e and g. This superposition prepares, as we have seen, an electric dipole that starts to rotate in the plane of the circular orbit at the frequency of the atomic transition *shifted* by the quantity $2N\Delta\nu$ when N photons are in the cavity (one of the levels is shifted by $+Nh\Delta\nu$ and the other by the opposite quantity $-Nh\Delta\nu$, whence the factor 2 in the expression of the frequency shift of the superposition).

The photon-number information is thus written in the phase of this rotating dipole. The phase accumulation induced by the N-photon field during the time T spent by the atom inside the cavity is $4\pi N\Delta\nu T$. The circular Rydberg atoms' sensitivity to the field is so great that the *phase shift per photon*, $4\pi\Delta\nu T$, can reach the value π (or 180°). To measure the phase of the superposition of states, a second microwave pulse is applied to the atom, at its exit from the cavity, in a second small auxiliary cavity R_2 identical to R_1. The atom's final state is then measured in an ionization detector placed behind R_2, the result being a binary unit of information indicating that the atom is found in *e* or *g*. The set of auxiliary cavities R_1 and R_2 surrounding the cavity that stores the photons is called a *Ramsey interferometer*, named after the American physicist, student of Rabi and mentor of David Wineland, who invented it in 1949 to measure precisely the quantum phase of a system oscillating between two levels.

Between its initial preparation in the *g* state and its final detection in *e* or *g*, the atom follows two paths whose probability amplitudes interfere in the device. If it is finally found in *e*, it is either because it switched from *g* to *e* in R_1 and remained in *e* while passing through R_2 or because it remained in g after the pulse R_1 and transited from g to e in R_2. In the first case it was inside the cavity in the e state; in the second, it crossed it in the g state. As the experiment cannot give any information on the state of the atom between R_1 and R_2, the probability amplitudes associated with these two paths add up and cause an interference phenomenon. To observe it, it is sufficient to scan around the atomic frequency ν_0 the frequency ν of the auxiliary microwave realizing the $\pi/2$ pulses in R_1 and R_2. The probability of finally finding the atom in e varies periodically between 0 and 1 when scanning ν, constituting what is called a *Ramsey fringe signal*. The more separated the two pulses R_1 and R_2 are in time, the narrower the fringes. The fringes' sensitivity to small variations of ν is thus greater the slower the atom is, taking more time to pass through the cavity from R_1 to R_2. If instead of recording the probability of finding the atom in e we measure the probability of detecting it in g, we find Ramsey fringes in phase opposition, the two fringe arrays simply exchanging their maxima and minima.

The appeal of this interferometric method lies in its sensitivity to the phase variation of the superposition of states being studied. When the phase difference between the two probability amplitudes changes from π, the Ramsey fringes translate by exactly half an interfringe, a maximum of probability becoming a minimum. By setting the frequency detuning of the Rydberg atom with the cavity so that the phase shift per photon $4\pi\Delta\nu T$ is equal to 180°, we obtain fringes in phase opposition

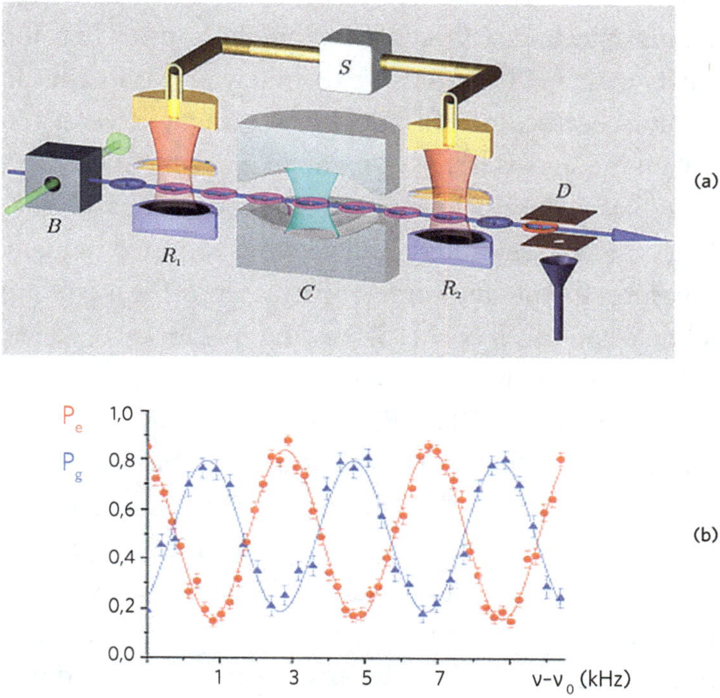

Figure VII.5. The Ramsey interferometer used to count photons without destroying them. (a) Sketch of the experimental apparatus: the circular atoms produced in box *B* pass through cavity *C* before being detected in *D* (ionization in an electric field selectively measuring states *e* and *g*). They are subjected before and after the crossing of C to two microwave π/2 pulses in the auxiliary cavities R_1 and R_2 that surround the main cavity C. These microwave pulses are produced by the coherent source *S* (per *Nature* 446 (2007), 297). (b) Probability of detecting the atoms in *e (red dots)* and in *g (blue dots)* as a function of the frequency *ν* of the microwave swept around the atomic frequency ν_0. The phase shift per photon of the atomic dipole is equal to π. By setting *ν* to adjust the interferometric signal to the maximum of one fringe, we detect atoms in one state or the other depending on whether the field contains an even or odd number of photons.

depending on whether the cavity contains 0 or 1 photon. Fixing ν to find with certainty the atom in e when the cavity contains 1 photon, we definitely find it in g if the cavity is empty. It is this ideal setting that allowed us to obtain the signals observed on September 11, 2006. Let's note that if the cavity contains a larger number of photons, the Ramsey interferometer thus adjusted actually measures the photon-number parity, since the atoms are all detected in e if $N = 1, 3, 5\ldots$ and in g if $N = 0, 2, 4\ldots$

The analogy between Ramsey's and Young's interferometers is striking. The Ramsey zones R_1 and R_2 play a similar role to that of Young's two slits. The inability to know in which zone, R_1 or R_2, the atom has undergone the transition from g to e maintains the quantum ambiguity that leads to the observation of the Ramsey fringes, just as the inability to know through which slit the particle passes in the slit experiment allows Young's interference fringes to build up statistically on the detection screen. Changing the phase accumulation of the atomic dipole between R_1 and R_2 by depositing a photon in the cavity is analogous to the interposition of a transparent plate behind one of Young's slits. By changing the path difference, this plate will move the fringes, just as the photon in the cavity moves the Ramsey fringes. Note the complementary roles of the atom and the field. Matter and field play the role of a dispersive plate, the atom changing the field's phase (and thus its frequency) while the photon modifies the phase of the atomic superposition that crosses the cavity.

The frequency shift $2\Delta\nu$ undergone by the atomic transition between the states e and g when the non-resonant cavity contains a photon is related, by the theory of the dressed atom, to the frequency $\nu_{R,0}$ of the atom's Rabi oscillation in the vacuum field. It is shown in a good approximation that $2\Delta\nu$ is equal to $\nu_{R,0}^2/\delta$. When the frequency detuning δ is on the order of $2\nu_{R,0}$, which is enough to make the probability of photon absorption by the atom negligible, the dispersive effect of the atom-field coupling thus produces on the atomic transition a frequency shift $2\Delta\nu$ on the order of $\nu_{R,0}/2 = 25$ kilohertz. The time T needed for the non-resonant atom to accumulate in the interferometer a phase shift of π radians per photon is on the same order as the time it takes, when resonant, to accomplish a complete Rabi oscillation in the cavity vacuum field. In both cases, the atom is able to acquire information about the photon in a very short time, less than a thousandth of the lifetime of the field in the cavity. The atom's extreme sensitivity to microwave fields proved essential to all the cavity electrodynamics experiments on photon counting and manipulation that we performed from 2006 onwards.

Back to Young's moving slit experiment

Developing our Ramsey interferometer led us, in the early 2000s, to reflect on the difference in nature between the Rabi oscillations induced by the vacuum in our cavity and the microwave pulses that prepared the superpositions of atomic states in the Ramsey zones surrounding it. The field that toggles the atom between *e* and *g* in these zones is classical. It acts on a Rydberg atom and prepares it in a superposition of states without itself being modified. The photons injected in these zones remain there only for a very short time and cannot record any information about the state of the atoms they influence. These zones thus behave like the fixed slit in Young's interferometer from the Einstein–Bohr thought experiment.

The field of the cavity oscillating between 0 and 1 photon when an atom passes through it, on the other hand, is analogous to the moving slit of the device in the thought experiment because its final state is entangled with the atom's. This difference gave us the idea of making an interferometer that combined the two pulse forms, the first mixing the states *e* and *g* in a quantum way, entangling the atom and the cavity field, and the second making the atom undergo a classical microwave pulse. We could thus perform an interference experiment very close to the one Einstein and Bohr had imagined with their moving slit.

By passing circular Rydberg atoms through this modified Ramsey device, we found that there were no fringes. Indeed, the first pulse left theoretically readable information about the path followed by the atom in the device. If the cavity field had been found in the vacuum, the atom would have crossed the interferometer in the *e* state, whereas it would have done so in the *g* state if a photon had been counted in the cavity. It was not necessary for the field to be detected for the fringes to disappear. The simple fact that this detection was possible was enough to erase them.

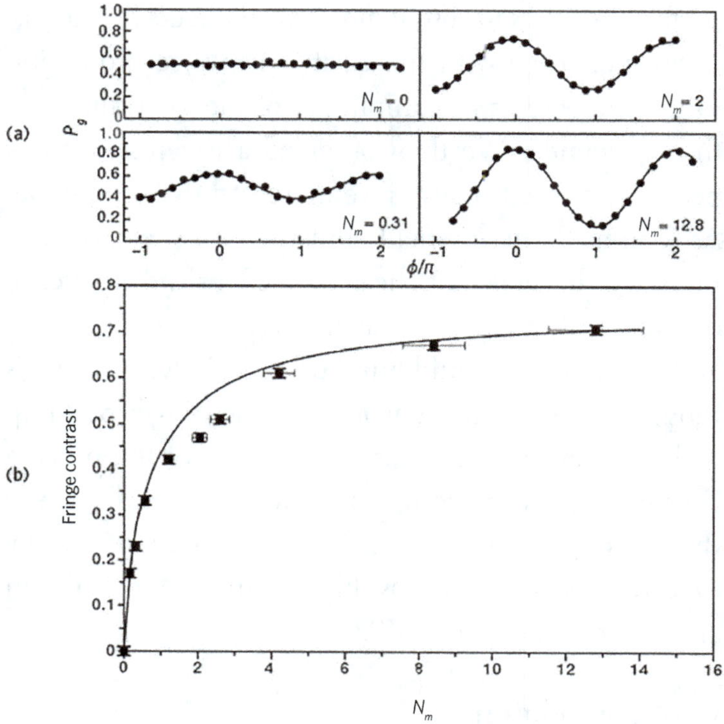

Figure VII.6. (a) Fringes of a Ramsey interferometer in which the first $\pi/2$ pulse is produced by a coherent field containing a very small average number N_m of photons, the second pulse being classical. This is the equivalent of Young's thought experiment with a small moving slit. We see that the contrast of the fringes, zero for $N_m = 0$, increases progressively as the field becomes larger and larger. (b) Fringe contrast as a function of N_m (experimental points and theoretical curve) (per *Nature* 411 (2001), 166).

To complete the experiment, we then executed the first pulse using a small coherent field of increasing amplitude injected into the cavity by coupling it for a brief moment with a conventional microwave source. To ensure that the pulse always caused the atoms to perform a $\pi/2$ Rabi oscillation, we reduced the atoms' interaction time with the cavity in inverse proportion to the increase in the amplitude of the quantum field thus generated. We then observed a progressive reappearance of the fringes as we increased this amplitude. As soon as the cavity field contained on average a few photons, the fact that the atom left or did not leave one more photon in the cavity ceased to be detectable and interferences could build up. Similarly, in Young's thought experiment, we would see the fringes build up with an increasing contrast as the mass of the moving slit increased to make it more and more classical. We thus illustrated the principle of complementarity by following as closely as possible the arguments Einstein and Bohr exchanged in 1927.

Counting photons and observing quantum jumps

Back to the experiments carried out with our supercavity. After observing single photons without destroying them, we generalized the method to non-destructively count photon numbers greater than one. As we will see, this experiment has analogies to Einstein's and Bohr's other famous thought experiment, that of the photon box.

To fully understand the experiment, we must specify the nature of the field we injected into the cavity. It was a small coherent field produced by a classical source, emitting a very short microwave pulse radiating inside our empty enclosure in a direction transverse to the mirrors' axis. A few photons scattered by the edges of the reflecting surfaces fall into the cavity mode and remain trapped there for about a hundred milliseconds. The preparation of the field is analogous to the excitation of a bell by a short percussion, producing a long vibration that is progressively damped. The field thus generated oscillates with a defined phase, that of the classical source that created it. In our experiments this source is strongly attenuated and the average number of photons injected into the cavity is on the order of a few units.

This number, as well as the field phase, shows significant fluctuations so as to respect the Heisenberg uncertainty relation discussed in Chapter 5. We can associate to this coherent field a complex number whose amplitude and phase are the averages of the fluctuating variables of the quantum field. This number is represented by a vector, a small arrow in the Fresnel plane of the radiation, as we saw in Chapter 3. The "fuzziness" associated with quantum fluctuations can be described as an uncertainty about the exact position of the tip of this vector.

Quantum fluctuations reflect the fact that the field of our classical source is prepared in a coherent superposition of states of different photon numbers. These superpositions and their properties were studied in detail by the American physicist Roy Glauber, hence the name often given them: Glauber states. The probability of finding by non-destructive counting the superposition's different *N* values is theoretically given by a bell curve centered around a mean value N_m, with a dispersion on the order of $\sqrt{N_m}$ around this value. The formula that describes this curve is called *Poisson's law*, named after the nineteenth-century mathematician, a contemporary of Fresnel, whom we met earlier in our history of light. Attached as he was to the corpuscular description of radiation, as we have seen, he doubted its undulatory nature. Ironic, then, to see his name two centuries later given to a law reconciling these two aspects, expressing how quantum physics associates the two faces of radiation to give it both an undulatory phase and a corpuscular granularity.

The coherent field thus injected into the cavity is now analogous to a transparent quantum plate "suspended" between several thicknesses, shifting the phase of the atomic dipole by several different angles at once. To count the number of photons—or, if we maintain the plate analogy, to measure its thickness—we send a succession of circular Rydberg atoms into the Ramsey interferometer. Each one gives us, by its detection in *e* or in *g*, a binary unit of information. After a few tens of atoms passing through the cavity in a period shorter than its damping time, the number of photons is reduced to a single value.

Without going into the details of the process of acquiring this information, let us simply review the principle. Suppose the Ramsey fringes are set such that the atom cannot be detected in the *e* state if there is a number N_1 of photons in the field. In other words, the phase of the Ramsey fringes associated with N_1 photons is set so that the detection of the atom in *e* has a zero probability to occur for this thickness of the quantum plate. If the atom is detected in this state, we can conclude only that there can be no N_1 photons, as this would have led to a result incompatible with the observation. The information provided by the detected atom in *e* thus allows us to remove N_1 from the initial photon-number distribution. Another atom crossing the interferometer with another fringe adjustment will then allow us to "decimate" another photon number, N_2, and so on until only one photon number N remains. Acquiring information *by elimination* in this way is an application of what is called *Bayes' law* after the English pastor, one of the fathers of probability theory in the eighteenth century, who introduced this type of powerful reasoning for the evaluation of *a priori* unknown quantities. The principle is to estimate the probability of the values of parameters that influence an observed phenomenon and thus to obtain information on the cause of an event when we know the law linking this cause to the probability of this or that measurement result.

Here we know the law governing the probabilities of detecting the atoms in *e* or in g for the various values of N, which in this experiment play the role of parameter to be measured. These probabilities, described by the Ramsey fringes corresponding to the different photon numbers, are sinusoidal functions of the frequency ν, whose phases are shifted by $4\pi\Delta\nu T$ when N increases by one unit. By judiciously choosing the phase of the fringes for each atom, we progressively decimate different photon numbers until we finally obtain the only number that has passed the tests to which the atoms successively crossing the cavity have subjected its field. At the end of this measurement, the field, whose initial energy was uncertain, becomes a state with a well-defined photon number and energy, called a *Fock state* in honor of the Russian physicist Vladimir Fock, a quantum field theory specialist who worked in the 1940s and 1950s.

The experiment is controlled by a computer that adapts the parameters of the interferometer and records in real time the successive detections of atoms in the *e* and *g* levels. From these data the computer reconstructs, as the atoms arrive, the histogram of probabilities of the different possible photon numbers, updated using Bayesian reasoning. The sequence of atomic states detected in a few milliseconds, well before the field has had time to dampen, appears as a sequence in the form *eegeggeeeg*... that determines, like a barcode, the final photon number created by the measurement.

Once this number is obtained, the computer continues to decipher the information provided by the atoms following each other in the interferometer.

These atoms confirm for a time the result initially found demonstrating that our counting method is indeed non-destructive. However, absorption of the field in the mirrors eventually occurs, causing the photon number to suddenly change from *N* to *N-1*, then to *N-2*... until a vacuum is reached. Each repetition of the measurement on the same initial field generates a signal in descending steps.

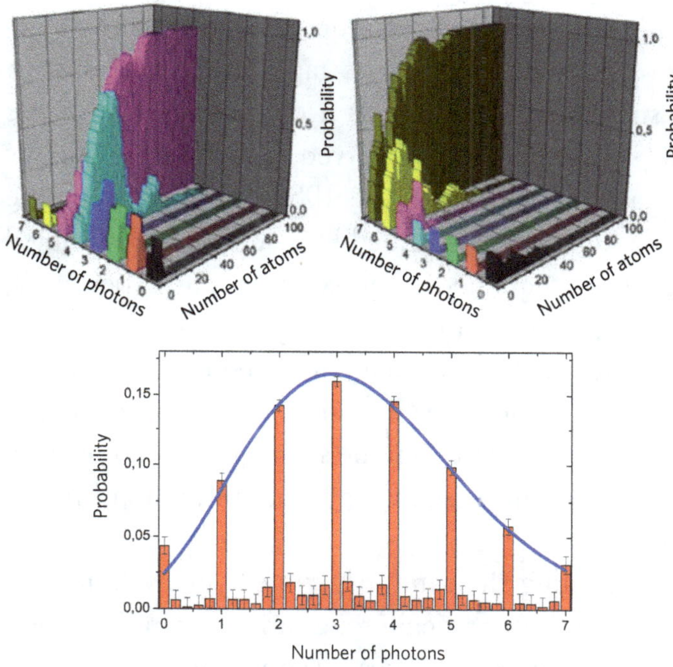

(a) (b)

Figure VII.7. (a) Evolution of histograms representing the probability that the cavity contains different numbers of photons, as a function of the number of atoms detected, for two executions of the counting procedure. Initially, the photon-number distribution is flat between 0 and 7 because the only information available is that this number is likely lower than 8 (the source that injected the field into the cavity is adjusted so that the average number of photons in the coherent field is close to 3). As atoms are detected in each sequence, the photon-number distribution is updated by Bayesian inference. Finally the field converges to 5 photons in the left execution and to 7 photons in the right one. (b) Histogram of photon-number distribution in the initial coherent field, reconstructed after two thousand independent counting sequences. The curve corresponds to the theoretical Poisson's law (per *Nature* 448 (2007), 889).

The starting stage and the duration of each step on this staircase changes randomly with each execution of the experiment. The rapid evolution between two successive steps signals a quantum jump in the field, similar to those observed on isolated atoms by ion trappers as early as the 1980s. In their case, the atomic jumps were revealed by a suddenly interrupted telegraphic sequence of photons. In our experiment, the photon jumps are signaled by the variation of the signal provided by a telegraphic sequence of binary results of atomic measurements. The roles of matter and light are reversed, but the same quantum properties are illustrated in these two experiments, one mirroring the other.

By repeating the same experiment many times, we reconstruct the statistical properties of the initial coherent field. The probabilities of counting N photons at the start of each stepped signal obey the theoretical Poisson's law. The duration of the N photon steps varies randomly from one execution to another, with an average duration τ_N equal to T_c/N where T_c is the damping time of the field amplitude in the cavity. This $1/N$ law reflects the increasing fragility of well-defined photon-number Fock states, which become harder and harder to maintain in the cavity as N increases.

The Ramsey interferometer, which plays an essential role in our experiments, is the device used in all microwave atomic clocks developed since the 1950s. This is particularly the case of cesium clocks, which define the second used by the International System of Units and are crucial instruments of the GPS navigation system. Cesium atoms in a horizontal atomic beam are subjected to two microwave $\pi/2$ pulses on the 9.2 gigahertz transition between the two hyperfine levels of this atom's ground state. The microwave frequency, locked at the top of the central Ramsey fringe, is the clock signal that can be stable at the level of one part in 10^{14}.

To narrow the fringes and further increase the clock's accuracy, a more modern version uses slow laser-cooled atoms launched vertically upward into the Earth's gravitational field. These atoms, whose speed is a few meters per second, form a fountain-like jet, following an ascending and then descending parabolic trajectory. They undergo the $\pi/2$ pulses in the same region of space, on the ascent and then on the descent. By detecting them optically after their fall, we obtain fringes a hundred times narrower than we get with the standard horizontal beam device and a clock stability on the order of 10^{-16}, or less than one second of imprecision in the measurement of time over a hundred million years. Our experimental device is in fact a special atomic clock whose ticking is sensitive to the number of photons stored in the cavity surrounded by the two Ramsey zones. Due to the Rydberg atoms' extreme sensitivity to the field, the presence of a single photon changes the period of the clock by about one ten-millionth, or about one second in three months. This small shift, obtained without light absorption, allows our clock to be sensitive to a single photon.

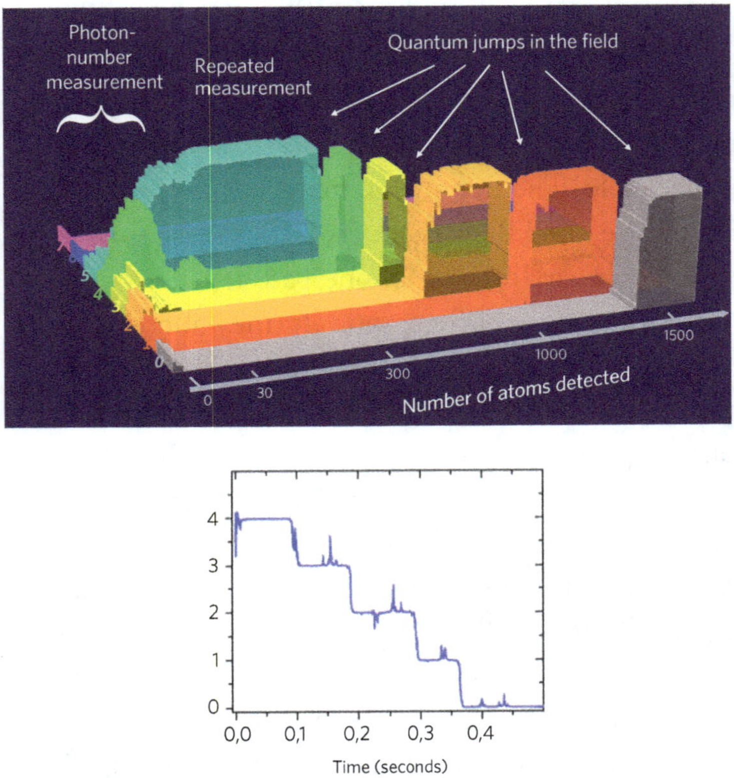

Figure VII.8. Observation of quantum jumps of the field. (a) Evolution of the histogram of the photon-number distribution in a long counting sequence. The photon number initially converges toward 5 in this execution. Repeated measurements after this convergence confirm this value before the cavity losses manifest themselves by successive quantum jumps, leading the photon number by steps toward 0. (b) Evolution of the photon number in another sequence, starting from a 4-photon Fock state. The stepwise decay of the field is clearly visible (per *Nature* 448 (2007), 889).

Even though our device is conceptually quite different from the Einstein–Bohr photon box, it has striking similarities. Both devices measure radiation energy without destroying it, the thought experiment using gravitation and ours the dispersive interaction between matter and light. In both experiments a clock is used to determine the moments when a photon leaves the box. In the device imagined by Einstein and Bohr, the clock activates a shutter. In ours, it detects the instant when a quantum leap occurs, signaling the disappearance of a photon.

Einstein's virtual box was designed to test Heisenberg's uncertainty relation between time and energy. We saw in Chapter 5 that this relation was confirmed after a subtle analysis of the box's operation. Does our experiment point in the same direction? It shows that an N-photon Fock state has a lifetime $\tau_N = T_c/N$, N times shorter than the mean time T_c of the exponential damping of the field energy in the cavity. Fourier analysis tells us that the uncertainty on the frequency of this field is $\Delta\nu = 1/2\pi T_c$. The result is that the N-photon energy is defined with an imprecision $\Delta E_N = Nh\Delta\nu = N\hbar/T_c$ and we verify the relationship $\Delta E_N \times \tau_N = \hbar$ in accordance with Heisenberg's uncertainty relation.

Our non-destructive photon-counting experiment illustrates the fundamental principles of quantum measurement. It shows how information about the observed quantum system (the field) is constructed from an interaction between this system and a measuring device (the succession of atoms passing through the Ramsey interferometer). Since each atom gives only one binary unit of information (a *bit* in computer language), it is necessary as soon as the number of photons exceeds 1 to extract the information from a sequence of atoms. The information is then built progressively and the computer controlling the experiment makes it possible to follow this progression, for the first time in a real quantum measurement. Before the irreversible measurement that creates the information (here the detection of each atom in *e* or *g*, a random process whose result can be predicted only statistically), the measured system and the apparatus are in an entangled state. If the field is in a superposition of states of different Ns, each of these states shifts the atomic dipoles by a different angle and the global system evolves in a superposition of correlated atomic and photonic states, a highly entangled state. It is the measurement of each atom that, destroying it after it has left the cavity, abolishes this entanglement and "forces" the field to undergo an irreversible transformation toward a state that depends on the information delivered by the result obtained (*e* or *g*). This irreversible evolution corresponds to the so-called *collapse* of the field's quantum state.

I have taken care, in describing this process, not to give the impression that the measurement reveals only a result that is *a priori* unknown but already present in hidden form in the cavity. Indeed, one could believe that the initial uncertainty about the photon number is of statistical origin, coming from imperfect knowledge of the system, and that the photon number is revealed by counting, like counting marbles in a box.

The quantum situation is quite different. The number of photons *does not exist* before measurement, as we have discussed in the framework of the thought experiments.

All that can be said is that the field has, in the initial Glauber state, probability amplitudes of containing different photon numbers. These amplitudes, which are complex numbers with a modulus and a phase, are likely to interfere with each other in experiments revealing the classical phase of the field. The number N that emerges from the measurement process is created by this process; it is not preexisting. Any attempt to give it a classical meaning before measurement leads to inconsistencies and paradoxes of the kind we described in Chapter 5.

This analysis of photon-number counting allows us to specify what the qualifier "non-destructive measurement" means in quantum physics. In general, during the information acquisition process, *the state of the system changes.* For example, it passes progressively from a Glauber state, superpositions of Fock states of different Ns, to a state whose photon number is well defined. The energy of this state, $Nh\nu$, generally differs from the average energy of the initial coherent state. This doesn't mean the measurement has changed the energy of the system in the classical sense. Indeed, the field energy has some initial blurring due to the dispersion of values of N in the initial superposition of states. This uncertainty is removed by the measurement, which after the collapse of the system state concentrates all the probability on a precise number of photons, somewhat like the way an optical instrument adjusted to focus the radiation passing through it makes a blurred image clear.

It may occur that the photon number thus indicated by the measurement differs by several units from the number of photons closest to N_m in the initial coherent field. These events reflect the existence of quantum fluctuations in the light intensity. When averaging the measurements over a large number of executions of the experiment, however, we always verify that the field energy is statistically conserved. Moreover, once the energy is set in this way, subsequent measurements confirm the result so long as the field is not subject to the inevitable damping effects. It is in this sense that the method can be called *non-destructive*.

Measuring the energy of a field does not exhaust all the information it contains. It focuses on determining the photon number and is not interested in the phase relations between the probability amplitudes associated with different photon numbers. These phase relations are in fact destroyed by the measurement of energy. Starting from a coherent field whose phase is defined, the measurement leads to a field in a Fock state, whose energy is known perfectly but whose phase is totally indeterminate. The acquisition of information about the system's energy completely muddles the acquisition of information about its phase, illustrating Bohr's principle of complementarity. The photon counter that reveals the field's corpuscular aspect is utterly insensitive to its wavelike aspect.

Quantum field radiography

Our experiment then presented another challenge. Could it, with a simple modification of the device, allow us to decode all the information contained in any quantum field—in other words, to reconstruct its complete quantum state? This reconstruction must result from the processing of a large number of measurements of different variables, sensitive either to the intensity of the field or to its phase. As a measurement on an observable erases any information on the complementary variable, the reconstruction cannot be done on a single execution of the experiment. It is necessary to have a large number of identical systems prepared in the same state, and to carry out averages of measurements on them.

This emphasizes an essential property of quantum physics, which says that a system's state is a statistical concept and that it is impossible to determine the wave function of an unknown system if we have only one realization of it. In other words, a system's quantum state or wave function is not a real object in the classical sense. It is a mathematical abstraction constituting the system's quantum identity, making it possible to determine the probabilities of the results of the measurements we could perform on it. This quantum identity cannot be copied by a duplicating device without erasing the information in the original system. This is called the *quantum no-cloning theorem*.

If quantum cloning were possible, it would indeed suffice to include a quantum copier in a measuring instrument that performed measurements of complementary variables on different subgroups of copies. The complete quantum state would thus be reconstructed from a single system, giving the wave function a status of individuality and reality contrary to the postulates of quantum physics. The only way to reconstruct a quantum state is to make measurements on a large number of specimens obtained by the repeated preparation of this state on identical physical systems.

In our experiment, we can generate in unlimited number sets of fields in various quantum states. These may be coherent states containing a few photons on average, prepared by repeatedly coupling the cavity to a classical source, as we have seen above, or Fock states obtained by non-destructive counting of the photons of these coherent states. Various quantum superpositions of the field can also be prepared in a reproducible way, as we will see later. A measurement is then performed on each of these realizations, whereupon the field is allowed to return to the photon vacuum and the sequence is restarted on a new preparation of the system in the same initial state, and so on.

The photon-counting experiment carried out on these sets is part of the work of state reconstruction. In order to obtain additional information on the field's phase, a coherent "probe" field of controlled amplitude and phase is injected into the cavity immediately after the preparation of the state to be reconstructed. The source of this tunable probe field, coupled to the Ramsey interferometer, then becomes an integral part of our measuring device. The field to be measured and the probe are added in the cavity and the number of photons in the field resulting from their interference is immediately counted. After a large number of repeated measurements, we obtain the histogram giving the probability distribution of the photon numbers in the resulting field. This distribution gives information on the state of interference between the initial field and the probe field. We then start over with probes of different amplitudes and phases. All this information, stored in the computer controlling the experiment, allows us to reconstruct the complete quantum state of the identical systems initially prepared. This method presents common points with holography, whose principle I recalled at the beginning of this chapter, as it consists in acquiring phase information on a light field by carrying out measurements of intensity after mixing this field with reference fields of known phase and amplitude.

This process is called *quantum tomography* because the information acquisition is analogous to the method bearing this name in radiology, which consists of detecting the intensity of X-rays transmitted by a patient's body by irradiating it from different angles so as to reconstruct a two- or three-dimensional image. In our case, combining the field under study with waves of varying amplitude and phase is equivalent to changing the angle at which the X-rays irradiate the patient's body in medical tomography, known as a *CT* (computer tomography) scan. The intensities collected by the X-ray scanner's detectors from the different directions of irradiation are raw data that a computer transforms into an image of the radiographed body. Similarly, in quantum tomography, the information contained in the histograms associated with combinations of the field to be measured with probes of various amplitudes and phases is encoded in numbers that the computer transforms into a three-dimensional image, a visual representation of the field's quantum state. By observing these images, we can deduce the properties of the quantum states thus reconstructed. As in medical radiography, this requires the trained eye of the specialist. I will simply describe here what these images represent.

In the case of a coherent field, the classical physicist is on familiar ground. The state of the field is indeed represented in the Fresnel plane, a scene we studied in detail in Chapter 3. A classical field is described there at a given time by a vector whose origin can be made to coincide with that of the coordinate axes. The tip of this vector, a point in the Fresnel plane, summarizes all the information about the amplitude and phase of this field. If we wish to add a third dimension to this description, we will represent a classical field with an infinitely fine vertical peak standing in the horizontal Fresnel plane at the tip of the Fresnel vector. When the quantum fluctuations of a coherent state are taken into account, this peak turns into a hill whose mathematical profile is given by a bell-shaped function called a *Gaussian function*, named after the German mathematician Carl Friedrich Gauss (mentioned in Chapter 3), who studied it and its important role in probability theory. This Gaussian hill is centered at a point at a distance of $\sqrt{N_m}$ from the origin and its circular base has an area equal to 1 so as to satisfy the uncertainty relation between the amplitude and phase of the quantum field.

The vacuum is a particular case of a coherent field, represented by a hill at the origin of the Fresnel plane's axes. The extension of its base then describes the fluctuations of the vacuum field. Coherent fields of finite amplitude are represented by identical Gaussian hills, simply translated in the Fresnel plane. When the mean number of photons becomes very large, the radius of the base of these hills (equal to 1) becomes negligible compared to their distance from the origin and the quantum effects fade away. These images can be generalized to any quantum field, represented in the Fresnel plane by a landscape of hills at positive altitudes above the plane and valleys at negative depths below it. This relief can be represented in 3D perspective or on a 2D image by coding, just as geographers code altitudes with colors.

The two-variable function of the Fresnel plane that I have just described qualitatively is called the *Wigner function* of the quantum field, named after the Hungarian-born American physicist who first introduced it to quantum physics. This is a kind of real-value wave function that contains all the information needed to calculate the probabilities of the results of any measurements made on the field. It is not necessary here to detail how this function is reconstructed from photon-counting measurements. Suffice it to say that in principle it is enough to extract from each histogram a number measuring the difference between the probabilities of finding an even number of photons and finding an odd number. This difference, a real number between +1 and -1, gives the value of the Wigner function at a point of the Fresnel plane; the set of measurements made with probes of different amplitudes and phases reconstructs the global Wigner function of the quantum field. In practice the situation is more complicated, as the measured signals are marred by a noise that somewhat blurs the ideal profile of the theoretical Wigner functions. I won't dwell on the classical methods of data evaluation and extraction that make it possible to minimize errors and obtain beautiful images of our quantum states of radiation.

After reconstructing the state of the vacuum and that of coherent fields of various amplitudes and phases and admiring their elegant Gaussian form, we studied the Wigner function of Fock states. These were prepared randomly by a photon-counting measurement of a coherent state. As soon as we obtained the desired state, we performed tomographic measurements before field damping occurred, accumulating the results over long hours of data acquisition during which all the parameters of the experiment had to remain stable. The Wigner functions of Fock states appear as circular waves centered at the origin of the Fresnel plane, resembling the waves created when a stone falls on the surface of a pond. A central peak—positive or negative depending on whether the number N of photons is even or odd—is surrounded by concentric ripples whose circular crests, which are in numbers $N/2$ or $(N+1)/2$ depending on the parity of N, lie within a circle of radius equal to $\sqrt{N}$.

The invariance of these functions by rotation around the origin clearly shows that the Fock states do not have a preferred phase—or, to put it another way, that their phase is completely indeterminate.

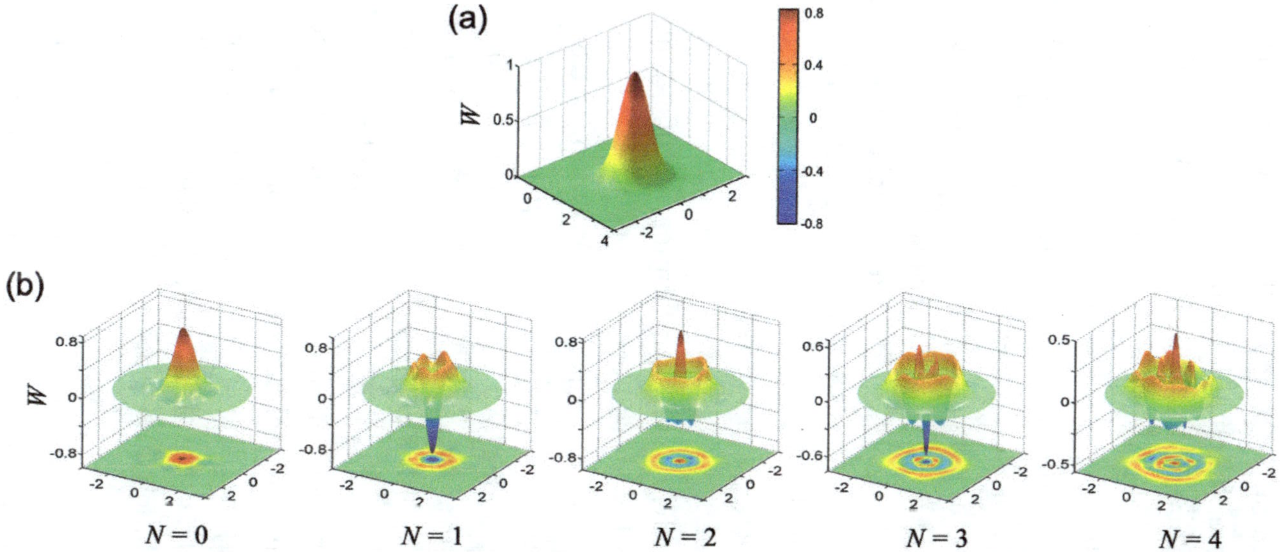

Figure VII.9. Experimental reconstruction of the quantum state of the field in the cavity. (a) The coherent field has a Gaussian Wigner function. (b) Fock states: the state N = 0 is represented by a Gaussian distribution centered at the origin of the Fresnel plane (radiation vacuum). The state N = 1 has a Wigner function in the form of a "sombrero" with a strong negative value in the center. Successive Fock states (N = 2, 3, 4…) have circular ripples with positive crests and negative valleys. (These Wigner functions are shown in 3D and, below, in a 2D version with a color code representing the "relief": positive values in red and negative ones in blue. The "altitude" 0 is green.) (per *Nature* 455 (2008), 510)

The presence of peaks and valleys between peaks where the Wigner function takes negative values indicates the Fock states' quantum character. When the Wigner function changes sign, its interpretation loses its meaning in terms of classical probabilities because it would imply the absurd existence of negative probabilities. The valleys of the Wigner function thus become the signature of the fundamentally non-classical properties of the field trapped in the cavity.

These non-classical aspects are associated with extremely fragile quantum interference effects, which quickly vanish under the slightest disturbance. Indeed, the Wigner function's ripples fade in a very short time compared to the damping time T_c of the energy of the field in the cavity, rapidly transforming the quantum states into statistical mixtures with a positive Wigner function to which a classical interpretation can be restored. This phenomenon, known as *quantum decoherence*, can be understood qualitatively by comparing the evolution over time of the Fock states' Wigner functions to that of the coherent states. The Gaussian peaks corresponding to the latter evolve continuously from their initial position toward the vacuum without changing shape, approaching the origin of the coordinates according to a law of decreasing exponentiality that describes the classical damping of the field's amplitude. This evolution takes place in an average time on the order of T_c.

The Wigner functions of N-photon Fock states, on the other hand, change their aspect very quickly, in a time on the order of T_c/N. The theory shows that the ripples of these functions blur and the negative valleys disappear, leading to a classical circular peak whose radius contracts in a time on the order of T_c. This erasure of the field's quantum characteristics is caused by the fact that the N-photon state undergoes, in a time on the order of T_c/N, a quantum jump to the *N-1* state, which itself rapidly cascades to N-2, and so on. The instants of these jumps vary from one execution to another of the initial field, and the quantum reconstruction describes a statistical average corresponding to all possible evolutions of the system. It is thus understood that the resulting Wigner function must evolve very rapidly toward that of a classical mixture of states, similar to that of a thermal field in which the probability amplitudes associated with the different photon numbers have no phase relations between them. The very rapid disappearance of these states' quantum signature explains why it gets harder and harder to reconstruct their Wigner function as soon as *N* exceeds a few units.

Schrödinger's cats of light

Beginning in 1996, our most spectacular experiments on non-classical field states have studied "photonic Schrödinger's cats." In the environment protected from the outside world by our superconducting mirrors, we were able to prepare superpositions of field states retracing the fate of the famous feline that the Austrian scientist imagined suspended between two classically incompatible states. We placed a field containing several photons in a quantum state that had two phases or two different amplitudes *at the same time*. By reconstructing these states and observing how they progressively lost their quantum characteristics, we were able to explore the boundary between the quantum and classical worlds, and the essential aspects of quantum decoherence.

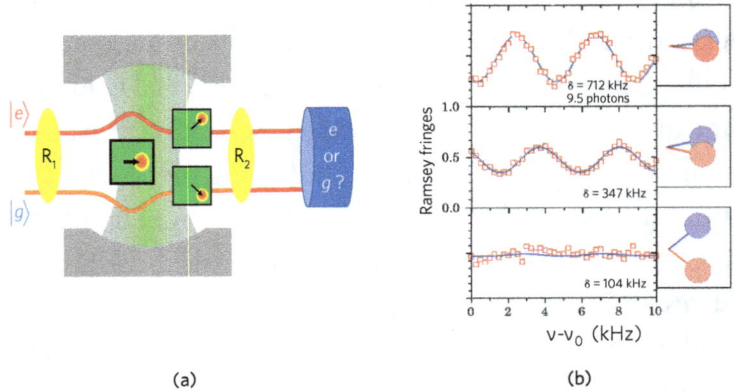

Figure VII.10. Preparation of a photonic Schrödinger's cat from a coherent field containing on average 9 photons. (a) Schematic of the experiment: an atom, prepared by a microwave $\pi/2$ pulse in R_1, crosses the cavity and imparts to the field two opposite phases at once. The atom then undergoes another $\pi/2$ pulse in R_2 before being detected in *e* or *g*. (b) Ramsey fringes detected on the atoms for three different values of the phase shift between the two components of the "cat." This phase shift is adjusted by changing the frequency detuning δ between the atom and the cavity. The upper trace shows well-contrasted fringes when the separation of cat components is small. As this separation increases, the contrast of the fringes decreases (*middle trace*). It fades completely (*bottom trace*) when the two components are well separated. This decrease in fringe contrast, as more and more information about the state of the atom when it crossed the cavity is inscribed on the field phase, illustrates the principle of complementarity (per *Physical Review Letters* 77 (1996), 4887).

Let it be said right away, to clear up any potential misunderstanding, that the name "cat" given to these states is purely metaphorical. Our systems contain only a few particles and are far from being as large and complex as the animal in Schrödinger's fable. The name has become popular in the scientific community and beyond because the questions of quantum entanglement, measurement, and decoherence studied with these systems are precisely those Schrödinger asked himself in an ironic, provocative way.

Our experiments were performed with the Ramsey interferometer used in the photon-counting experiment, whose information we analyze in a complementary way. Instead of studying how the phase of an atomic dipole is altered by the field, we are now interested in the reciprocal effect, the modification of the field phase by the atom. As in the photon-counting experiment, we begin by preparing a field in a Glauber state whose mean photon number can be varied continuously from zero to a few units. This field is represented in the Fresnel plane by a Gaussian peak centered at the tip of the Fresnel vector associated with the classical field of the same phase and amplitude. We have seen that a single atom shifts the frequency v of the field by $+\Delta v$ or $-\Delta v$, depending on whether it is in the *e* or *g* level.

Before it enters the cavity, the atom is prepared by crossing the R_1 zone in an equal-weight superposition of these two levels. The field will then start to oscillate *simultaneously* with both frequencies, one smaller than v, the other larger. When the atom exits the cavity after having interacted with the field for a time T, the field has accumulated two different phases, one in advance of $\Delta\nu T$, the other lagging by the same amount. The Fresnel vector associated with the field points *in two directions at once*.

This "two-headed" field is entangled with the atom. The global state of the atom-field system is the sum of two terms, one corresponding to the atom in state e in the presence of a field with a Fresnel vector in phase advance of $\Delta\nu T$, the other to the atom in state g with a Fresnel vector field in phase lag of the same quantity. The situation in which our field is found is thus reminiscent of the living and dead cat entangled with an atom that is excited and de-excited *at once*. We can also say that in this experiment the cavity field is the instrument that measures the atom's energy. It is represented by an arrow in the Fresnel plane, similar to the needle of a dial pointing in two different directions depending on whether the atom has one energy or the other. Here we illustrate the first stage of a quantum measurement, the one where the measuring device (the field) interacts with the measured system (the atom) so that the information about the observable being measured can pass from one system to the other.

Note that at this stage there is no difference from the photon-counting experiment described above. At the moment when the interaction between matter and light ceases, we may say equally rightly that it's the atom that measures the field's energy (by bringing to its dipole information about the number of photons) or that it's the field that measures the atom (by rotating in the Fresnel plane at an angle dependent on the atomic energy). However, there is a difference that breaks this symmetry. The atom is a microscopic particle that can evolve between only two states in this experiment, whereas the field is an object that is small, certainly, but whose number of states can be increased at will, simply by injecting more and more photons into the cavity. In an experiment involving only one atom interacting with a field containing several photons, it is therefore more natural to consider the field to be the measuring device. By increasing the number of photons, we are thus able to move continuously from a situation where the measuring instrument, initially described as a quantum system, becomes an increasingly classical object.

Let's continue to describe the story of our atom that became entangled with the field as it passed through the cavity, letting it finish its journey in the interferometer after having undergone a second microwave $\pi/2$ pulse in the Ramsey R_2 zone. When it is finally detected, what can we say about the probability of finding it in *e* or in *g*? Is it the result of a quantum interference effect between the amplitudes associated with the two paths followed by the atom in the interferometer, depending on whether it passed through the cavity in *e* or in *g*? The answer to this question depends on the field state. If the cavity is empty, there is interference because the atom has left no trace of its passage. Repeating the experiment a large number of times by scanning the frequency ν of the $\pi/2$ pulses in R_1 and R_2, we record well-contrasted Ramsey fringes. If, on the other hand, the cavity initially contains a coherent field of several photons, the contrast of the fringes decreases and may even fade. We actually performed this experiment, well before the complete reconstruction of the states, because it requires only the detection of a single atom in each preparation of the cat, and can therefore be done with a less efficient cavity than the one used in 2006.

This experiment, which we performed in 1996, is in fact a test of complementarity. If the information about the path followed by the atom is inscribed on an element of the interferometer, the quantum ambiguity is removed and the fringes disappear. In our interferometer, the greater the distance between the two cavity-field Gaussian components, the more information is transmitted to the field. We can vary this distance, with a constant average number of photons, by changing the phase shift of the field induced by the atom. To do this it is sufficient to modify the frequency detuning between the atom and the cavity. At a large detuning, the phase shift is small and the two Gaussian components of the field overlap, maintaining ambiguity about the state of the atom in the cavity and resulting in good fringe visibility. As the atom-cavity detuning is reduced, the field components separate more and the fringe contrast decreases, canceling when the two Gaussian components of the cat cease to overlap completely. As in the aforementioned experiment with the modified interferometric setup where one of the Ramsey pulses was replaced by a quantum field, we again illustrate the complementarity between the distinguishability of the paths and the visibility of the interferences.

Let us note again that it is unnecessary to perform an explicit measurement on the field for the interference to disappear. The simple fact that this field was able to record the path information by interacting with the atom suffices to remove the quantum ambiguity. By performing this experiment ten years before we were able to measure the field, then, we knew simply by observing the fringes' disappearance that we had prepared a Schrödinger's cat–like state in the cavity.

We were able to confirm this cat's existence by repeating this experiment in 2007, once we had a cavity with a very long damping time. Let's return to the story of our atom at the moment before its detection. It is entangled in a non-local way with the field it left behind in the cavity, a situation similar to the one Einstein described in his famous EPR paper. The final measurement of the atom will then destroy the entanglement and provoke the collapse of the field state, under the spooky action at a distance that troubled Einstein so much. Because the atom underwent a mixing of its states while crossing R_2, its detection in the e or g state doesn't make it possible to know which state it was in when it crossed the cavity. The R_2 zone thus plays the role of a *quantum eraser* that suppresses the path information acquired by the field in its interaction with the atom. Thus the quantum ambiguity on the field phase survives this detection. The measurement destroys the entanglement of atom and field, but the field remains suspended between two states, in a superposition of two coherent fields of different phases. Depending on whether the atom is detected in e or in g, the relative phase between the two probability amplitudes changes from π. For a reason I will explain later, these two states are called *even cat* and *odd cat*.

We reconstructed these states with quantum tomography, using a succession of atoms crossing the interferometer and injections of probe fields interfering with the photonic cat. The operations, including the preparation of the cat and its measurement, followed one another in each execution of the experiment within a few milliseconds, a very short interval compared to the damping time of the field. After generating a large number of identical cats, we reconstructed their Wigner function. This showed two well-separated Gaussian peaks, corresponding to the two coherent states of advanced and delayed phase. Between these peaks, ripples take alternately positive and negative values. These are quantum interferences that indicate the coherent character of the superposition between the two states. Even and odd cats show the same Gaussian peaks and differ from each other only by the signs of their ripples. Adding them erases the interference and leads to a Wigner function consisting simply of the two peaks, representing a classical statistical situation. This is the situation that would be obtained if the measurements had been made on two sets of executions in which the field was initially prepared with one classical phase *or* the other.

The "cats" in our experiments contained a handful of photons at most. Beyond that, a circular Rydberg atom was no longer able to act globally on the quantum field without saturation effects and strong distortion of the Wigner functions altering the ideal situation I have just described.

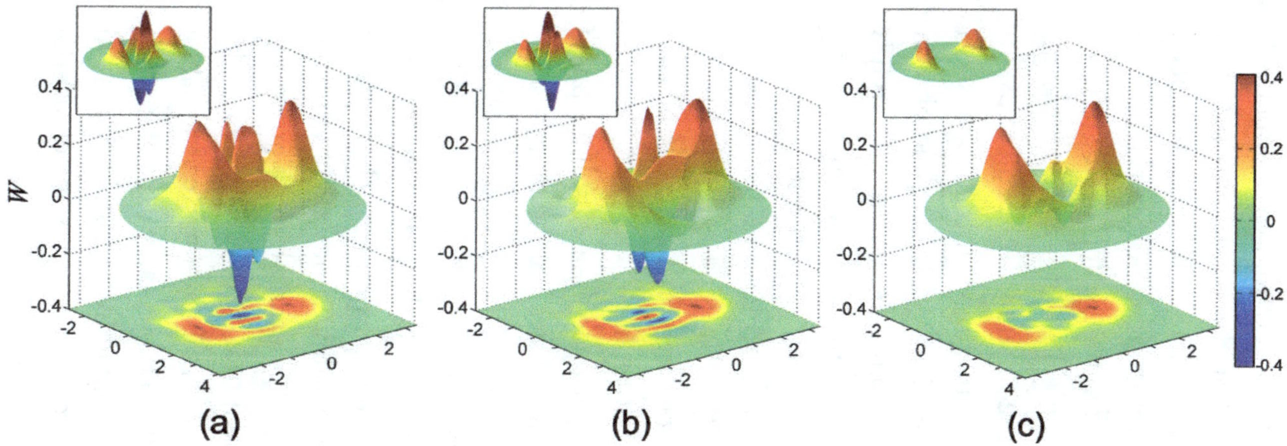

Figure VII.11. Experimentally reconstructed Wigner functions of an even (a) and odd (b) photonic Schrödinger's cat prepared from a coherent field containing on average 3 photons. These cats are obtained by detecting atoms in the *e* (odd cat) or *g* (even cat) state. We can clearly distinguish the two Gaussian peaks associated with the two components of the cats and the interference fringes between the two, taking alternately positive and negative values. These interferences have opposite signs in the even and odd states. Not distinguishing the two atomic states is the equivalent of summing the Wigner functions of both cats. We then obtain the function represented in (c), which describes a classical mixture in which the interference fringes have disappeared. The field in the cavity oscillates with one phase or the other and no longer with both phases "at once." The figures in the small frames at the top show the theoretical Wigner functions (per *Nature* 455 (2008), 510).

The phase difference between the two field components was 135°. It is easier to describe what happens when this difference is 180°, a situation we have since been able to achieve by performing the same experiment with slower atoms, giving the field a greater phase shift.

Let's now analyze this ideal situation. It corresponds to the phase setting we used in our first counting experiment of 0 or 1 photon: an atom shifts the phase of the field by + or -90° and, conversely, a single photon changes the phase of the atomic dipole by 180°. In this case we have seen that the atom is detected in the *e* or *g* state, depending on whether the photon number is odd or even. In other words, the photon counter measures the photon-number parity (identified with the photon number itself if it cannot exceed one, as it was the case in our first counting experiment).

This justifies the names given to these cats. If the phase shift between the two coherent states of the superposition is exactly π, the cat for which both probability amplitudes have the same sign contains only even numbers of photons, while the cat for which these amplitudes have opposite signs contains only odd numbers. The photon distributions of these two cats thus show one N value out of two. The absence of odd or even photon numbers in these histograms is a quantum interference effect, the equivalent of dark fringes on an interferometer, indicating the non-classical character of these states. This property is related to the existence of ripples with negative valleys in the Wigner function of even and odd cats, which cannot be interpreted in terms of classical probabilities.

The generation of cats as the first step in a non-destructive photon-counting process illustrates once again the principle of complementarity. By measuring the field's energy using successive atoms, we progressively destroy the information about its phase. The preparation of a two-component cat represents the first stage of this destruction. While the initial field had a well-defined phase (give or take the quantum uncertainty), splitting its phase causes it to lose precision. After the second atom, each component will again be split into two, the splitting process continuing until the phase is uniformly distributed over 2π, leading to the Wigner function of a Fock state with a completely indeterminate phase. In this sense, a well-defined photon-number state is a multi-component Schrödinger's cat, suspended between a large number of coherent states with phases equally distributed between 0 and 2π.

Let's pause for a moment to dwell on the strangeness of the two-component states we have thus prepared and reconstructed. We might naively think that this situation is identical to that which would be obtained by injecting two fields of different phases into the cavity simultaneously or successively. It's easy to see, however, that the two experiments don't lead to the same result at all. In the case of two injections, a classical interference experiment would be performed by adding two Fresnel plane vectors. We would then simply obtain a coherent field whose amplitude was the sum, as defined by the vectors, of the amplitude of the two fields, a result very different from the cat generated by our experiment. In the case of two fields of opposite phases, the addition would simply bring the field back into the vacuum, a trivial result. The energy of the field would be that of the vacuum's fluctuations, whereas the cat we generated has the energy of several photons. To prepare Schrödinger's cats, a classical manipulation of coherent fields is not sufficient. It is necessary to perform a quantum operation that first generates a superposition of atomic states and then entangles it with the field state.

If the notion of a field suspended between two different phase states doesn't resonate with the reader, she may consider an equivalent but more surprising Schrödinger's cat. After the generation of a superposition of coherent states of the same amplitude and opposite phases, a field identical to one of the components of this cat is injected into the cavity. This is added to this component, doubling its value, and is subtracted from the other, bringing this other component of the superposition to the vacuum. We have thus transformed our phase cat into an amplitude cat. We now have a cavity that is both empty *and* filled with a coherent field containing, on average, up to ten photons.

The Wigner function of this cat is deduced from that of the phase cat by a translation that brings one of its peaks back to the origin of the Fresnel plane. The quantum strangeness of this superposition is described by Wigner function ripples identical to that of the initial phase cat, except for this translation. We thus create a classically absurd situation, a box that is both empty and full, with the possibility of observing interference phenomena between the two terms of this choice. To take up Schrödinger's feline metaphor, we have created a situation where the box is empty *and* at the same time imprisons a cat, with quantum amplitudes of equal amplitude. To probe this strange state, we could send a *quantum mouse* into the box, where it will be devoured by the cat in the first alternative term and remain alive in the second.

Exploring the quantum-to-classical boundary

We tested the fragility of a phase cat by reconstructing its Wigner function as a function of time, letting the field evolve in the cavity before probing it by tomography. By taking pictures of this Wigner function corresponding to increasing evolution times, we observed that the quantum ripples disappear rapidly. The Wigner function quickly becomes a simple superposition of two Gaussian peaks which then converge slowly toward the vacuum. We can thus distinguish in this evolution a fast time constant, on the order of T_c/N, measuring the average lifetime of the quantum coherence, and a slow time constant, T_c, associated with the classical decay of the energy of the field in the cavity. These features are similar to the one we have analyzed in the evolution of Fock-state Wigner functions.

How to explain the speed of the Schrödinger's cat decoherence? Here again we can invoke the principle of complementarity. As soon as a photon disappears into the environment, it carries with it information about the phase of the trapped field, which destroys quantum coherence. If the field contains on average N_m photons whose energy dissipates in an average time T_c, the first of these photons will disappear in a time on the order of T_c/N_m and quantum coherence will not be able to survive any longer.

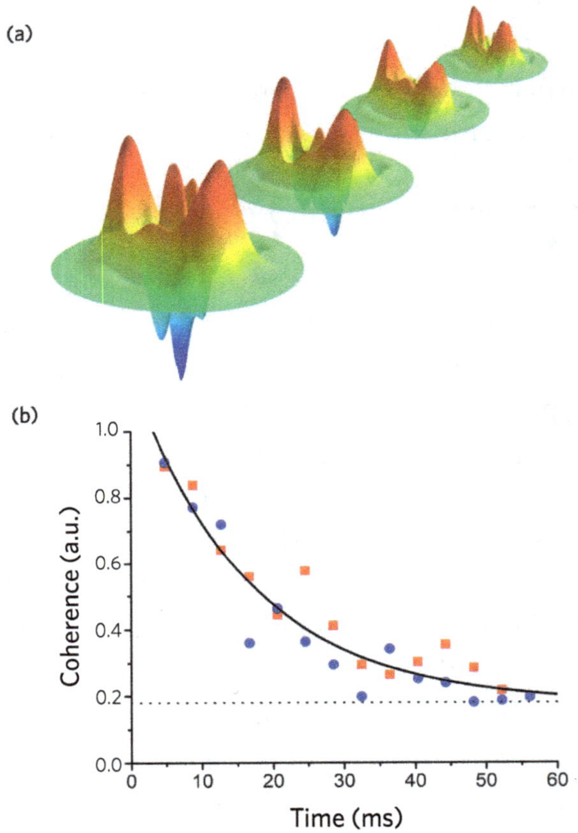

Figure VII.12. Decoherence of a Schrödinger's cat. (a) Wigner functions of the cat shown at four successive times. We observe the progressive disappearance of interference, while the position of the Gaussian peaks has changed very little. We see that the decoherence time is much shorter than the damping time of the field in the cavity. (b) Measurement of the decay, as a function of time, of the coherence term for even (*red dots*) and odd (*blue dots*) cats in arbitrary units (a.u.). From these curves the exponential decay law of cat coherence is established, with a decoherence time varying as $1/N_m$ (per *Nature* 455 (2008), 510).

This decoherence is easily understood in the case of a cat whose components have opposite phases. For a cat with an even number of photons, the loss of the first photon will transform it into an odd cat, which will in turn result in an even cat again after the loss of two photons, and so on.

Since the moments of photon loss are not measured by tomography, the Wigner function of the reconstructed cat becomes, after a time on the order of T_c/N_m, the sum of the functions of even and odd cats, in which the term of quantum interference has disappeared. In other words, the loss of information on photon-number parity destroys the quantum ambiguity and reduces it to a simple classical uncertainty. The cat no longer has one phase *and* the other, merely one phase *or* the other.

Instead of corresponding to a single result (the cat is found alive or dead), the Wigner function gives equal statistical weight to both possibilities, because the reconstruction reports measurements performed on a large number of identical systems. Decoherence leads to a classical situation where the cat is dead (or alive) in 50 percent of the cases. This analysis is generalized to any superposition of two coherent states, even if they are not in phase opposition. Their decoherence occurs in a time inversely proportional to the square of the distance of the Gaussian peaks associated to the components of this state.

This experiment illustrates the very different fate of two families of field states trapped in the cavity when they are allowed to evolve freely under the effect of mirror losses. The coherent states, which describe a well-localized field in the Fresnel plane, keep a Wigner function of invariant form, continuously evolving toward the Gaussian peak representing the vacuum. Their superpositions, on the other hand, are very unstable and lose their quantum coherence faster the more distant their components are in the Fresnel plane.

Continuously coupling the field to the microwave source that created it can compensate for losses in the cavity and maintain a stationary coherent state. With the exception of uncertainty relations, which impose a small Gaussian blur on the field, the coherent states describe a classical situation that would have surprised neither Young nor Fresnel. On the other hand, counteracting the damping of the field by maintaining a continuous coupling of cavity and field source does not allow us to keep a Schrödinger's cat coherently suspended between two states represented by distinct peaks. The disappearance of the non-classical ripples of the Wigner function of photonic Schrödinger's cats is irreversible and cannot be compensated for by classical methods.

Why are coherent states stable and their superpositions so fragile? The fundamental reason lies in their different manner of coupling with the environment and letting information escape from it. Let's try to analyze this difference. The field of the cavity is damped by the effect of its scattering on small rough surface irregularities and residual absorption by the mirrors. The details of these phenomena are unimportant. They can be modeled by generally assuming that the field leaking from the cavity excites small oscillators in the external environment, be they modes of the electromagnetic field scattered laterally outside the cavity or electron oscillations in the mirrors. If the cavity field is coherent, we can imagine it as a small source, a radio station losing and radiating energy with a well-defined phase, coupling weakly to all oscillators in the environment. Each behaves like a small independent antenna picking up the signal from the cavity and oscillating in phase with it.

These antennas all receive the same information indicating that the cavity field oscillates with a given phase and amplitude. In this case, there is no entanglement between the field and the external oscillators that detect it. The state of the overall system is simply the product of the field state and that of its environment. This impermeability to entanglement is a fundamental property of coherent radiation states, which explains their great stability.

If, on the contrary, the cavity field is in a superposition of coherent states, each component of the cat will correlate to a particular phase and amplitude state of each of the small antennas spying on it. This will very quickly induce an entanglement between the cat and its environment, in the sense used by Einstein in his EPR article. The global state of the field and the surrounding antennas will have two components, coherently superimposed. In each, the cavity field will have a well-defined phase, correlated to that of all the antennas. A measurement of one of these microscopic antennas in the environment, whose state is very close to the vacuum, provides only very partial information on the state of the cat. But these "spy" antennas are quite numerous, so very quickly the measurement of their global state reveals that of the cavity field. Moreover, it's unnecessary for this measurement by the environment to really be carried out. That this measurement is possible in principle suffices to destroy all interference effects related to the cat's coherence. We always find the same complementarity argument.

There is a remarkable analogy between the states of the field in the cavity and those of a particle—electron or ion—that evolves like a small mass oscillating linearly under the effect of the restoring force of the trap in which it's confined. We can classically represent the state of this particle at a given instant by a point in a plane analogous to the Fresnel plane. The x and y coordinates of this point describe the particle's position and velocity. Heisenberg's uncertainty relation replaces this discrete point with a small region in which the probability of finding the particle's position and velocity is expressed by a Gaussian law. We thus represent a coherent state of the oscillator, whose mathematical description is identical to that of a coherent state of the field in our cavity.

David Wineland and his group were able to prepare an ion trapped in a superposition of two states in which it oscillates with two different phases at once, a state that has the same characteristics as our photonic cats. They were able to study these ionic Schrödinger's cats and their decoherence in experiments similar to ours, reaching the same conclusions. The states that describe a particle in two distinct points at once with two different velocities see their quantum strangeness disappear faster the more distant these points are from one another.

These experiments, carried out on photons or ions, allow us to shed light on a fundamental aspect of quantum measurement and the transition between the quantum and classical worlds. The systems most resistant to decoherence are those that are well localized, both in terms of their position in real space (in the case of ions) and in the Fresnel plane space (for photons). Information on this localization is disseminated in the environment without entanglement. For example, an ion will give information about its position by scattering laser photons that locate it to within a fraction of the wavelength of light. Any observer reading the environment will obtain the same information on the position of the system, which gives this variable a classical character. The same goes for a field in a coherent state. The signals it emits in the environment all give unambiguous information on its position in the Fresnel plane. We can thus say that a particle's position and velocity, or a field's amplitude and phase, are *semi-classical* quantities, to which—except for Heisenberg's uncertainties—we can associate a character of objective reality, since independent observers, each taking measurements on distinct parts of the environment, will agree on what they see.

We call these objectifiable states *pointer states* because they ideally represent the tip of a needle coupled to a quantum system pointing in one direction and giving unambiguous information about an observable of this system. In the case of our quantum field, the coherent state points in a direction that allows us to measure the energy of the atom crossing the cavity.

It's a different story with the superposition of coherent field states that are almost instantaneously entangled with the environment and very quickly become statistical mixtures. Confirming the preceding qualitative analysis, the precise theory of decoherence generally shows that the mean time T_D during which these superpositions remain in a suspended state between different classical realities is equal to $2T_c/D^2$ where D is the distance of their Gaussian peaks in the Fresnel plane. The greater D is, the faster the decoherence. For even or odd Schrödinger's cats, superpositions of fields of opposite phases containing on average N_m photons, D is equal to $2\sqrt{N_m}$ and the decoherence time $T_D = T_c/2N_m$ is $2N_m$ times shorter than the time T_c of damping of the field energy. The decoherence time T_D of the Schrödinger's cat states of an atomic oscillator is given by the same expression, where T_c is the damping time of its energy and N_m is the mean number of its vibration quanta.

The decoherence theory thus summarized leaves one parameter free. It says that the characteristic decoherence time is inversely proportional to the square of the separation between the cat's components, but the constant of this law of proportionality depends on the speed with which the system's energy is lost in the environment. In the case of our experiments, this constant is the damping time T_c of our cavity, which depends on the quality of its mirrors. This is a technical parameter that measures the quality of protection we can provide to the quantum coherence of our Schrödinger's cat states. There is no obvious theoretical limit to this level of quantum isolation. Our record cavity had a damping time of 130 milliseconds, but one can imagine Q factors ten times longer, which would allow us to maintain cats containing ten times more photons for time intervals on the same order. Thus, there is no clear boundary between the classical and quantum worlds. Pushing this limit by making Schrödinger's cats bigger and bigger is more a problem of technology than one of fundamental physics. It is nevertheless an essential problem if we wish to exploit quantum logic for practical applications in information processing.

Returning to the image of Schrödinger's cat, we now understand why it's practically impossible to prepare a real animal in a coherent superposition of living and dead states. Such a superposition would imply that it was in a well-determined quantum state at the beginning of the experiment. However, our feline is a complex system coupled to a vast environment of molecules and thermal photons indispensable to its survival. It is therefore permanently entangled in an environment in which information on everything that happens to it is recorded immediately. In a nutshell, decoherence theory tells us that the cat is, from the very beginning of our experiment, a classical object that can only be alive *or* dead, never alive *and* dead. If it is enclosed in a box with a radioactive atom, it is the cat, the classical system, that will act as a measuring instrument to determine the state of the atom. By passing from life to death, it will determine the moment of the atom's quantum jump. The Schrödinger's cat fable has only served us as a metaphor to describe the non-classical superposition states of a few particles studied in cavity-photon or trapped-ion experiments.

What drove us above all in our research was curiosity, as well as the passionate desire to highlight in the most demonstrative possible way the strange laws of quantum physics—without seeking any specific application. In this sense we have followed in the footsteps of the theory's founding fathers, with the advantage of having a technology they could not even have dreamed of. Our work has always mixed theoretical and experimental aspects. While we were putting together our apparatus, we were thinking in parallel about the experiments we could do. In a series of theoretical papers, we proposed, before we carried them out, the methods permitting the atom and photon manipulations described above.

We were accompanied in this theoretical work by two Brazilian researchers, Luiz Davidovich and Nicim Zagury, with whom I had become friends at our first meeting, a French-Brazilian physics symposium organized in Rio de Janeiro at the end of 1983, when the country was emerging from long years of dictatorship. Luiz and Nicim subsequently visited us in Paris many times in the 1980s and 1990s, and it was with them that we wrote our preparatory papers on the non-destructive photon-measurement experiments. It was while writing these papers that we also realized that this counting method naturally involved the preparation of photonic cats whose decoherence could be studied experimentally. This decoherence had been analyzed theoretically in a general way by the American researcher Wojciech Zurek. Inspired by his papers, we showed how we could adapt our cavity QED setup to study decoherence experimentally.

These theoretical works have been the subject of several of my courses at the Collège de France. I was appointed there in 2001, on the recommendation of Claude Cohen-Tannoudji and Pierre-Gilles de Gennes, two instructors in the quantum physics graduate program I attended in the 1960s, both of whom had since pursued exceptional scientific careers crowned by Nobel Prizes. Our theoretical work on Schrödinger's cats, non-destructive photon counting, and decoherence were analyzed in detail in a book I wrote with Jean-Michel Raimond between 2001 and 2006. Ironically, the book in English, *Exploring the Quantum: Atoms, Cavities, and Photons,* was completed in the summer of 2006, a few weeks before the experiments we described theoretically therein became feasible in the laboratory. This timeline left us with a sense of incompleteness, and we often considered writing a sequel to describe the experiments we've done since then. This book, less technical than that one, is partly a response to that desire. I hope I have been able to present to a wide audience the experiments that have occupied us for so many years, and to share our fascination with quantum physics and its paradoxes.

Toward a quantum computer: utopia or future reality?

Every time I present the results of our work in a seminar or at a conference, I am inevitably asked when they will lead to the construction of a true quantum computer. The question is a bit frustrating, as our research wasn't done with this goal in mind, but I still have to attempt to answer it. So I must touch on this still-mythical device. In a "classical" computer, the elementary operations are performed by logic gates that couple together binary electrical signals taking two values that can be coded as classical bits, 0 or 1. At each gate, a bit called the *control*, which remains unchanged during the operation, determines what happens to another bit called the *target*. For example, nothing happens to the target bit if the control is 0; if the control is 1, the target bit switches from one state to the other. Any calculation by the computer is performed with a very large number of such operations performed on a very large number of bits.

In a quantum computer, the bits would be quantum objects called *qubits*, which could be prepared and evolve in a coherent superposition of 0 and 1. Gates would operate on the same principle as classical gates, but the bits that feed them would live in the space of the superposed states of quantum physics. Such gates were realized as early as the 1990s with trapped ions, in experiments by David Wineland's group following a proposal by the quantum optics theorists Peter Zoller and Ignacio Cirac. My group showed at the same time that it was possible to operate a quantum gate in which the cavity field containing 0 or 1 photon was the control qubit and the Rydberg atom the target one. Numerous experiments, on a limited number of qubits, showed around the same time that such gates could be used to create and detect entanglement on demand.

The operation of successive gates on a large set of qubits would create a large entanglement, and the machine would very quickly find itself in a massive state of superposition in which the evolution of bits would be suspended between not a few but an enormous number of different classical paths. It has been shown that this superposition speeds up the execution of certain calculations far beyond what a classical computer can do following only one path at a time. At the end of the quantum computation, the result sought would be revealed by a measurement sensitive to an interference effect between all the paths the machine followed in parallel.

An example of a calculation that would benefit from quantum acceleration is one that allows a number containing hundreds of digits to be decomposed into prime factors. A conventional computer is unable to accomplish this task in a physically acceptable amount of time, whereas an algorithm invented by the American mathematician Peter Shor would allow it. This possibility is worrisome for the world economy: the system of cryptographic keys used to ensure the secrecy of banking communications and protect all credit cards is based on the extreme difficulty of finding the prime factors of a very large, publicly accessible number. The search for the quantum computer thus has considerable economic stakes. The arrival in this field in recent years of the major internet companies—Google, Microsoft, and IBM—is a sign of the strategic interest of this hypothetical machine.

Building a quantum computer, however, presents an immense difficulty to overcome. What we are trying to build would be a Schrödinger's cat supercomputer that could perform logical operations leading to massive entanglement between its bits without being affected by decoherence, *i.e.* without information about the computation leaking into the environment. Beyond a certain system size, decoherence becomes unavoidable. While it is possible, at the cost of a great deal of effort, to operate a few dozen elementary quantum gates acting on a few dozen qubits, it is impossible to passively protect the bits from decoherence beyond these orders of magnitude, which don't allow for useful calculations. We are thus very far from running a machine that can compute quickly and more effectively than a classical computer.

One way to solve this difficulty would be to actively maintain the machine's coherence by detecting disturbances produced by the environment and correcting their effects as they occur. This is called *quantum error correction*. An approach in this direction consists in detecting random quantum jumps associated with information losses in the environment and retroacting on the system to restore it to the state prior to the jump. We have shown in our experiments that it is possible to maintain a non-classical state—a radiation Fock state—for an indefinite time in our cavity. A sequence of atoms continuously measures the number of photons, and when this number changes other "correcting" atoms are injected into the cavity to add or remove a photon. However, this kind of method remains unable to protect a quantum machine from the effects of decoherence.

Subtler methods have been implemented to maintain the coherence of even or odd Schrödinger's cat states in a cavity. These experiments are performed in a new experimental field of quantum information called *circuit quantum electrodynamics*, or circuit QED. Our Rydberg atoms are replaced by artificial quantum systems realized by superconducting circuits. These small circuits have quantized levels between which they can evolve by absorbing or emitting radiofrequency photons. These photons can be stored in superconducting cavities whose geometry is different from that of our experiments. The physics, however, is very close to that of cavity QED. Superconducting circuits are qubits bound together either by direct electrical contacts or by coupling to the photons in a microwave cavity, following mechanisms similar to those of our experiments.

These systems evolve faster because the artificial qubits, macroscopic in size, have a much stronger coupling to microwaves than our Rydberg atoms do. The time constants of operations in circuit electrodynamics are measured in nanoseconds, not the microseconds of our experiments. Decoherence is also faster, which means the number of operations that can be performed without error correction is still limited.

A clever method developed in these experiments to combat decoherence consists in encoding bits 0 and 1 in even and odd photonic Schrödinger's cats with one or two photons on average, stored in superconducting cavities. The logical operations of the gates are performed by coupling these fields to superconducting qubits. The decoherence is manifested by the loss of photons that disrupts the cats' parity. This parity change is detected by superconducting qubits measuring the field parity non-destructively, as we have done in our experiments. Errors manifested by cat parity jumps are corrected by feedback on the field without destroying the qubit coherence. The error detections and corrections involve operations on fields and superconducting qubits that I won't explain in detail here. These error correction operations, developed at Yale University by Robert Schoelkopf, Michel Devoret, and Steven Girvin, remain for the moment in the phase of proof of principle, and it's difficult to extend them to systems containing more than a few qubits.

Feynman's dream: quantum simulation

While the quantum computer remains in limbo, other applications of the manipulation of individual quantum systems are more realistic in the short to medium term. Quantum simulation consists in mimicking the particle arrangements of condensed matter physics in one-, two-, or three-dimensional artificial qubit architectures. Cold atoms can be placed in an array of wells of optical potentials realized by the intersection of interfering laser beams. We can also trap sets of ions that arrange themselves regularly in space under the effect of the attractive force of the trap and their Coulombic repulsion forces. A set of superconducting qubits printed on an electronic chip can also be coupled together. By adjusting the interactions between these qubits, we seek to reproduce on a different scale those between atoms in the real system we are imitating. Today's computers are unable to calculate the exact behavior of these sets of atoms as soon as their number exceeds a few dozen; quantum superpositions then become too numerous and the Schrödinger equation to be solved contains too many variables. By emulating in the laboratory the real system with individually controlled and measurable qubits, we can observe the evolution of the artificial system and deduce the behavior of the real one under similar circumstances, which a classical computer could not do.

Whereas in a molecule or a solid atoms are separated by distances on the order of an angstrom, the separations between the qubits of cold-atom quantum simulators are on the order of microns. In an ion trap, the distances between particles are of the same order of magnitude. For both atoms and ions, the interaction energies are always much smaller than in condensed matter, where the interatomic distance is one thousand to ten thousand times smaller. Laser-cooled atomic systems are particularly fragile. To stay ordered, they must be maintained at extremely low temperatures, on the order of one microkelvin or less. However, quantum physics is the same despite the difference in scale and temperature, and what happens in simulators can be extrapolated to conditions in much denser and hotter real matter.

These simulators, whose emergence was announced by Richard Feynman as early as the 1980s, are adapted to the specific problems they are asked to solve: for instance, studying the different phases or configurations atomic or electronic systems can take according to the nature of the particles (bosons or fermions), the range of interactions between them, the dimension of the network in which they are confined (a line, a surface, or a volume), or the temperature. By doing these experiments, we hope to find new atomic configurations that could have interesting properties for various applications.

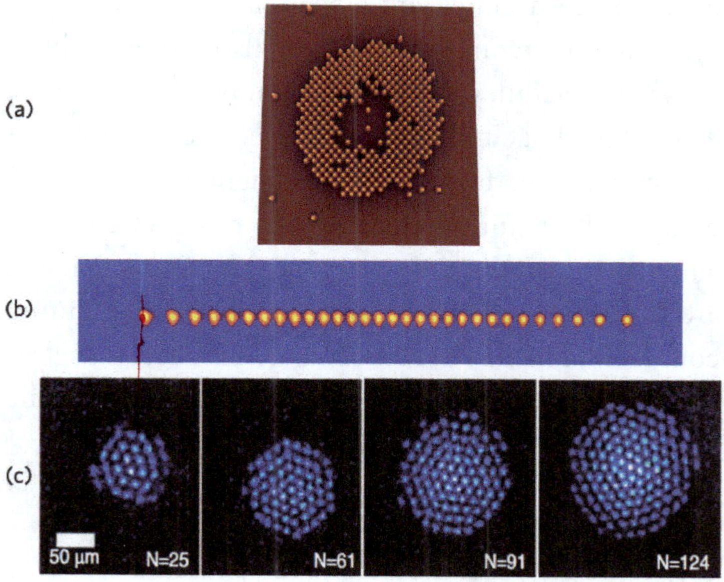

Figure VII.13. Examples of quantum simulators. (a) Ultracold rubidium atoms trapped in a two-dimensional optical lattice. Each point is an atom scattering light in a very high-resolution microscope (atom separation: 0.53 micron) (Immanuel Bloch, Munich). (b) Chain of thirty calcium ions in a one-dimensional trap. The ions are in equilibrium due to their attraction toward the center of the trap and their Coulombic repulsion (distance between the ions: 3 microns). Each ion is a qubit evolving between two states, one scattering light, the other not. The coupling of the qubit's states with a collective vibration mode, realized by laser pulses, makes it possible to create controlled interactions between the ions (Rainer Blatt, Innsbruck). (c) Beryllium ion crystals in a trap (the number N of ions is shown in each figure). The ions are arranged in an ordered lattice under the combined effects of their attraction at the center of the trap and their Coulombic repulsion (per J. Bollinger, NIST, *Science* 352 (2016), 1297).

These quantum simulations could, for example, open up new avenues for the manufacture of high-temperature superconducting materials or the synthesis of large molecules with novel therapeutic properties. These targeted simulations to solve specific problems can support a certain level of decoherence, which also exists in the natural material we're trying to emulate. Thus they don't require systematic error correction and are much less difficult to implement than a quantum computer. The first encouraging achievements suggest that this field of research will expand rapidly in the coming years.

Spooky action at a distance, quantum cryptography, and quantum teleportation

Quantum communication is another very active area of quantum information, allowing two distant partners, our proverbial Alice and Bob, to exchange messages in secret. Located in remote stations, they share pairs of entangled photons propagating either in fibers or in the atmosphere, using satellites as relays. Measurement of the state of these photons by the two partners gives random results, haphazardly distributed but perfectly identical sequences of 0s and 1s which can serve as tamper-proof coding and decoding keys for the messages to be exchanged. Let's briefly explain how this works. Alice starts by translating her message into a long sequence of 0 and 1 bits. Since there are twenty-six letters in the Roman alphabet, plus ten digits and five or six punctuation marks (space, period, comma, colon, semicolon, quotation marks…), all of these can be encoded with six qubits, yielding $2^6 = 64$ possibilities. Two supplementary qubits per letter make it possible to choose between upper and lower case and write in italics. With $2^8 = 256$ sequences of 0 and 1, Alice can thus encode the basic symbols required to write a text. One page of this book, which contains roughly two thousand characters (spaces included), is thus encoded in a sequence of $2{,}000 \times 2^8 = 512{,}000$ bits; the whole book would be about 250 million bits long.

To make this text totally undecipherable, Alice adds to each bit of the text a bit of the random key she shares with Bob. This key must be as long as the message itself, to avoid encoding repetitions that would allow a clever spy to use a statistical linguistic analysis to guess the encoding rule. To encode a bit of the key, Alice performs an addition modulo 2 according to the rules: 0+0=0; 0+1=1+0=1; 1+1=0. This addition is in fact realized by a gate coupling the two bits acting as control and target. The resulting text looks like complete gibberish, and is sent by Alice to Bob over a classical channel (a simple telephone link open to interception will do). At the end of the line, Bob performs the decoding operation opposite to Alice's, subtracting bit by bit the randomized message from the code. The spooky action at a distance that so displeased Einstein thus makes it possible to secretly transmit information between different stations. This is called *quantum cryptography*. Any attempt to intercept the keys made of entangled photons leads to a detectable destruction of the entanglement, making it possible to suspend communication before it's used to transmit a message. To ensure the security of the bits' sharing, Alice and Bob need only to choose a short sequence of the key (about a hundred bits will do) and verify publicly that they find the same results for this sequence. The probability that this will happen by chance after the keys have been tampered with is only 1 in 2^{100}, an utterly negligible possibility.

In order to make quantum communication via optical fiber possible over long distances, we must combat the attenuation of photons that are absorbed as soon as their path through the most transparent glasses exceeds a few tens of kilometers. This requires the realization of *quantum repeaters* connecting fiber segments in order to propagate the entanglement over long distances. These devices are much more delicate to develop than the classical repeaters that amplify signals in the fiber networks that carry the optical signals of the classical internet. The quantum information carried by the photons of one fiber segment must be transmitted to the next segment by a connector made of atoms or artificial qubits. Information exchanges between light and matter can be done by processes similar to those demonstrated in cavity QED experiments. The development of these devices is the subject of numerous studies in different laboratories.

I have chosen here to describe the procedure of quantum key distribution using qubit entanglement, first proposed by Artur Ekert in the early 1990s. Other methods, based on the exploitation of the Heisenberg uncertainty principle, are also possible. In all versions of key distribution, secrecy is ensured by the principles of quantum physics, which imply the impossibilities of performing a perturbation-free measurement and of quantum cloning.

Having pairs of entangled qubits allows Alice and Bob, separated by a large distance, to transmit to each other the state |ψ> of a qubit, the superimposition of states |0> and |1> affected by arbitrary probability amplitudes. The principle of this operation, called *quantum teleportation,* was conceived in the 1990s by a group of quantum information theorists led by the Canadian Gilles Brassard and the American Charles Bennett, who had proposed the first cryptographic key exchange protocol a few years earlier. To achieve teleportation, Alice makes the particle in the state |ψ> interact with the qubit of an entangled pair shared with Bob. When she performs a measurement on this system, the qubit in Bob's possession is projected into a state depending on the result obtained. Alice communicates this result to Bob, who can use this classical information to apply an operation to his qubit that brings it to the state |ψ>. Neither Alice nor Bob knows the state they have exchanged. Contrary to a classical fax machine, which allows the sender to keep a copy of the transmitted information, teleportation destroys the teleported state in Alice's possession, in accordance with the quantum no cloning theorem that prohibits the duplication of a quantum state.

Demonstrations of teleportation of qubits carried by photons have been performed since the 1990s by the groups of Anton Zeilinger, Francesco De Martini, and Nicolas Gisin in Europe, and by the American Jeffrey Kimble in California. An experiment of teleportation of the state of a photon at great distance between a ground station and a satellite was recently carried out in China by Jianwei Pan's group. If the quantum computer ever becomes a reality, the teleportation of quantum information between remote stations will make it possible, by exchanging qubits, to connect networked quantum machines. This would create what we can call a quantum internet.

Quantum metrology and optical clocks

Finally, let's talk about quantum metrology. An individual quantum system that is precisely controlled and manipulated can be an extremely sensitive probe for measuring the physical parameters on which its evolution depends. In an ideal classical situation, the system, prepared in a pointer state, behaves like the needle of a measuring instrument oriented in different directions according to the value of the measured parameter. Heisenberg's quantum uncertainties impose a fuzziness on the direction of this needle and thus a fundamental imprecision on the measurement, called the *standard quantum limit*. In some cases, it's possible to exceed this limit by preparing the probe system in a non-classical state. For example, the quantum needle can be prepared in a Schrödinger's cat–like state in which it points in two different directions at once. By observing a quantum interference effect associated with this superposition, we obtain information on the measured parameter whose accuracy will exceed the standard limit.

We can intuit this remarkable result by observing the Wigner function of a photonic Schrödinger's cat containing an average of N_m photons. The Gaussian peaks of this function have a width on the order of 1 in the Fresnel plane, while the ripples between peaks are separated by only $1/\sqrt{N_m}$. We can see, then, that if we disturb the system (for example by adding a very small coherent field in the cavity), the narrow coherent ripples of the cat state will be more sensitive to this disturbance than the wider Gaussian peaks. The implementation of these ultraprecise measurement methods is tricky. In particular, it requires operating in a time that is short compared to that of decoherence. My research group has demonstrated these possibilities in recent experiments on photonic and atomic Schrödinger's cat states.

But it's in the field of time measurement that quantum metrology has made the most spectacular progress in recent years. We've seen how studies on light have been closely linked over the centuries to the development of increasingly precise clocks. Concluding with an evocation of the latest scientific advances in this field will help us appreciate the immense recent progress made in the measurement of time thanks to lasers.

Huygens's clocks, which played such an essential role in the beginnings of modern science, gave way a century later to Harrison's mechanical chronometers, which, mounted on ships, allowed the first precise measurements of longitudes. At the beginning of the twentieth century, quartz clocks replaced the counting of the periods of mechanical oscillators with that of electrical vibrations excited in a crystal oscillating like a small tuning fork. In three centuries, the inaccuracy of time measurement has decreased from about ten seconds to one millisecond per day, an improvement of four orders of magnitude that essentially reflects the increase in the frequency of the oscillations counted (one per second for the pendulum, several tens of thousands for the quartz).

The advent, in the middle of the last century, of atomic clocks that count the periods of microwaves absorbed and emitted by cesium atoms allowed a six–order of magnitude jump in the uncertainty of clocks, reduced in a few years to one nanosecond per day (about one part in 10^{14}). This jump still reflects the increase in clock frequency to about 9.2 gigahertz for cesium. We've seen that the use of cold atoms, by making it possible to lengthen the atomic interrogation time, has further increased this already extraordinary precision by another two orders of magnitude, bringing it to about ten picoseconds per day (ten one-thousandths of a billionth of a second, *i.e.* a relative uncertainty of 10^{-16}).

A last gigantic leap was made in the early 2000s, when it became possible to count the much faster oscillations of light emitted or absorbed by atoms on optical transitions between the ground state and a very long-lived excited level of an atom or ion. The frequency of the oscillations is then on the order of hundreds of terahertz, between 10^{14} and 10^{15} hertz. The imprecision of the most recent research clocks is now reduced to a few tens of femtoseconds per day (ten one-millionths of a billionth of a second), corresponding to an uncertainty of 3×10^{-19}. Knowing that the age of the universe is about 10^{18} (one billion billion) seconds, this means two optical clocks that were synchronized at the beginning of time wouldn't deviate from one another today by more than two or three tenths of a second! This extraordinary leap in precision was achieved when optical frequencies were counted by dividing them by a factor of about a hundred thousand and brought back to frequencies that could be measured directly by electronic circuits.

The instrument that allowed this division is the frequency comb, mentioned in the previous chapter. This is a laser oscillating simultaneously in about a hundred thousand equidistant modes, covering an octave between infrared and ultraviolet. Its modes are precisely phase-locked and periodically reach their oscillation maxima at the same time. This locking is the result of a coupling among all the modes, due to the fact that they are all powered by the same amplifying medium. Huygens discovered a similar phenomenon in mechanics when,

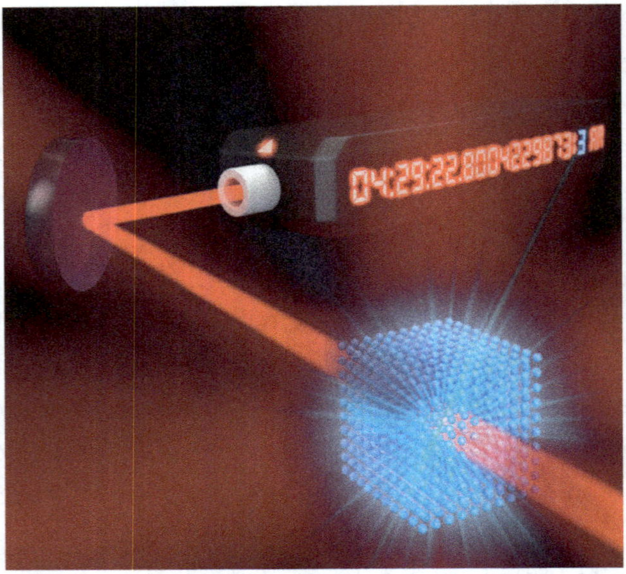

Figure VII.14. Simplified view of the optical clock with ultracold strontium atoms. The frequency of a very stable laser is locked to that of an optical transition of strontium atoms trapped in a three-dimensional optical lattice. This laser in turn stabilizes a frequency comb laser (*not shown*) that transfers the optical frequency into the microwave range. The uncertainty of this clock, 3×10^{-19}, corresponds to an error of less than one second in the age of the universe. It allows us to define the frequency of the strontium transition to seventeen significant digits (ν = 429,228,004,229,873.13 hertz). This precision is limited by that of the best cesium clocks that define the second today. Eventually the second will be related to the frequency of an optical clock whose exact value will be fixed by convention (Jun Ye, JILA).

while ill, he observed the oscillations of two pendulums hanging on a wall in front of his bed. As a result of the vibrations they induced in the wall, the pendulums would be locked in phase and oscillate in unison.

The history of science is full of these striking connections that illustrate its unity and connectedness across the centuries. It is striking that the great watchmaker of the seventeenth century, by discovering the phenomenon of phase locking of clocks, provided the key to an effect that would, three and a half centuries later, lead his descendants, the atomic watchmakers of the twenty-first century, to dramatically improve the precision of their instruments.

The locking of the laser modes regularly spaced in frequency leads to a periodic constructive interference of all waves, resulting in a sequence of light pulses following each other with a period equal to the light's round-trip time between the laser mirrors. We thus obtain a clock whose ticking is materialized by the clicks produced by the successive light pulses escaping from the cavity. These clocks actually realize the thought experiment (described in Chapter 4) that allowed Einstein to illustrate the phenomenon of time dilation.

To ensure the accuracy of these clocks, an extremely frequency-stable laser probes the ultranarrow absorption line of a trapped aluminum ion or a small set of ultra-cold strontium or ytterbium atoms confined in an optical lattice. The frequency of this probe laser is servo-controlled at the center of the ion or atomic line, then compared to that of one of the modes of the frequency comb laser, which is in turn servo-controlled by finely adjusting the separation of its mirrors. The interval between two consecutive modes of the comb is then a precise divider of the optical frequency.

A quartz oscillator whose frequency is read directly by an electronic digital counter is locked to this interval frequency. Calibration of the device is done by comparison with a cesium clock.

The accuracy thus obtained far exceeds that of the microwave clock. The study of the stability of these optical devices can only be done by comparing them with one another. Eventually the cesium clock will disappear and in a few years the *new second* will be defined as a multiple of the oscillation period of one of these fantastically precise optical clocks. Their precision is such that one can measure in the laboratory the variation of time with the altitude predicted by general relativity. The effect, about 10^{-16} per meter, was demonstrated about ten years ago by David Wineland with two aluminum ion clocks, one of which was on a table resting on a hydraulic jack that allowed him to raise it by an adjustable height relative to the other. He then observed, for a difference in altitude of one foot (31 centimeters), a shift of 3×10^{-17} between the two clocks. The progress made over the last ten years should make it possible today to detect the difference in the times indicated by two clocks whose altitudes differ by only a few millimeters.

This extraordinarily precise watchmaking owes everything to the lasers that play an essential role in the device's various elements. It is a laser that cools the reference ion in the aluminum clock by bringing it into the ground state of vibration around its equilibrium position, thus eliminating any Doppler effect that would be detrimental to the oscillator's stability. It is lasers that create, in neutral strontium atom clocks, the optical lattice that holds the atoms in a perfect artificial crystal, and it is other lasers that cool the atoms in these optical lattices. An ultra-stable laser is added to probe the reference transition and transfers the information to the last laser, the frequency comb.

We can see this extraordinary contribution of light to the measurement of time as the culmination of a cycle that illustrates science's multidisciplinary character. It was the first mechanical clocks of the modern era that made it possible to determine the speed of light, paving the way for other fundamental discoveries in optics. In return, after several centuries, light offers the science of time measurement the means to improve itself to a point the watchmakers of the seventeenth century could not have imagined.

The story doesn't end here. Quantum metrology opens up ways to further improve the precision of these clocks. In today's devices that interrogate trapped cold atoms, particles respond to the laser that probes them independently of one another. By entangling the atoms with each other, we could achieve a situation where the atomic assembly solicited by the laser's optical field would evolve in a Schrödinger's cat–like state where all the atoms are both in the ground state *and* the excited state of the probed transition. The quantum coherence of this non-classical state would evolve in time much faster than that of the individual atoms, increasing the time sensitivity of the device. If we could place a hundred atoms in such a state, the already extraordinary precision of optical clocks would be increased by another order of magnitude.

What can we do with clocks that are so sensitive to relativistic effects? Combined with the information provided by the current GPS system, they will undoubtedly make it possible one day to map the Earth's geoid, the equal-gravity surface of the Earth, with a precision of a few millimeters. Changes related to the variation of ocean levels or the thickness of ice caps will be measurable with increased precision, as will continental drift. The detection of minute variations in gravity, precursor signals of earthquakes, may become possible.

Here again, this science brings us back to the past. The extremely precise measurements of the shape of the Earth that we can glimpse ahead of us recall the adventures of La Condamine and Maupertuis. It's no longer in kilometers but in millimeters that we evaluate the precision with which the shape of the Earth is determined. If the orders of magnitude are very different, the spirit of the research is the same. Scientists do these studies because they want to know. And, like Maupertuis in the eighteenth century, they believe their thirst for knowledge will one day prove useful.

The futurology exercise I'm attempting here has its limits. Recall the postcards of 1900 that tried to predict what the world would be like in the year 2000. The repercussions of today's fundamental science in tomorrow's applications remain largely unpredictable—such is the charm of research. The scientists with whom I have worked or exchanged ideas in my life have all been driven by the desire to push the field of knowledge further and further—not so much because it was useful as because it satisfied their curiosity and their need to know. I have had the privilege of participating with them in this great adventure that began four centuries ago, whose guiding thread is light. The story is not over. It will continue to deepen our knowledge and deliver new instruments, no doubt more unexpected and surprising than the ones we can imagine today.

POSTFACE

Science and truth

I have tried to illustrate, through the history of light, what it means to search for truth in science. The step-by-step process that led to an ever deeper understanding of light has given humanity fundamental knowledge about the universe and the world we live in. Participating in this history over the last half-century has been an exhilarating adventure, one that has allowed me, through thought, to live alongside the great minds that preceded us, to rub shoulders with the scientists near and far who have carried out this research through to the present day. In my own research I have experienced the particular joy that accompanies the first observation of a phenomenon revealing a hidden aspect of nature. I have also had the privilege of witnessing great discoveries made by others, a pleasure of a different nature but no less profound.

Scientists need time and trust to carry out their research. Time is necessary because nature doesn't unveil its secrets easily. It often leads us down false paths, and sometimes tests our patience and resolve. Trust, on the other hand, has many faces. It is first and foremost the trust we must have in ourselves, in our ability to analyze, to understand, to imagine new approaches when unexpected situations arise. It is also the profound conviction that natural phenomena obey rational laws, that the models we build of the world form a coherent web woven by science day after day, discovery after discovery.

The trust we need is also that of the institutions to which we belong, which must provide us with material and moral support. Finally and above all, it is that of society, which must share our thirst for knowledge and the belief that it is an essential element of our culture and civilization. I have personally worked in an environment where all of these conditions were met. I have not lacked for time or for trust. For a long time I felt my situation was common, that scientists were generally understood and supported by the public and by the powers that be.

My optimism, however, is tested today. Economic difficulties have reduced the resources allocated to research in France and in many other countries, making working conditions very difficult for young researchers in particular. But there are more serious problems. At a time when science has never been richer in discoveries that broaden our vision of the world, when it offers us means of action and control over nature that were unimaginable only a few decades ago, it is paradoxically misunderstood, even denigrated and attacked. Anti-scientific currents have always existed, but they are now taking an especially pernicious turn with the rise of *fake news* and *alternative facts*.

Science is not the only target of this tide of misinformation and lies, but it is particularly vulnerable. Conspiracy theories are based on an insidious form of doubt, in total opposition to the rational, constructive doubt of the scientific method. By parodying an essential element of the scientific process in this way, science's detractors elaborate a strategy that is as perverse as it is effective.

Why is it that scientists must, four centuries after the advent of modern science, defend themselves against lies? There are reasons linked to the way societies evolve in a world subject to profound crises, to the psychology of individuals who feel increasingly isolated and tend to attach themselves in a tribal manner to the cultures or the religious beliefs that seem to reassure them. Globalization of the economy and the market has left many people by the wayside, unprotected. The fear engendered by this situation makes any global activity that carries in it universal values no group can appropriate—like science—appear to be a threat.

This anti-scientific tribalism leads to misunderstandings on the part of a public that, because it fails to grasp the scientific process, is easily swayed. The perception of certain fields in physics, chemistry, biology, and medicine is thus distorted, manipulated consciously or unconsciously by ideological passions or financial interests. We hear, for example, that global warming is a Chinese invention meant to economically weaken the West, that genetically modified organisms poison our crops, that vaccination is a danger to our children, or that the COVID pandemic is being exploited by dark forces aiming to control the planet. The impression conveyed by these visions of conspiracy is that scientists seek to impose their occult power on society.

Such untruths, propagated frighteningly effectively on the internet in particular, reduce scientific theories to the status of opinions that can be dismissed without proof and put on a level with the beliefs of various traditions. We see in this a resurgence of cultural relativism as advocated by certain currents of sociology and anthropology. If science is merely an activity whose results depend on the sociocultural conditions in which it is practiced, why not put scientific theories on the same pedestal as opinions exempt from the need to be supported by evidence?

Cultural relativism is, if not the cause of the difficulties facing science today, undoubtedly a related factor. In order to defend science and its values, it is necessary to study the deep sociological and psychological origins of these lies, and to analyze in particular the conditions in which social networks function by favoring the confinement of communities of internet users within delusions they sustain among themselves. The effects of permanent and uncontrolled access to an ever-accelerating flow of information also have a role in the crisis we are seeing. This is a subject I have discussed often with Claudine, whose work as a sociologist has led her in recent years to consider these issues.

As a scientist, I thought the best defense of science I could provide was to explain to the public, especially a non-scientific one, the power and beauty of the scientific approach. That is what I have tried to do in this book while talking about light. I have tried to explain what truth means in science, to describe the way it evolves and is constructed, patiently, in a permanent back-and-forth among observation, experimentation, and theory. I have evoked the doubts and interrogations inherent in the scientific method, which constantly calls into question the very models it constructs as it subjects them to increasingly precise and demanding tests.

I have also spoken of the force with which physics has made its mark as an explanation and interpretation of the world, giving us the means to act on and control nature. I have evoked its reductionist character and its "connectedness," implying that questioning one aspect of scientific truth can have repercussions on our whole description of the world. The story I have outlined also illustrates the universality of science—which, contrary to the claims of cultural relativism, knows no boundaries. Finally, I have analyzed the difficulties researchers have had to overcome throughout history to combat the prejudices and dispel the illusions that have long blinded us, preventing us from seeing and understanding a nature that increasingly challenges our intuition.

The history of light has been rich and complex and full of surprises: clouds obscuring our view and lightning bolts suddenly opening new horizons, as happened a century ago with the advent of relativity and quantum physics. Today we live in another crucial period where light will undoubtedly guide us even further in our knowledge of the world. It is not irrelevant that some of the deepest questions facing science still concern the notion of light. When we talk about dark matter or black holes, we are still implicitly talking about fundamental properties of light, its inability to interact with certain forms of hypothetical matter or to escape from intense gravitational fields. When we speak of the as yet unconsummated marriage of general relativity and quantum physics, we seek an ultimate unification of the laws of physics, one that may be inspired by those that preceded it, in which light played an essential role.

When we speak of the promise of quantum information, it is always in light, vehicle of information and instrument for the control and manipulation of quantum matter, that we place our hopes. There is still much to be discovered and invented. It pleases me to imagine the astonishment that would seize Galileo, Newton, Fresnel, Maxwell, or Einstein if they came back among us today and learned what the researchers who followed in their footsteps have understood and accomplished by juggling with photons. And I would love, like Langevin's twin, to return to Earth, if only for a moment, in fifty or a hundred years, to learn what the generations of researchers after me have found. But that is an impossible dream. No rocket will be fast enough to bring me into the future. And even this certainty, given to me by the theory of relativity, comes from light.

BIBLIOGRAPHY

Some works to complement the reading of this book

This list, far from exhaustive, includes works that have inspired me, and others in which the reader will find details, sometimes fictionalized, about characters or episodes in the history of science related in this book. Three personal works round out this bibliography. The first, *Physique quantique*, is the text of my inaugural lesson at the Collège de France (2001). The second, *Exploring the Quantum: Atoms, Cavities, and Photons*, written with Jean-Michel Raimond, is a detailed presentation of cavity quantum electrodynamics and its relation to other fields in modern quantum physics. The last is the text of the lecture I gave in Stockholm when I was awarded the Nobel Prize in Physics.

BIOGRAPHIES AND AUTOBIOGRAPHIES OF SCIENTISTS CITED

Anatole Abragam, *De la physique avant toute chose?* Odile Jacob, 2000.

C. D. Andriesse, *Huygens: The Man Behind the Principle.* Cambridge University Press, 2011.

Bernard Cagnac, *Alfred Kastler*. Éditions Rue d'Ulm, 2013.

Claude Cohen-Tannoudji, *Sous le signe de la lumière*. Odile Jacob, 2019.

Niccolò Guicciardini, *Isaac Newton and Natural Philosophy*. Reaktion Books, 2018.

Alan W. Hirshfeld, *The Electric Life of Michael Faraday*. Walker Publishing Company, 2006.

James Lequeux, *Le Verrier, savant magnifique et détesté*. EDP Sciences, 2009.

James Lequeux, *Hippolyte Fizeau, physicien de la lumière*. EDP Sciences, 2014.

Thomas Levenson, *Einstein in Berlin*. Bantam, 2004.

Walter J. Moore, *Schrödinger: Life and Thought*. Cambridge University Press, 1992.

Abraham Pais, *Subtle Is the Lord: The Science and the Life of Albert Einstein*. Oxford University Press, 2005.

Abraham Pais, *Niels Bohr's Times*. Clarendon Press, 1991.

Dava Sobel, *Galileo's Daughter*. Penguin, 2000.

WORKS ON THE HISTORY OF OPTICS AND ELECTROMAGNETISM

Jed Z. Buchwald, *The Rise of the Wave Theory of Light: Optical Theory and Experiment in the Early Nineteenth Century*. University of Chicago Press, 1984.

Olivier Darrigol, *A History of Optics: From Greek Antiquity to the Nineteenth Century*. Oxford University Press, 2012.

Olivier Darrigol, *Electrodynamics from Ampère to Einstein*. Oxford University Press, 2000.

ON THE HISTORY OF NAVIGATION AND MEASUREMENTS OF THE EARTH

Ken Alder, *The Measure of All Things: The Seven-Year Odyssey and Hidden Error That Transformed the World*. Free Press (reprint edition), 2003.

Peter Galison, *Einstein's Clocks, Poincaré's Maps*. Norton & Company, 2003.

Allan W. Hirshfeld, *Parallax: The Race to Measure the Cosmos*. Palgrave Macmillan, 2001.

Dava Sobel, *Longitude: The True Story of the Lone Genius Who Solved the Greatest Scientific Problem of His Time*. Bloomsbury, 2007.

Florence Trystram, *L'Épopée du méridien terrestre*. J'ai Lu, 1979.

ON THE THEORY OF RELATIVITY AND THE TESTS VERIFYING IT

Albert Einstein, *Relativity: The Special and the General Theory*, 100th anniversary edition. Princeton University Press, 2015.

Hanoch Gutfreund and Jürgen Renn, *The Road to Relativity: The History and Meaning of Einstein's "The Foundation of General Relativity."* Princeton University Press, 2015.

John Waller, *Einstein's Luck: The Truth Behind Some of the Greatest Scientific Discoveries*. Oxford University

Press, 2002 (in particular the third chapter of part I, on Eddington's observation of a solar eclipse in 1919).

Richard Wolfson, *Simply Einstein: Relativity Demystified*. Norton & Company, 2003.

ON THE FOUNDATIONS OF QUANTUM PHYSICS AND ITS PHILOSOPHICAL IMPLICATIONS

Niels Bohr, *Atomic Physics and Human Knowledge*. Dover Books on Physics, 2010.

Roland Omnès, *Les Indispensables de la mécanique quantique*. Odile Jacob, 2006.

Abraham Pais, *Inward Bound: Of Matter and Forces in the Physical World*. Clarendon Press, 1986.

Erwin Schrödinger, *Physique quantique et représentation du monde*. Seuil, 1992 (French translation of two books by Schrödinger: *Science and Humanism* (Cambridge University Press, 1951) and *Die gegenwärtige Situation in der Quantenmechanik* (Naturwissenschaften, 1935)).

ON EINSTEIN'S ESSENTIAL CONTRIBUTIONS TO THE BIRTH OF QUANTUM PHYSICS AND HIS OPPOSITION TO THE "COPENHAGEN INTERPRETATION"

Edmund Blair Bolles, *Einstein Defiant*. Joseph Henry Press, 2004.

Albert Einstein and Max Born, *The Born–Einstein Letters, 1916–1955*. Palgrave Macmillan, 2004.

A. Douglas Stone, *Einstein and the Quantum: The Quest of the Valiant Swabian*. Princeton University Press, 2013.

ON RECENT ADVANCES IN QUANTUM OPTICS AND ATOMIC PHYSICS THANKS TO THE LASER

Claude Cohen-Tannoudji and David Guéry-Odelin, *Advances in Atomic Physics: An Overview*. World Scientific Publishing Company, 2011.

Nicolas Gisin, *L'Impensable Hasard. Non-localité, télé-portation et autres merveilles quantiques*. Odile Jacob, 2012.

Anton Zeilinger, *Dance of the Photons: From Einstein to Quantum Teleportation*. Farrar, Straus and Giroux, 2010.

BY THE AUTHOR

Serge Haroche, *Physique quantique*, inaugural lecture at Collège de France. Fayard, 2005.

Serge Haroche and Jean-Michel Raimond, *Exploring the Quantum: Atoms, Cavities, and Photons*. Oxford University Press, 2006.

Serge Haroche, *Nobel Lecture: Controlling Photons in a Box and Exploring the Quantum to Classical Boundary*. https://www.nobelprize.org/uploads/2018/06/haroche-lecture.pdf.

Index of cited scientists

Acknowledgments

This book owes much to the students, colleagues, and postdoctoral visitors with whom I have sought over the years to deepen our knowledge of atoms and light. I was fortunate to work in a group where scientific curiosity could be exercised without constraint, in an atmosphere of trust and friendship. I would not have been able to accomplish anything had these conditions not been met. My gratitude goes first to Jean-Michel Raimond and Michel Brune, who accompanied me throughout the adventure that led us to tame photons. All those who have come from all over the world and passed through our group have made essential contributions to the successive stages of our research. Most of them went on to pursue brilliant careers in France and abroad. I can't list them all here, but I want them to know I am happy and proud to have shared with them some of the exciting moments of this adventure.

Having reached baton-passing age, I am now free of administrative and teaching duties. I still have the privilege of following the research carried out in my group by the generation that took over. Sébastien Gleyzes, Igor Dotsenko, and Clément Sayrin, working with Michel and Jean-Michel in an administrative and financial context much harder than the one I have known, make discoveries year after year that lead their research in promising new directions. I admire their inventiveness and dynamism, and I am grateful to them for maintaining, with their students, that passion for knowledge without which no research is possible.

Conscious of the fact that each generation of scientists is a link in a chain of knowledge assembled over the course of time, I have tried in this book to trace the chain back to its anchor in the seventeenth century, the origin of modern scientific thought. To brush up this epic on the science of light, I have relied on the works of science historians. I had the pleasure of talking with one of them, Olivier Darrigol, who described to me the subtle and paradoxical representations the scientists of past centuries made of light. I thank him for his insights, which greatly helped me develop Chapters 2 and 3.

I would also like to thank two particularly demanding readers: Jean-Michel Raimond, who studied the manuscript with the attentive eye of the expert and allowed me to correct certain inaccuracies, and my son-in-law, Thomas Peugeot, who read these pages with the eye of a passionate non-specialist. Their opinions and encouragement were precious to me. Patrick Imbert, from the Collège de France's photo service, helped me format several illustrations, for which I am grateful to him.

I owe great thanks to Odile Jacob, who suggested that I write this book and had the patience to wait until circumstances gave me the time to do so. The whole team of her publishing house has helped me greatly in solving the many material problems that accompany the preparation of such a book. I also enjoyed my interactions with Daniel Levin Becker, who translated this book into English.

Finally, I must say it is my wife, Claudine, whom I owe the most. She has been with me since the beginning of my career and has constantly helped and supported me. I cannot dissociate my scientific life from the richness of our personal life, and I cannot imagine what I would have accomplished without her. To thank her for all of this would be paltry. I can at least say thank you to her for putting up with me during the long months when my attention was taken up by the writing of these pages.

Paris, February 2020–April 2021.

This book was typeset
in Electra and Whitney
by Nord Compo
in Villeneuve-d'Ascq (France).

Publication number: 4150-0003-Y
Copyright registration: March 2022

www.ingramcontent.com/pod-product-compliance
Lightning Source LLC
LaVergne TN
LVHW050050061026
849532LV00001BA/29
9782415000035